POLITICS OF CLIMATE CHANGE

Climate policy is a matter of real politics. Every industrialized nation has signed up to the UN Framework Convention on Climate Change. This initially commits each country to reduce climate warming gases to the same level in 2000 as in 1990. The Convention exists to ensure that industrialized countries will cut their CO_2 emissions to the point where there is no threat to global survival. To do this will mean significant changes in energy use and transportation, and in human behaviour.

Politics of Climate Change provides a critical analysis of the political, moral and legal response to climate change, in the midst of significant socio-economic policy shifts. Evolving from original EC commissioned research, it examines how climate change was put on the policy agenda with the evolution of the United Nations Framework Convention and the subsequent Conference of Parties, and considers the uncertainties of climate futures.

Examining the policies of climate change in Europe from the perspective of both practical politics and institutional redesign, this book contains important analysis of the significance of institutional change for international agreements, and of the role of international law in determining national policies.

Tim O'Riordan is Professor of Environmental Sciences and Associate Director of CSERGE at the University of East Anglia; **Jill Jäger** was in the Climate Division of the Wuppertal Institute, Germany. She is now Deputy Director of Programs at the International Institute of Applied Systems Analysis in Laxenburg, Austria.

GLOBAL ENVIRONMENTAL CHANGE SERIES
Edited by Michael Redclift, Wye College, University of London, Martin Parry, University of London, Timothy O'Riordan, University of East Anglia, Robin Grove-White, University of Lancaster and Brian Robson, University of Manchester.

The *Global Environmental Change Series*, published in association with the ESRC Global Environmental Change Programme, emphasizes the way that human aspirations, choices and everyday behaviour influence changes in the global environment. In the aftermath of UNCED and Agenda 21, this series helps crystallize the contribution of social science thinking to global change and explores the impact of global changes on the development of social sciences.

POLITICS OF CLIMATE CHANGE

A European Perspective

Edited by
Tim O'Riordan and Jill Jäger

Global Environmental Change Programme

London and New York

First published 1996
by Routledge
11 New Fetter Lane, London EC4P 4EE

Simultaneously published in the USA and Canada
by Routledge
29 West 35th Street, New York, NY 10001

Routledge is an Independent Thomson Publishing Company

Typeset in Garamond by BC Typesetting, Bristol

Printed and bound in Great Britain by
A Printer

British Library Cataloguing in Publication Data
A catalogue record for this book is available from the British Library

Library of Congress Cataloging in Publication Data
Politics of climate change: a European perspective/edited by Tim
O'Riordan and Jill Jäger.
    p.   cm. – (Global environmental change series)
Includes bibliographical references and index.
ISBN 0–415–12573–1 (cloth). – ISBN 0–415–12574–X (pbk.)
    1. Climate changes–Environmental aspects. 2. Climatic changes–
–Political aspects. 3. Greenhouse gases–Environmental aspects.
I. O'Riordan, Timonthy. II. Jäger, Jill. III. Series.
QC981.8.C5P658 1996
363.73′87–dc20   95-42459
                                                                         CIP

ISBN 0–415–12573–1 (hbk)
        0–415–12574–X (pbk)

CONTENTS

LIST OF FIGURES

LIST OF TABLES

NOTES ON CONTRIBUTORS

Jill Jäger   was in the Climate Policy Division, at the Wuppertal Institute for Climate, Energy and Environment, Wuppertal, Germany. She is now Deputy Director for Programs, International Institute for Applied Systems Analysis, Laxenburg, Austria.

Tim O'Riordan   is Professor of Environmental Sciences and Associate Director of the Centre for Social and Economic Research on the Global Environment at the University of East Anglia, Norwich, UK.

Elizabeth J. Rowbotham   is an Articled Student at Russell and Du Moulin, Barristers and Solicitors, in Vancouver, Canada.

Susan Subak   is a Senior Research Associate at the Centre for Social and Economic Research on the Global Environment, University of East Anglia, Norwich, UK.

Andrew Jordan   is a Research Associate at the Centre for Social and Economic Research on the Global Environment, University of East Anglia, Norwich, UK.

Richard Macrory   is Professor of Environmental Law at the Centre for Environment and Technology, Imperial College, London, UK.

Martin Hession   is a Lecturer in Environmental Law at the Centre for Environment and Technology, Imperial College, London, UK.

Nigel Haigh   is Director of the Institute for European Environment Policy, London, UK.

Christiane Beuermann   is a Research Associate at the Wuppertal Institute for Climate, Environment and Energy, Wuppertal, Germany.

Anne Kristin Sydnes is Director of the Climate and Energy Programme, Fridtjof Nansen Institute, Lysaker, Oslo, Norway.

Alessandra Marchetti is an Environmental Consultant in Florence, Italy.

Konrad von Moltke is Director of the Centre for Global Governance at Dartmouth College, Hanover, New Hampshire, USA.

Atiq Rahman is Director of the Bangladesh Centre for Advanced Studies, Dhaka, Bangladesh.

PREFACE

The literature on climate change is congested and voluminous. So why are we adding to it? Well, to begin with this is not a book about climate change as such. It is more a reflection on how institutional adaptation takes place between global, national and local scales in the name of climate protection. So the focus is the application of the theory outlined in Chapter 4, namely the ever augmenting interpretation of what institutions are, how they function, and how they help in the social accommodation to a collective threat. To provide case material for this focus we selected climate change because it is the object of enormous political bargaining as well as intensive scientific research and integrating socio-economic assessment.

The climate change issue illustrates the interconnections between politics, science and social response. It also challenges democracy in that it seeks policy shifts well in advance of its consequences being evident. In short, the processes of both conceptualizing climate change, evaluating responses, and subsequently carrying out new or additional policy measures, are all part of institutional mechanisms. The analysis that follows tries to set institutional adjustment to climate change 'in the round'. This is why we do not ultimately concentrate on climate change on its own, but largely in terms of its influence on a host of organizational structures and relationships, and intermediate policy measures.

The point of departure is the evolving science–policy synergy that is nowadays climate change. In Chapter 1 we review how the early period of scientific investigation has become transformed into a politically manipulated process where 'the science' of climate change is revealed as being embedded in deeply controversial interest group biases. This was always the case; now it is more evident. Chapter 3 examines how the scientific devices for comparing the atmospheric warming effects of various gases are themselves contentious, as also are the monitoring and reporting mechanisms. Yet the latter must be broadly accepted, otherwise the performance of national commitment cannot form the basis of collective agreement.

The UN Framework Convention on Climate Change provides another organizing framework for the study that follows. It not only emerged from an enormously arduous process of multilateral bargaining. It also revealed how international negotiations reflect differences of ideology within a common purpose that has at its core the aim of human survival on this planet. Chapter 2 reviews the legalities of the Convention, and its subsequent political interpretation.

The research itself was financed by the Research Directorate General (DG XII) of the Commission of the European Communities with the following aims:

- to assess the significance of the UN Framework Convention on Climate Change in coordinating policy and in creating mechanisms to ensure its compliance;
- to investigate the effectiveness of the Convention in terms of new organizational arrangements and policy outlooks in selected countries;
- to review the significance of the first phase of reactions as a guide to possible further responses to further reductions in greenhouse gases; and
- to consider the scope for integrating more effectively the social, economic transport and industrial dimensions of climate policy both nationally and internationally.

The research team looked at the European Community as a whole, as well as four selected countries, namely the UK, Germany, Norway and Italy. The selection of the European Community as a separate legal and political entity was deliberate in that no full assessment had heretofore been made of the legalities of the powers of the Community to bring Member States into line in order to meet such a collectively agreed purpose. Chapter 5 provides the theory and legal argument for this external competence. The material is very complex, but the authors have managed to make sense of an area little understood beyond the freemasonry of European Community lawyers. The political implications and the manner in which the Community went about its climate change politics is discussed in Chapter 6. By a whisker, the Community will probably meet its objectives of bringing its greenhouse gas emissions in 2000 to 1990 levels, but more by luck than by purposeful intent. How far it can further reduce its collective emissions remains a matter for conjecture. That conjecture is provided in Chapter 12.

Chapter 4 looks at what institutions are, how they are conceptualized, in various disciplines, and how they adapt to threat and opportunity. It is intended to provide an overview of a theme that is much in need of simplifying and amalgamating. It seeks to show that institutions do play a vital role in the conduct of highly complex and wide ranging policy issues such as climate change, and the dynamics of institutional adjustment can be mapped. Its theoretical conclusions are tested in Chapters 7–10, and reflected upon in Chapter 12.

The four national case studies form the basis of Chapters 7 to 10. These chart the evolution of national responses, the trials and tribulations of the political processes involved, and the reactions of the key stakeholders. Each chapter concludes with some observations of the principal institutional innovations. Chapter 11 provides a perspective from beyond Europe. The US experience is not a happy one, and this is explained. The Third World is by no means a homogeneous entity, but there is a Third World view which makes uncomfortable reading. Climate change politics have a long way to go before they satisfy the legitimate aspirations of NGOs in developing countries. Chapter 12 suggests how this evolution might be charted and why, given the theory outlined in Chapter 4. This touches on issues raised towards the end of Chapter 1, so hopefully you, the reader, are given a sense of continuity and cohesion in what is really a very amorphous process. It is largely because responding to climate change is so policy interconnective that the multidimentional character of this response can both be judged successful and a failure at the same time.

This book depended on the cooperation of the research team, their informants and the support of their funders, notably DG XII. We owe a great debt to them all. The final text also benefited enormously from the editorial support of Sarah Lloyd at Routledge and the secretarial gifts of Pauline Seeley, Ann Dixon and Rosie Cullington. The whole exercise has been hard work but fun. We hope we can capture that spirit in the text that follows.

Tim O'Riordan and Jill Jäger

1

THE HISTORY OF CLIMATE CHANGE SCIENCE AND POLITICS

Jill Jäger and Tim O'Riordan

OVERVIEW

The prospect of a planet, intricately and marvellously attuned to almost any calamity, being stressed by the hand of humans in the twinkling of an evolutionary eye, became a high level political issue in many countries in the late 1980s. Most people reading this book will be familiar with the result, namely increased investment in climate change science, the emergence of a global commitment to collective action via convoluted political and legal arrangements, the uneasy and distancing relationship between science and policy (though the two are never separable), and the continuing suspicion of key players. These include the 37 or so small island nations who claim they face oblivion and want much prompter and more concerted action; oil producing states and fossil fuel industrial cartels who strenuously seek to delay action in the foreseeable future; big emitters who are unwilling to mend their ways; would-be emitting developing nations who will not play ball until the established emitters either shut up or pay up; and the enormous breadth and representatives of non-governmental organizations who are beating on the negotiators' tables with a host of conflicting demands. These aspects of climate change science and policy are addressed in an enormous literature which is readily accessible (see for example Clayton, 1994; Read, 1994; Victor and Salt, 1994). This chapter provides a summary of the main points, charts the evolution of climate change science and politics, evaluates the first responses to the UN Framework Convention on Climate Change (UN FCCC) and sets the scene for the chapters that follow.

To help the reader anticipate the overall direction of this text, here is a basic summary of this chapter. (For a full summary of the whole book, please read Chapter 12.)

1 The role of climate change science is changing from networks of collaborating, but separately operating, natural and social scientists, to integrated modelling and the emergence of cross-scientific interdisciplinary assessments.

1

2 It is now widely recognized that to be fully effective in the policy process, this emerging interdisciplinary science must become an integral part of the policy machinery and not try unsuccessfully and mistakenly to remain aloof from it.

3 The complexities of legal arrangements in the international forums around climate change science and policy response mean that the politics and economics of policy options will have to become fused with the legal agreements and associated institutional arrangements. This will make future climate change negotiations extremely convoluted, involving fluctuating combinations of interests and negotiating positions. Over time all of the science, no matter how interdisciplinary, will become enmeshed in this engulfing process of bargaining and evolutionary compromise.

4 As climate change issues become more and more immersed in wider social, political and economic policy arenas, notably tax policy, social redistribution issues, developmental pathways and north–south financial and technological deals, so the 'pure' climate change negotiations, complicated as they have already become, will enter an even more demanding phase of multilateral and multipolicy bargaining and horsetrading. Over time any distinctive climate change politics that remain may continue to disappear in this stormy sea of international diplomacy and local action.

Climate change science is essentially the product of an amazing period of interdisciplinary networking and modelling by international groups of scientists working imaginatively and creatively over a period of approximately 20 years. Arguably the development of high level computer technology, and easy international air travel, have revolutionized this form of scientific endeavour. Huge teams of highly skilled people can review each other's work, perform integrated assessments, and generate ideas that far exceed the aggregation of their particular knowledge. In both the natural and the social sciences this is an impressive and exciting development with enormous implications for the future of international and interdisciplinary endeavour. Global environmental threats generated the need for a wide ranging assessment of danger and the socio-economic and political consequences of various possible responses. A steady diet of fresh scientific perspectives helps to maintain regular doses of funding, helped in turn by an endless round of conferences.

The fresh evidence from the natural science community is not producing much that is really new. To keep the pot hot, more attention is now being given to sophisticated integrated modelling of the likely outcomes of possible policy options, ranging from pricing the causes of the trouble to evaluating programmes of joint investment between rich and poor nations aimed at improving their economies while reducing the likelihood of climate change (see Dowlatabadi, 1994). These models by no means fully integrate all aspects of the natural and social sciences, they do not take into account the subtleties of how different groups in different societies generate demands for natural resources, including fossil fuels, cattle and rice production and timber products, nor do they yet appreciate the socially divisive aspects of brutal politics that worsen the plight of those already economically and/or environmentally vulnerable. So the new round of models will have to be more interactive, recognizing that certain policy options, aimed at reducing either the causes or consequences of climate change, may actually produce perverse outcomes, such as more widespread removal of surface vegetation, that add to climate change, as well as creating even more human misery and environmental stress.

To be even moderately successful this modelling exercise will have to involve a much closer integration between natural and social scientists. To date this particular intellectual linkage has been notable for its failures. These have in part been caused by a serious difficulty in understanding the disciplinary cores and peripheries that even established practitioners in each discipline involved find troublesome to comprehend. But it is also a feature of the enormous range of theories and perspectives that beset the social science community when dealing with the climate change issue. For example, students of the sociology of science (for example, Jasanoff, 1990; Wynne, 1994) assert that climate change science itself is a social construction that cannot be disentangled from political biases, interpretations and expectations of funders and regulators, with a well meaning but futile wish to *appear* neutral, objective and separate from policy making.

The very *image* of a science separate from the policy process is a feature of the politics of modern science. Professor Bert Bolin (1994, 29), a distinguished climate change scientist, noted that the role of science should be 'to delineate a range of future opportunities, and analyse what the implications of development along one course or another might be . . . [but] not to recommend one or the other'.

Indeed, Bolin goes as far as to clarify his position in a diagram reproduced here as Figure 1.1. He characterizes the pure research effort of the big global networks, here summarized as the World Climate Research Programme (WCRP), the International Geosphere Biosphere Programme (IGBP) and the Human Dimensions Programme (HDP) with major monitoring efforts, such as the Global Climate Observing System (GCOS), the Global Terrestrial Observing System (GTOS), the Global Ocean Observing System (GOOS), the World Weather Watch (WWW) and the Global Atmospheric Watch (GAW). These, he sees as largely independent of the policy process, but linked to that process by the Intergovernmental Panel on Climate Change (IPCC) which Bolin chairs, and of which more later in the chapter. IPCC undertakes the assessments for the political process which is discussed below.

This separation of responsibilities may be an expressed wish by the head of this most prestigious climate change scientific panel. But it is frankly

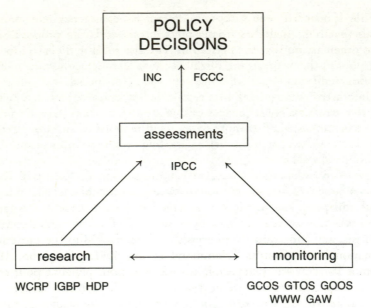

INC	Intergovernmental negotiating committees
FCCC	Framework Convention on Climate Change
IPCC	Intergovernmental Panel on Climate Change
WCRP	World Climate Research Programme
IGBP	International Geosphere Biosphere Programme
HDP	Human Dimensions Programme
GCOS	Global Climate Observing System
GTOS	Global Terrestrial Observing System
GOOS	Global Ocean Observing System
WWW	World Weather Watch
GAW	Global Atmospheric Watch

Figure 1.1 A scientist's perception of the role of science to policy. Research and monitoring are regarded as independent activities that feed into integrated, but still politically detached, assessments that in turn inform policy makers. In practice the whole process is structured by patterns of perception and expectation that are shaped by social judgements and political biases

Source: Bolin (1994, 27)

impossible for such a panel actually to remain aloof from the political processes that both shape its existence and respond to its propositions — if for no other reason that the pattern of response itself will determine how the scientific effort will proceed, and how it will influence future political and legal agreements.

Another arena where intellectual conflict upsets the cause of inter-disciplinary climate change science relates to the political implications of certain judgements which appear value neutral but which contain enor-mously political implications. This matter is addressed further in Atiq Rahman's viewpoint in Chapter 11. As an example, there is a long running dispute over the role of the source of any particular greenhouse gas discharge as a factor in determining global warming and its relative contribution to the total emission. The Intergovernmental Panel on Climate Change, for example, tried to produce a comparative table of global warming potential for each gas, and subsequently to produce a league table of the most signi-ficant national contributors (Hammond et al., 1991; Houghton, 1992). But from a Third World perspective, such calculations penalize poor coun-tries, who assert, for instance, that deforestation is a matter of desperation not choice, and subsistence animals who provide a variety of useful func-tions cannot be compared with surplus overfed livestock in rich countries (Agarwal and Narain, 1992; Rahman et al., 1993). The answer, claim the critics, lies in alleviating poverty and social oppression, not in technical fixes and population control. This is a heated dispute that colours some of the scientific assessments and political negotiations between north and south. Well intentioned as it is, Bolin's diagram founders at the base of the triangle. Both scientific research and monitoring, let alone the integrated assessment process, are tinged by political values and ideological conflict.

To illustrate this point further, economists have sought to calculate the total value of social and ecological damage arising from human-induced climate change. Even without the uncertainties as to how much temperatures and sea levels will rise, these cost benefit calculations have also sparked off a huge row. The trouble lies in the estimate of death and injury induced by the specific climate change, over and above all other causes of death and injury, and in the valuation of a life. The economists who are responsible for providing the IPCC assessments seek to do so in US dollar equivalents, and in terms of the discounted value of earnings for a prematurely killed indivi-dual. Critics argue that dollar equivalence does not fully take into account how local currency can only be translated into dollars via linkages that have more to do with international financial markets and trading arrangements than currency equivalence. The response is that sensitivity analysis can adjust for such factors. One can very quickly see how tendentious all this becomes. And the value of a life can also become a matter of heated debate if it is linked to such currency assessments and the earning power of people in poor or rich economies.

5

Economists calculate that the value of a 'northern life' may be up to ten times higher than the value of a 'southern life', and estimate that the total global damage from climate-change-induced events is around $362 billion, or approximately 1.5–2.5 per cent of gross world economic product (see Fankhauser, 1995). The analysts are, however, adamant that this comparison is a rule of thumb that can be adjusted via various statistical weightings. If, however, the value of a life is made equal in money terms, and the purchasing parity of local currency is taken into account, then, so argues the Global Commons Institute (Meyer, 1995), the 'true' damages would amount to $720 billion or 3.2 per cent of GWP. This would justify, so the advocates claim, much more expensive ameliorating measures now, targeted especially at the Third World. The Global Commons Institute figures are just as politicized as the IPCC economist calculations. Frankly there is no value neutral analysis of the costs and benefits of climate change, or of accommodation to the effects of climate change. To prove the point, the Global Commons Institute analysis, astutely packaged and circulated, has been taken up by the Indian environment minister on behalf of many developing nations. The result is that any scientific assessment is thrown into the arms of politics and international relations. One can readily see how the 'science' loses its significance in such disputes. Yet these disputes are very real and very serious for those at the centre of the negotiations.

These two examples illustrate the problem that climate change science and politics cannot be dissociated either from the political process, or from a wider set of policy issues within which all climate change analyses must be interpreted. These relate to levels of consumption between north and south, luxury US subsistence needs, international trading arrangements, debt responsibilities, aid relationships, population growth, human rights and forms of democratic political cultures in different countries, and the unanswerable theme of who or what is ultimately responsible for inequality and poverty.

THE PROBLEM OF ANTHROPOGENIC CLIMATE CHANGE

For more than a hundred years scientists have been considering the possibility that human activities may change the earth's climate. Over the past 20 years concern has concentrated on the fact that the concentrations of certain gases in the atmosphere are increasing and that this may lead to a warming of the earth's surface and to climatic changes. The basis of this concern is the *enhanced greenhouse effect*.

There is no scientific dispute over the fact that certain gases exert a so-called greenhouse effect on the earth's surface and lower atmosphere. Basically these gases, which are generally referred to as *greenhouse gases* or GHGs for short, do not have much, if any, influence on the incoming, short wave radiation as it passes through the earth's atmosphere. This radiation is absorbed at the earth's surface and is radiated away as long wave, infrared radiation. The GHGs in the atmosphere absorb and reradiate this long wave radiation from the earth's surface. Therefore, if the amount of one or more GHGs in the atmosphere is increased, the amount of long wave radiation absorbed and reradiated increases. The additional GHGs are acting like a blanket keeping more heat close to the earth's surface – the analogy with a greenhouse is not quite correct.

The greenhouse gases which have given rise to concern include carbon dioxide (CO_2), water vapour (H_2O), methane (CH_4), nitrous oxide (N_2O) and chlorofluorocarbons (CFCs). The first of these to receive attention was CO_2, while, as will be discussed in more detail below, the latter three were brought into the scientific debate in the mid-1970s. All of these gases except for the CFCs are natural constituents of the atmosphere. That is, there is a natural greenhouse effect. These naturally occurring greenhouse gases keep the earth's surface and lower atmosphere about 33 °C warmer than it would be without them. Without the natural greenhouse effect, the surface of the earth would be inhospitable.

The 'enhanced greenhouse effect' is a result of increases of the concentration of GHGs in the atmosphere. Observations have shown that concentrations have been increasing since the industrial revolution or even longer. The CO_2 level in the atmosphere increased from about 280 ppmv (parts per million by volume) in the pre-industrial period to about 355 ppmv in 1992. The concentration of CH_4 increased from 700 ppbv to 1714 ppbv over the same time span. Such increases are a result of human activities which add the gases to the atmosphere. Studies during recent years have confirmed that the following human activities add greenhouse gases to the atmosphere:

CO_2 combustion of fossil fuel, deforestation, desertification and agriculture;

CH_4 rice paddies, cattle rearing, biomass burning, coal mining, ventilation of natural gas, landfill, deforestation and wood fuel use;

N_2O agriculture, fossil fuel combustion, use of catalytic converters in cars;

CFCs   used in a wide range of products, such as insulation, air conditioning, and solvents. The phase-out of CFCs is regulated by the Montreal Protocol and its successive Amendments as a result of their calculated role in the destruction of stratospheric ozone.

The greenhouse gases do not contribute equally to increased radiative forcing. Various estimates have been made of the contributions in the 1980s. Table 1.1 shows the estimates made by the IPCC in 1994 and indicates that more than half of the increased radiative forcing was from CO_2. Subsequent work has revised these figures slightly, but did not change the finding that CO_2 has made the largest individual contribution to the

Table 1.1 A summary of key greenhouse gases affected by human activities

	CO_2	CH_4	N_2O	CFC-12	HCFC-22 (a CFC substitute)	CF_4 (a perfluoro-carbon)
Pre-industrial concentration	280 ppmv	700 ppbv	275 ppbv	0	0	0
Concentration in 1992	355 ppmv	1714 ppbv	311 ppbv	503 pptv	105 pptv	70 pptv
Recent rate of concentration change per year (over 1980s)	1.5 ppmv/yr 0.4%/yr	13 ppbv/yr 0.8%/yr	0.75 ppbv/yr 0.25%/yr	18–20 pptv/yr 4%/yr	7–8 pptv/yr 7%/yr	1.1–1.3 pptv/yr 2%/yr
Atmospheric lifetime (years)	(50–200)*	(12–17)†	120	102	13.3	50000

* No single lifetime for CO_2 can be defined because of the different rates of uptake by different sink processes.
† This has been defined as an adjustment time which takes into account the indirect effect of methane on its own lifetime.
1 pptv = 1 part per trillion (million million) by volume.
Source: IPPC (1994b, 17)

increased radiative forcing, while CH_4 and CFCs have contributed roughly equal but smaller amounts.

In order to estimate future possible changes in radiative forcing, scenarios of future emissions are developed and used to calculate future concentrations of greenhouse gases. From the estimates of concentrations, it is possible to derive estimates of the change in radiative forcing.

The interest of many societal actors is, however, not in the potential change of radiative forcing but in the amount of warming at the earth's surface and in resulting climatic changes. As we have already noted, such estimates are difficult to make and thus more controversial. The effect on climate can be estimated, up to a point, using computer models ranging from simple models to complex, time-dependent three-dimensional models of the atmosphere and oceans. On the basis of model results, scientists generally agree that a doubling of the atmospheric CO_2 concentration (or an increase in the concentrations of other GHGs with the same total radiative effect as doubling the CO_2 concentration) would give a globally averaged surface temperature increase of 1.5–4.5 °C, with a 'best guess' of 2.5 °C (see Figure 1.2).

Since CO_2 has made the largest individual contribution to the enhanced greenhouse effect, discussion, as reflected in other chapters of this book, has focused on controlling CO_2 emissions. In order to understand some of the institutional developments in response to the problem of the enhanced

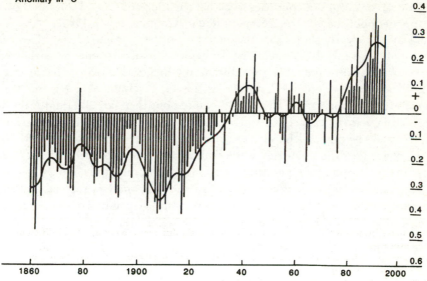

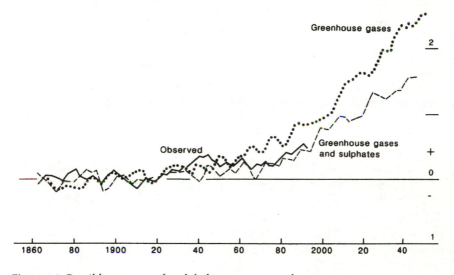

Figure 1.2 Possible outcomes for global temperature and various assumptions, particularly the so-called 'dampening' effect of sulphate aerosols. The chart above shows the pattern of global temperature changes from 1860 to the present day

Source: © *The Economist* 22 April 1995, 10, published from Hadley Centre data

greenhouse effect, it is useful to look at the estimated shares of different countries or groups of countries to global emissions of CO_2. Table 1.2 shows estimates published by the German Enquete Commission in 1990.

Table 1.2 clearly shows that the emissions of CO_2 from the energy sector are quite unevenly distributed. The per capita emissions of the industrialized countries listed in the table are substantially higher than the global average of 4.1 and of developing countries (e.g. China, 1.9; India, 0.7; Nigeria, 0.3). However, as Subak and Clark (1990) pointed out there are different ways of presenting emissions accounts. Their analysis looked at carbon emissions from land clearing and fossil fuel consumption between 1860 and 1986 for 132 countries. The countries were ranked according to CO_2 releases per nation, per capita and per land area. Moreover, the study looked at the cumulative emissions between 1860 and 1986. This analysis showed that the ranking of the countries varied quite considerably depending on the accounting system used. The top three countries for each system were:

Current mean CO_2 release per nation	United States, Soviet Union, China
Current mean CO_2 release per capita	United States, Ivory Coast, East Germany
Current mean CO_2 release per land area	Netherlands, East Germany, West Germany
Cumulative CO_2 release per nation	United States, Soviet Union, United Kingdom
Cumulative CO_2 release per capita	Canada, Australia, East Germany
Cumulative CO_2 release per land area	Belgium, United Kingdom, West Germany

Table 1.2 Energy-related CO_2 emissions, shares of global CO_2 emissions and per capita emissions for selected countries and the EC in 1986

Country	CO_2 emissions from energy sector (million tonnes)	Share of global-energy-related CO_2 emissions (%)	Per capita CO_2 emissions from energy sector (tonnes)
USA	4766	23.8	19.7
Germany*	1067	5.3	13.7
UK	676	3.4	11.9
Italy	365	1.8	6.4
Norway	31	0.2	7.5
EC*	3187	15.9	9.4

*Including the former German Democratic Republic

Source: Enquete Commission (1990)

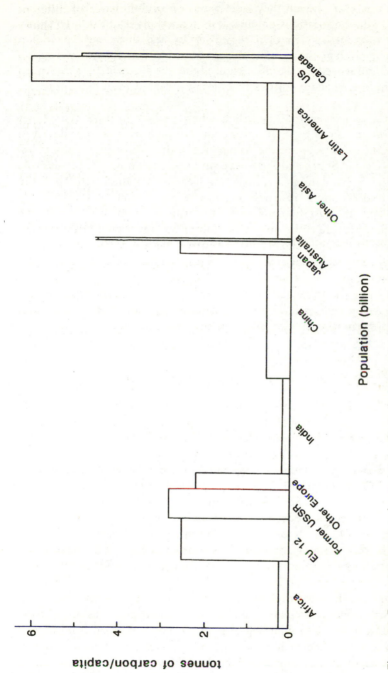

Figure 1.3 Per capita emissions by country and population numbers. This diagram makes it clear that the United States, Europe and the former USSR are the main contributors, but that China and India are waiting in the wings. Any indication of a move towards serious trading of emissions between these blocs would transform future climate change negotiations.

Source: Grubb (1995, 42)

These rankings, even though they were based on preliminary data, showed quite clearly that negotiations on emissions reductions could be affected by how different countries view their responsibility and this view depends on the accounting system chosen. The negotiations for the climate convention were indeed influenced by the discussion about the historical emissions, in which the industrialized countries had clearly released more CO_2 into the atmosphere.

These data look even more startling when a chart is produced of per capita carbon emissions by country or group of countries as reproduced in Figure 1.3. One can readily see the dominance of the rich northern countries, notably the United States and the European Union nations. The precise role of the former Soviet Union is a little less clear as emission data are more unreliable, and economic restructuring has reduced some of the formerly very subsidized and wasteful emissions. From this table one can see why the potentially large future emissions of the poorer but developing countries are such a source of dispute. The growing countries want room for manoeuvre, and unconditional technical assistance and trade aid to enable them to grow more efficiently in GHG emission terms. Above all, they want to see real reductions from the north before they commit their peoples to a more restricted development path. This has enormous implications for the politics of joint implementation as we shall see below.

THE DEVELOPMENT OF SCIENTIFIC UNDERSTANDING

As already mentioned, scientific concern over the possibility that human activities could change the global climate has been expressed at various intervals over the past hundred years. However, since the beginning of the 1970s the concern received increasing attention in national and international institutions. At first analysts concentrated on the potential climate impacts of the increasing CO_2 concentration. Starting in 1958 in Mauna Loa, Hawaii, Charles Keeling of the Scripps Institute of Oceanography at La Jolla, California, kept accurate observations of the atmospheric concentration of CO_2. By the mid-1970s almost 20 years' worth of data were available and these showed that the concentration was steadily increasing (see Wang *et al.*, 1976). At the same time atmospheric scientists intensified their efforts to estimate the impact that an increasing CO_2 concentration would have on global climate. At first with relatively simple models, but by the end of the 1970s also with general circulation models of the atmosphere, estimates were made of the response to a doubling of the CO_2 concentration from its pre-industrial value. Lanchberry and Victor (1995) provide an excellent summary of the main scientific developments since the 1970s.

Figure 1.4 shows the evolution of climate change science and politics. This suggests that the scientific work began to increase its efforts in the mid-1950s

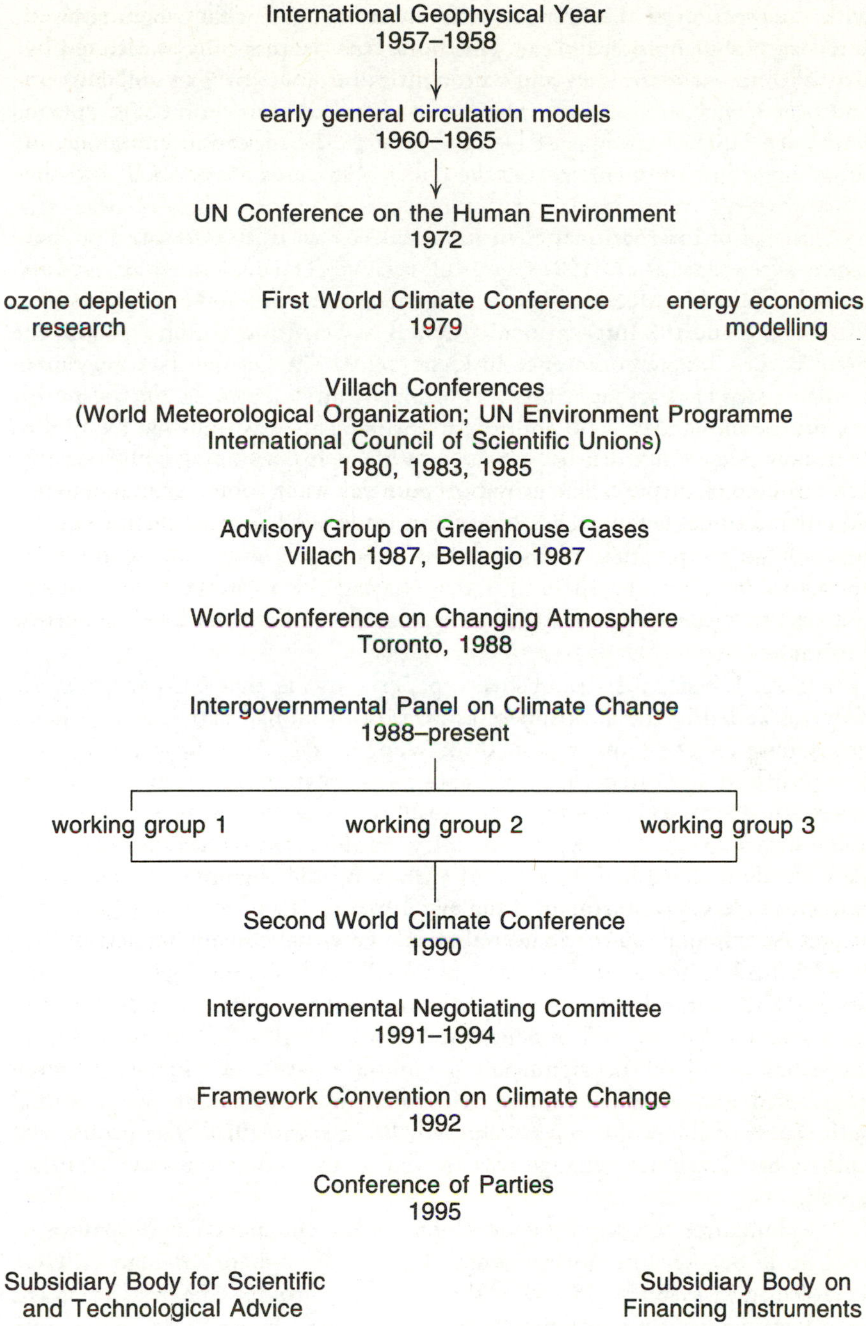

International Geophysical Year
1957–1958

↓

early general circulation models
1960–1965

↓

UN Conference on the Human Environment
1972

ozone depletion
research

First World Climate Conference
1979

energy economics
modelling

Villach Conferences
(World Meteorological Organization; UN Environment Programme
International Council of Scientific Unions)
1980, 1983, 1985

Advisory Group on Greenhouse Gases
Villach 1987, Bellagio 1987

World Conference on Changing Atmosphere
Toronto, 1988

Intergovernmental Panel on Climate Change
1988–present

working group 1   working group 2   working group 3

Second World Climate Conference
1990

Intergovernmental Negotiating Committee
1991–1994

Framework Convention on Climate Change
1992

Conference of Parties
1995

Subsidiary Body for Scientific
and Technological Advice

Subsidiary Body on
Financing Instruments

Figure 1.4 The evolution and pathway of climate change science and politics,
1950–1995

with the creation of the first major international scientific programme, the International Geophysical Year. This began the process of connecting scientists of different disciplines and nationalities into networks of collaboration and peer review. It also began the era of global circulation models, spurred on by the US Department of Defense. The 1972 Stockholm conference on the Human Environment created the political impetus for national responsibility towards transboundary environmental deterioration, and began the modern era of international environmental law with its associated political regimes (see Haas *et al.* (1993) for a full review). The evidence was sufficient for the World Meteorological Organization, the UN Environment Programme and the International Council of Scientific Unions to hold the first World Climate conference in Geneva in 1979. Unlike its successor in the same city 11 years later, this was primarily attended by scientists, and, at the time, commanded little political attention. This was followed by a series of conferences and workshops held in Villach in Austria. In 1979 the US National Academy of Sciences published a report in which they concluded that the doubling of the CO_2 concentration would lead to an increase of the average temperature of the earth's surface of between 1.5 and 4.5 °C. In the following years up until and including the 1994 IPCC assessment, the scientific consensus was that there was no reason to widen or narrow this range.

In 1975, US scientists published a paper showing that CO_2 was not the only gas added to the atmosphere as a result of human activities that has a greenhouse effect (Ramanathan, 1975; Wang *et al.*, 1976). Research quickly followed that indicated that only about half of the enhanced greenhouse effect was due to CO_2, the rest was a result of gases such as methane, CFCs, and nitrous oxide. However, the policy implications of this finding were only slowly realized and the issue of anthropogenic climate change was still framed as a 'CO_2 problem' until the mid-1980s.

In 1985 a major international review of the issue formed the basis of the third Villach conference organized by the World Meteorological Organization (WMO), the United Nations Environment Programme (UNEP) and the International Council of Scientific Unions (ICSU). Scenarios for future emissions of all of the significant greenhouse gases, not just CO_2, were considered and an international scientific consensus about the potential seriousness of the problem was achieved (Bolin *et al.*, 1986). The problem of anthropogenic climate change was moved at this point onto the political agenda.

A culmination of scientific development in climate change analysis is the work of the Intergovernmental Panel on Climate Change (IPCC). Established in 1988 the IPCC had three working groups. The first was asked to assess available scientific evidence on climate change. The second was to assess environmental and socio-economic impacts of climate change. The third was to formulate response strategies. The three working groups published reports in 1990. The report of Working Group 1 had the most impact (Houghton *et al.*, 1990). The executive summary of the report was clearly worded and pointed out that the group was *certain* that there is a natural greenhouse effect and that human activities are increasing the atmospheric concentrations of greenhouse gases. Furthermore, the group 'calculate[d] with confidence that' CO_2 was responsible for over half of the enhanced greenhouse effect in the past and is likely to remain so in the future. The group also made a number of predictions based on current model results, including predictions of future rates and amounts of change of global average temperature and rough estimates of the kinds of regional climate changes that could occur.

The 1990 report introduced the idea of 'Global Warming Potential' (GWP) as a tool for translating scientific information into policy-relevant material. Essentially the GWP allowed the relative warming effect of different gases to be compared. Alternatively, GWPs could be used to calculate the effect of reducing CO_2 emissions by a certain amount compared with reducing CH_4 emissions, for a specified time horizon. In the negotiations of the Montreal Protocol and its Amendments for protection of the stratospheric ozone layer a similar concept, the 'Ozone Depleting Potential' (ODP), was regarded as very useful. However, as we noted earlier, the use of GWPs was more controversial, since the calculation requires information on the atmospheric lifetime of the gases as well as knowledge of the fate of the gas. Both of these are not well known. In the 1994 report the GWPs were revised. It was concluded that the typical uncertainty of the GWP values is ± 35 per cent relative to the CO_2 reference.

For a hundred year time horizon the IPCC 1994 report gave the following GWPs referenced to the absolute CO_2 GWP:

methane	24.5
nitrous oxide	290
CFC-11	4000
CFC-12	8500
HCFC-22	1700

By multiplying the GWP by the estimated annual emissions typical of the 1980s the future warming potential of current emissions can be estimated. Such calculations show that on a 20 year time horizon CH_4 emissions are comparable with those of CO_2. Current emissions of HFCs are small, making their contribution to future radiative forcing modest at present, but if these emissions were to increase in the future, their contribution to future radiative forcing would become more important.

On the basis of the strong statements made by IPCC Working Group 1, the issue of climate change was given even more priority on the political agenda and negotiations for the climate convention began. The other two working group reports were less persuasive. This was blamed in part on the fact that the working groups had worked in parallel, i.e. the climate scenarios were not readily available for the kinds of impacts assessments that Working Group 2 was doing and for the development of response strategies in Working Group 3.

The success of WG1 was partly because it was regarded as 'scientific' and hence detached from political bias, and in part because the late 1980s saw a series of much publicized abnormal climate events across the globe. Notable was the very hot and dry summer of 1988 in the United States, an event that led to Congressional hearings and much political attention to scientific predictions. But WG1 was also weak because it operated on the basis of consensus, it was therefore inherently conservative scientifically speaking, and it sought to detach itself from the negotiations taking place in the all important Intergovernmental Negotiating Committees that led to the Rio Convention. As Lanchberry and Victor (1995, 33) observe, 'although non-controversial conservatism may be a good feature in a body designed to reflect the balance of scientific thought, it is not necessarily a good feature in a body called on to inform a negotiating process'.

The other working groups were far less successful, partly because it is undesirable to separate adjustment from impacts, and also to isolate analysis from policy prescriptions. The fact that the Russians chaired WG2 and the Americans chaired WG3 was also politically dictated, and virtually guaranteed that both groups would be ineffective in informing the all important negotiation process. It is also a sad reflection on the state of the social sciences at present, that their practitioners cannot produce a coherent view of what causes climate change in terms of human needs and wants and associated economic and technological 'drivers', what should be done about these, and what would be the social, political and economic consequences. Compared with the consensus-oriented format of the natural scientists, the social scientists have behaved in a more disorganized and non-credible manner.

The work of the IPCC has been criticized for representing rich world agendas and interests and for limiting the extent to which the industrialized world needs critically to re-examine and reconstruct its own deep commitments in order adequately to address the contemporary debate. Wynne (1994) has questioned whether the deeper meanings on which the scientific knowledge is built serve to constrain the vision of what is at stake, socially and culturally. Furthermore, he points out that the acceptance of the precautionary principle implies that scientific uncertainty can eventually be made acceptably certain, which means that policy makers are encouraged to limit the pain for a period during which they hope that the scientific mists should clear. This is one reason amongst many why delay is such a refuge for the big emitters and the fossil fuel owners. It is also a reason that is often assisted by statements from the scientific community that general circulation models can be made sufficiently precise to allow for more reliable prediction.

After an update in 1992, it was proposed that the IPCC produce a full review in late 1995. However, to help negotiations at the first conference of Parties to the Convention in Berlin in 1995, some parts of the 1995 assessment were published after lengthy discussions in 1994. Discussions focused on rules of procedure, requiring that governments have adequate time to review reports before they are published. The executive summaries of the reports were discussed 'line-by-line' in plenary sessions and politics continued to dominate some of these discussions. Oil producing countries and some industrial organizations regularly entered these 'scientific' debates to protect their own interests.

At the time of writing, the 1995 IPCC summary for policy makers was still in the final draft form (July 1995). However, the final version is unlikely to depart very far from the wording that

> the observed increase (in global temperature) over the last century (0.3%–0.6%) is unlikely to be entirely due to natural causes and that a pattern of climatic response to human activities is identifiable in the climate record.
> (IPCC, 1995, 6, 1)

This is the closest the Panel will go to claiming that human-induced global warming has started. This is a significant statement, given the collective unwillingness to be so precise before. But it is still unlikely to move the political machinery, at least for a few years yet, into any significant new initiatives over greenhouse gas reduction.

The writing teams for the IPCC 1995 reports included more participants from developing countries, to counter criticism that the first IPCC reports were dominated by scientists from developed countries. Although the IPCC is seen as an inherently conservative institution in which it is difficult to bring in extreme views, the consensus documents have played a major role in influencing the policy process. Its future is now somewhat uncertain. In many ways, the conventions of 'science' have run their course. The next phase of climate change 'science' may be far more overtly politicized, and more integrally locked into the CoP mechanisms. Most will depend on the effectiveness of the CoP secretariat and the role played by its subsidiary assessment panels.

CLIMATE CHANGE POLITICS

As described in Figure 1.4, the issue of climate change was moved onto the political agenda as a result of the international conference held in Villach in 1985. At this conference the invited scientists reached a consensus about the seriousness of the problem of the enhanced greenhouse effect and suggested that it was time for them to begin talking to policy makers. The conference also recommended that an advisory group be set up to provide regularly updated evaluations of the issue to the heads of WMO, UNEP and ICSU.

This group, consisting of six experts and called the Advisory Group on Greenhouse Gases (AGGG), was established and met for the first time in July 1986. One member of the AGGG, Gordon Goodman, then of the Beijer Institute in Stockholm, was interested in pursuing the goal of the 1985 Villach conference to establish a dialogue between scientists and policy makers on the issue of climate change. It was agreed at the meeting in July 1986 that Goodman, with the help of an international steering committee, should organize two workshops to be held under the auspices of the AGGG in 1987. These workshops were held in Villach (September 1987) and Bellagio (November 1987). They produced an assessment of the potential impacts of the enhanced greenhouse effect and a prioritized list of steps to be taken in response to the issue of climate change (Jäger, 1988).

A large proportion of those involved in the organization of the Villach and Bellagio workshops got involved in the organization of the Toronto conference held in mid-1988. The government of Canada had offered to host a conference on 'The Changing Atmosphere' during a hearing for the Brundtland Commission. The conference played a major role in raising international awareness of the issue of climate change and stimulated action in a number of countries. It is of significance, however, that the Toronto conference was not a governmental conference. The participants were invited in their own right and not as delegates of countries. As we have noted, the conference happened to coincide with a major drought in the US Midwest. Hearings were being held in Washington about the drought and a respected atmospheric scientist testified that this was evidence of global warming. This became headline news and media attention to the Toronto conference increased. The large media attention and the presence of a pro-active group of representatives of environmental non-governmental organizations were two important influences on the outcome of the conference. The conference statement began by pointing out that the threat of anthropogenic climate change was second only to the threat of nuclear war. The statement called on all developed countries to cut their CO_2 emissions by 20 per cent from the 1987 levels by the year 2005. This 'Toronto goal' was subsequently discussed in many countries and implemented in some, for example Austria. After the Toronto conference, a study team in Germany was asked to investigate whether the 'Toronto goal' was achievable in Germany. It concluded that an even more ambitious reduction was achievable and the Enquete Commission recommended a 30 per cent reduction and the Cabinet agreed to a 25–30 per cent reduction goal (see Chapter 7).

A few months after the Toronto conference the Intergovernmental Panel on Climate Change (IPCC) was established. The discussions about the character and role of such a panel began in mid-1987. One of the reasons given for setting up the panel was that the issue of climate change must be dealt with by governments and not just by a group of interested scientists. The three working groups of IPCC reported on their results just before the

Second World Climate conference (SWCC), held in Geneva in November 1990. The strong statements in Working Group 1, the 'science' group, were endorsed by the SWCC. However, since the SWCC was split into two parts, a scientific and a ministerial conference, the conference statement of the scientific part called for serious policy responses to the issue, while the Ministerial Declaration was much less committed to action.

Shortly after the SWCC but not as a direct result of the conference, the United Nations General Assembly called for the establishment of an Inter-governmental Negotiating Committee (INC) for a Framework Convention on Climate Change (FCCC). The INC/FCCC met for the first time in February 1991. The aim was to negotiate a convention for signature at the UN conference on Environment and Development in Rio de Janeiro in June 1992. The negotiations began slowly with a series of procedural questions. The basic idea was to negotiate a framework convention, as had been done for long range transboundary pollution and stratospheric ozone. The European Community and the small island states favoured the establish-ment of a target and timetable for limiting the emissions of greenhouse gases, while the United States and the oil producing states were against this procedure. Shortly before the last session of the INC in April 1992 the United States and the UK brought in the compromise formulation which is now in Article 4 of the FCCC with the non-binding call for an attempt to return to 1990 emissions of CO_2 and other greenhouse gases not controlled by the Montreal Protocol. The second controversial point in the INC negotiations was on financial issues, with differences of opinion between developed and developing countries on who should be in charge of the financial mechanism set up to support developing countries in imple-mentation of the convention.

The UN FCCC is discussed in more detail in Chapter 2. It is important to note here that the Convention was signed by more than 150 states and the EC at the Rio conference in June 1992. It had been negotiated in a remark-ably short period of time due to the pressure to have something to be signed at the UNCED. Although it is called a 'Framework Convention', the UN FCCC contains more than other similar framework conventions for long range transboundary air pollution and protection of the stratospheric ozone layer. The UN FCCC has more extensive commitments than the other conventions: it sets up a number of new institutions and provides financial assistance and technology transfer for developing countries. The institutions established by the UN FCCC include the Conference of Parties (CoP), subsidiary bodies on science and implementation, reporting and review procedure. The Convention also states that '[t]he Parties should take pre-cautionary measures to anticipate, prevent or minimize the causes of climate change and mitigate its adverse effects', thus establishing the precautionary principle as a guiding principle in the international response to climatic change.

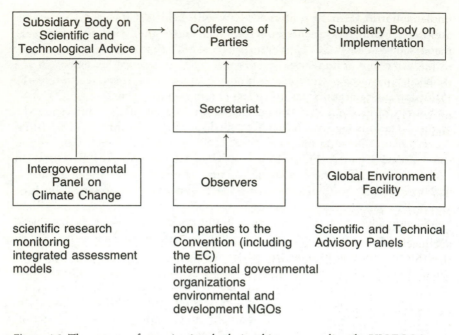

Figure 1.5 The pattern of organizational relationships surrounding the UN FCCC process

Figure 1.5 summarizes the relationships between the various organizations associated with the UN FCCC process. As noted above, the future of IPCC remains slightly unclear, though it is likely to continue in place, with much restructuring of its assessment procedures. That will be aided by better development of integrated modelling, and by a closer infusion of political negotiations in the policy evaluative component of the IPCC's work. Many will regret what they see as 'interference', but the outcome could result in a more effective IPCC role, if the transition is skilfully managed.

After the signing of the Convention in Rio, several steps were foreseen. Fifty ratifications or similar approvals by signatories of the Convention were required for it to become legally binding. This was set to take place three months after the minimum number of ratifications had been deposited. Thereafter, the first CoP had to be held within a year. At the Rio conference the Federal German Chancellor Kohl invited the Parties to hold the first CoP in Germany. After the 50th ratification was deposited on 21 December 1993 the Convention came into force on 21 March 1994. The first CoP was held in Berlin from 28 March to 7 April 1995.

As described in more detail in Chapter 2, the Convention sets out a substantial part of the agenda for the first CoP, including the discussion of the adequacy of the commitments made in the FCCC and the criteria for joint implementation. However, as the INC negotiations for the FCCC came to a close, a number of countries agreed that rather than lose time between the signing of the Convention and the first CoP, the INC should continue to meet and negotiate in preparation for the first conference. The pressure to reach an agreement before the UN Conference on Environment and Development in Rio meant that the Intergovernmental Negotiating Committee (INC) for the UN FCCC avoided a number of contentious issues and left them to be dealt with by the CoP.

FROM RIO TO BERLIN

The UN FCCC, as discussed in Chapter 2, commits developed countries and those with economies in transition to aim to reduce their CO_2 emissions to 1990 levels by the year 2000. The CoP was given the task of reviewing the adequacy of this initial agreement, summarized in Table 1.3. Meetings of the INC in 1994 indicated that parties from developed countries did not generally feel that the commitments would be adequate to meet the ultimate goal of the Convention. This led to a general discussion of the necessary next steps, ranging from strengthening the commitment made in the FCCC to agreement on a protocol with targets and timetables for emissions reductions. Six months before the first CoP, the Association of Small Island States (AOSIS) submitted a draft protocol for consideration at the first CoP and Germany submitted a document containing elements of a protocol. However, as the CoP approached, it was clear that the most likely scenario was that the first CoP would, at most, call for the negotiation of a protocol by a certain date. The unease with a target and a timetable was too great to ignore. The cards remained firmly in the hands of the Americans and the oil producing states.

A second controversial topic dealt with by the INC between Rio and Berlin was 'joint implementation' (JI). Included in the UN FCCC, JI can allow one country to help another to implement a project or change a policy with the result that the mutual sum of greenhouse gas emissions is lower than it would otherwise have been. Some countries and actors (especially some environmental NGOs) expressed serious concerns at the use of JI. On the one hand JI appears to offer market-oriented countries the chance to implement emissions controls where they are cheapest. Developing countries and some NGOs saw JI as an effort by the rich countries to avoid making substantive commitments and undergo necessary structural changes. As Victor and Salt (1994) paraphrase the view of opponents: 'JI is a loophole through which [the rich countries] will double count existing flows of aid and international investments'. After two years of debate about JI, some developing countries began to change their position towards limited acceptance of the concept. The CoP was given the job of deciding on the issue of

Table 1.3 Stronger emission curbs needed to meet Climate Convention target

Most countries must adopt additional measures to cut emissions of carbon dioxide and other greenhouse gases in order to meet the Climate Convention target of returning output of these gases to 1990 levels by the year 2000, according to a new report released by the Climate Action Network. The report is based on independent evaluations of the national climate plans of 21 advanced industrialized countries. The report emphasizes that emissions from cars and trucks are rising particularly rapidly, and that much more effort is needed to increase vehicle efficiency and shift to other means of transportation. While the national plans are found to be very disappointing overall, there are a number of promising specific policy initiatives that have been implemented in one or more countries.

Replicating the best initiatives from each country could begin to remedy the deficiencies revealed in the first route. The following chart is a brief summary of the findings of the NGO evaluations of OECD countries' National Plans for Climate Change Mitigation.

Country	Target	Will national target be achieved with current measures?	Projections beyond 2000	Unique measures
Australia	'Stabilize GHG emissions based on 1988 levels, by the year 2000 and to reduce these emissions by 20% by the year 2005 …'	No. The best available projection is for emissions to increase by 6.6% between 1990 and 2000	No	
Austria	Toronto target	No, government report states that could stabilize Austria's CO_2 emissions at the 1990 level by 2000–2005 will bring in further measures to implement Toronto target	2005	
Belgium	5% CO_2 reduction by 2000 (1990 baseline)	Only if EU energy/CO_2 tax is implemented (now unlikely)	No	
Canada	Federal target: Toronto target	No	No	Supply of renewables (but underfunded)
Denmark	Toronto target, and on EU level a 5% CO_2 emissions reduction by 2000 to 1990 levels, will implement further measures to implement Toronto target	No, but 18% reduction will be met	2005	Integrated combination of market and non-market measures; CO_2/energy taxes; application of a CO_2/energy tax on the commercial sectors without adverse effects for the competitiveness of Danish products; cogeneration
France	2 tonnes of CO_2 per inhabitant by 2000	No	2010	
Germany	25–30% CO_2 reduction by 2005	No, but will achieve a 10–20% reduction	2020	Kerosene/mineral oil tax for air traffic in Europe
Ireland	20% increase over figures for 1990 in accordance with the EU burden-sharing arrangement under the EU stabilization target	Yes, emissions will rise	No	Tentative steps in wind, wave and hydro power
Italy	Stabilization of CO_2 emissions at 1990 levels by 2000	No	No	

Country	Target/objective	Achievement	Year	Measures
Luxembourg	Stabilization of CO_2 emissions by 2000 and Toronto target	Yes, will achieve Toronto target	No	Alternatives to HFC use, payment for power from small CHP plant
Japan	Stabilization of per capita CO_2 emissions at approximately the 1990 level by 2000	No	No	
Netherlands	CO_2 3–5% reduction in the year 2000, 1989–1990 base; CH_4: 10% reduction in 2000, 1990 base; N_2O: Stabilize in 2000, 1990 base; NO_X: 55% reduction in 2000, 1988 base; NMVCOs: 60% reduction in 2000, 1988 base; CO: 50% reduction in 2000, 1990 base	CO_2: Yes, if energy tax implemented; CH_4: Yes, 25%; N_2O: No, 5% increase; NO_X: No, 30% reduction; CO: not predicted; NMVOCs: not predicted	2010	Planning, research and public awareness; possible national CO_2/energy tax
New Zealand	Return net anthropogenic emissions of CO_2 by 2000 to 1990 levels; Toronto target as objective	If net emissions approach not allowed, then no	2005	
Norway	Stabilization of CO_2 emissions by 2000, 1989 base	No, 12% increase	No	CO_2 tax, including emissions from oil and gas extraction. Study the possible shifting the burden of taxation from labour to taxation of polluting activities and resource use
Portugal	40% CO_2 emissions increase by 2000, 1990 base in accordance with the EU burden-sharing arrangement under the EU stabilization target	Yes, emissions will increase	2010	
Spain	25% CO_2 emissions increase by 2000, 1990 base in accordance with the EU burden-sharing arrangement under the EU stabilization target	Yes, emissions will increase	No	
Sweden	Stabilization on CO_2 emissions from fossil fuels by 2000, and decline after that, 1990 base. 30% reduction of methane emissions from disposal of wastes by 2000, 1990 base	No	2005	CO_2 and sulphur tax; Energy efficiency procurement programme
Switzerland	'Stabilization, by the year 2000 of the consumption of fossil fuels and of resulting CO_2 emissions at their 1990 level, followed by a gradual reduction'	Yes	2005	Constitutional requirement to shift transalpine freight transport from road to rail within the next 20 years; immediate stabilization of transalpine road capacity; Energy 2000
United Kingdom	To return each of the main GHGs to 1990 levels by the year 2000	Maybe	No	Energy Savings Trust
United States	To reduce net emissions of GHGs to 1990 levels in the year 2000	No	2010	Market aggregation/transformation measures (superefficient refrigerator programme), least-cost planning

JI but even a well-monitored pilot phase will clearly only postpone serious debate on its ultimate role within the UN FCCC.

The other main debate with regard to the UN FCCC concerned financial and technological transfers. The FCCC commits OECD countries and the EU to provide 'new and additional financial resources to meet the agreed full costs incurred by developing country Parties' for national reports and some other activities but not for emissions controls by developing countries. Some financial transfers have been made but little has been done to transfer technology. Progress in negotiations on transfer questions was slow and dominated by major north–south differences in position.

The matter of financing climate protection strategies illustrates clearly the degree of commitment between the major players to the whole ideal of stabilizing climate futures through international cooperation and mutual assistance. Countries simply see this matter very differently. Many of the poor nations, as cited in Rahman *et al.* (1993), are unconcerned about climate change. Their greenhouse gas emissions are very modest on a per capita basis, and they cannot clearly visualize how future climate change will stress their vulnerable peoples any more than a host of other causative factors, many of which, they believe, are due to imbalances of trade, debt and subservience to international financing institutions. For them the key to the future lies in massive technology transfer on a 'know-how' basis, carried out at local level through NGOs and regional community action.

India and China have yet to declare their full hand on this difficult matter. Both are edging towards some form of bilateral carbon 'trade' based on clearly defined packages of technical and financial assistance for their economic restructuring. Donor nations are wary that their gifts will not be used to reduce greenhouse gas emissions, so want time and actual experience to see how far an effective regime that is independently verifiable can be put in place.

The one mechanism that is established to do the job of post-Rio financing is the Global Environment Facility (GEF) (see Jordan (1994) for a comprehensive assessment). This is a creature of the World Bank, the UN Environment Programme and the UN Development Programme. Its purpose, established before Rio but greatly strengthened afterwards, is to provide a financing mechanism to aid those countries who wish to meet global goals but would not do so purely on their own initiative. This may be because they lack the resources, and more likely it is because the costs involved provide a global gain; then there is an element of additionality to their level of investment. For example, an investment in a gas conversion technology replacing inefficient coal burning would clearly be a global as well as a local gain. Some of this conversion would be a national financing matter, possibly with aid. The rest should be a GEF investment. One can guess that the precise relationship between the two is a matter of some dispute. This is why the

GEF has a panel on technical and scientific advice, including a number of economists.

GEF operates to its own financing rules, while UN FCCC is mainly the concern of environmental and/or foreign ministries. This is not a happy relationship, but the heavy administrative cost and bureaucracy of GEF mean that it will stay in place, attached to the UN FCCC. In recent years the GEF has put more emphasis on small grants to NGOs so it looks as if there will be greater flexibility in this critical area.

The Berlin CoP meeting was reminiscent of Rio. Some felt that it ducked all the key issues, others were satisfied that renewed commitments were made, with a promise of further reductions of greenhouse gases beyond 2000. The small island states were bitterly disappointed, as were the climate action networks. Realistically, as we note below, climate change has passed the novelty stage; it no longer claims broad public attention. The serious consequences appear far away, and recession, unemployment and excessive national public expenditure beleaguer most governments. There is not the political stomach for tough precautionary action while real energy prices fall, and the cold war is over. Russia has masses of fossil fuel reserves it is willing to sell for hard currency, and the enticing promise of joint implementation offers a lifeline for trade, aid and job creation in both the north and the south. Yet a significant majority of the CoP did want further development of the UN FCCC.

What did Berlin accomplish? Six outcomes stand out

1   A 'Berlin Mandate', or declaration of intent, to create a formal protocol under the Convention that would lead to GHG reductions according to an agreed timetable (see Appendix 2 for the final text). This does not involve any new obligations for the developing countries. This process was helped by the formation of a 'green group' of EU and developing countries, coupled to major international NGOs. The Mandate itself will involve a sophisticated apparatus of monitoring and reporting (see Chapters 3 and 6), as well as substantive new measures. The evolution of this phase will be important since the 1990–2000 stabilization programmes have virtually avoided any serious restriction on business as usual in the north. What that target will be is at present not certain. The Toronto target of 20 per cent reduction on 1988 emission levels by 2005 seems remote. This will be carried forward by a 'process' involving an *ad hoc* working group aimed at producing 'qualified reduction objectives within specified time-frames'.

2   A pilot phase for joint implementation, until 2000, but not involving any credits to donor nations. This is designed to test out the practicability of agreements and suitable verification procedures. In theory, developing countries have much to gain, via transfers of technology, and management skills, but donor nations benefit little unless they hope (or expect) that in

the final protocol these experimental schemes will be translated, if only in part, into GHG 'offsets'.

3 A resolution on technology transfer, according to which the Parties must provide information on environmentally friendly technology, and the Secretariat (see 4 below) must produce an inventory. Furthermore the Parties must report regularly on their own initiatives at technology transfer to developing countries. The 23 OECD Member States, plus the EU, announced a climate protection technology initiative.

4 A Climate Change Secretariat will be established in Bonn. This is what the Germans really wanted, namely the prestige of an international organization of high profile on their own soil. The patronage of a key, and committed, big emitter, with considerable resources of cash and other support for the Secretariat, should not be underestimated.

5 Voting procedures were left to the next meeting of CoP – in Montevideo in 1996 – to sort out. Initially the OPEC states insisted on unanimity for all agreements. Normally two-thirds majorities are required for major international agreements. A delayed outcome on this tricky area allowed important decisions to be made at Berlin, and tempers to cool. Right now, the provisional rules of procedure that permit agreement by broad consensus hold.

6 The NGOs remain important players, especially the business community. There is reason to believe that the major multinationals are willing to accept a protocol because they prefer internationally agreed rules, clear mandates, and an open negotiating process in which they can exert leverage. In any case most are undergoing various kinds of eco-audits and environmental management procedures. These require that they set energy efficiency targets, take climate change into account in all their activities, and seek to ensure that their suppliers and distributors are doing the same.

THE ISSUE ATTENTION CYCLE AND REGIME EFFECTIVENESS

In terms of the environmental issue attention cycle discussed in Chapter 4, the debate by informed insiders on the issue of the enhanced greenhouse effect and its impacts was ongoing in the 1970s and 1980s. By the mid-1980s, however, a concerted attempt was made to assess the state of knowledge. The international assessment that provided the background material for the 1985 Villach conference was broadly based and subject to an international scientific peer review. The scientific consensus reached at the Villach conference about the potential seriousness of the problem can be seen as the phase of 'discovery' of the issue. The rise of media attention began shortly thereafter. For example, in Germany the number of articles per year published in a selection of two magazines and one weekly newspaper increased from less than 20 in 1985 and 1986 to almost 100 in 1990 (see Chapter 7).

By the time that the IPCC released its reports in the middle of 1990 the media interest was very high and stimulated by the strong statements made by IPCC Working Group 1. Public opinion polls undertaken for the Gallop organization worldwide, and the Eurobarometer programme for Europe in particular (Hofrichter and Klein, 1991; Rudig, 1995) show great differences in the level of anxiety over climate change as well as ozone depletion, chemical pollution generally, and local degradation of environmental amenities. Figure 1.6 suggests that concern over global warming is generally less great than anxieties about the depletion of stratospheric ozone, but that there is great variation from one country to another. More to the point, the evidence from these polls shows that there is both considerable confusion in the public mind as to what global warming is, and what should be done about it (Rudig, 1995). This in turn means that the public is ambivalent over policy measures, especially new taxation. Willingness to accept a carbon tax rises if the public can be assured that this will be offset by reductions in other taxes, such as income tax or consumer taxes, and that at least part of the revenue is geared to ameliorating the effects of climate change (Rudig, 1995, 23–24). There is also evidence that the public may be sceptical that any single policy initiative would have any real effect, but that a concerted series of coordinated measures was thought to be politically unlikely.

Between 1990 and 1994 there were no major scientific developments which would substantially alter the conclusions of the IPCC Working Group 1. The Summary for Policymakers of the 1994 Working Group 1 report 'affirmed the basic understanding of climate change contained in the 1990 report' and stated that 'new findings add detail to our knowledge but do not substantially change the essential results concerning radiative forcing of climate which appeared in the 1990 or the 1992 IPCC scientific assessments'. Much debate between 1990 and 1994 centred on the costs and benefits of climate change responses. On the one hand, some model calculations have shown an ambivalent effect on atmospheric greenhouse gas concentrations from immediate, drastic, emissions reductions. As Figure 1.2 suggests, there might even be a cooling effect from human-caused sulphate pollutants, enough to reduce possible warming by as much as a degree centigrade. Therefore, it is possible that climate warming has been masked during recent years by sulphate particles. Measures to reduce acid deposition will thus have an effect on the climate change issue. All in all in the 1990s doubts remain about the seriousness and timing of climate change effects, doubts sufficient to stay the hand of the concerned, and provide succour to the disbelievers (see, for example, Beckerman, 1995).

There is thus no doubt that by the mid-1990s attention to the issue of climate change had moved into the phase of 'counting the costs'. Media attention to the issue generally dropped after the Rio conference. The hot

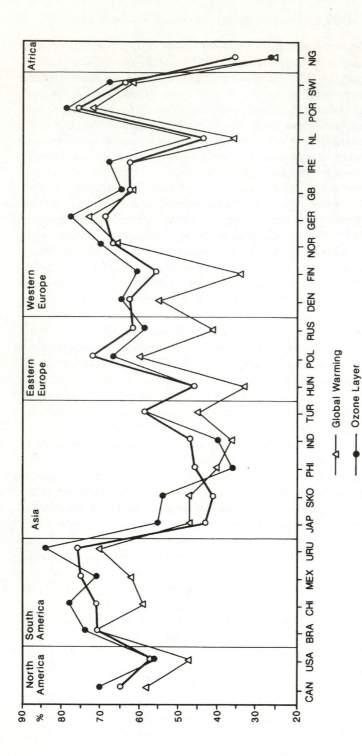

Figure 1.6 Concern over global warming and ozone depletion compared with an average of concern over a range of environmental problems

Source: Rudig (1995, 17)

summer in Europe in 1994 was discussed more with reference to the associated 'summer smog' (health-affecting ozone levels near to the ground) than as a forewarning of summers to come as a result of the increasing greenhouse effect.

As long as attention was high and scientific issues were the focus, it was easier to reach consensus, even on weak targets. However, negotiations on policies and measures are more difficult in international and national arenas. As soon as measures such as carbon or energy taxes are discussed, arguments about industrial competitiveness are raised. During the period of economic recession such views are taken seriously in industrialized countries.

Whilst a measure of international and national interest in climate change seems to have lessened since the signing of the UN FCCC, the case studies that follow in this book show that local initiatives have developed and that in a number of countries it became clear that climate protection policy cannot be pursued for its own sake. Achievement of climate protection goals appears more likely if policies are formulated within the implementation of other policy agendas with respect to issues such as economic restructuring, fiscal reform and local sustainability initiatives. Yet, there is also need for some kind of global organizing theme around which all nations and peoples can sense solidarity, commitment and the joy of working alongside everyone else to retain the habitability of this planet. The UN FCCC seems to provide that vital role. Imperfect it undoubtedly is, both as a legal document and as a call to action. But without it, there would be no collective commitment to responding to climate change, its causes, its consequences, and its politics. For better or for worse, that is an extraordinarily important beginning along a road that is bound to be long, bumpy, deafening with the din of dispute and awash with the tears of frustrated dreams.

Yet the irony of all this is that any international convention sets only a very limited target for initial action, with usually only modest amendments in the following renegotiations, unless the science becomes much more bold, and public concern more politically articulated. To a degree that happened with the removal of stratospheric ozone depleting chemicals under the Montreal Protocol. For climate change, with so much more at stake, and decisions embedded in the heart of government and the economy, such a future is far less likely. Simply because modest but agreed targets are in place usually means that there is almost no political incentive to go beyond them, even when to do so might be both politically painless and economically efficient. Such is the entrapment of global convention.

REFERENCES

Agarwal, A. and Narain, S. (1991) *Global Warming in an Unequal World: A Case of Environmental Colonialism*, New Delhi: Centre for Science and Technology.
Beckerman, W. (1995) *Small is Stupid*, London: Duckworth.

Bolin, B. (1994) 'Science and policy making', *Ambio* 23(1), 25–29.

Bolin, B., Döös, B. R., Jäger, J. and Warrick, R.A. (eds) (1986) *The Greenhouse Effect, Climatic Change and Ecosystems*, SCOPE 29, Chichester: Wiley.

Clayton, K.M. (1994) 'The threat of global warming', pp. 110–130 in T. O'Riordan (ed.) *Environmental Science for Environmental Management*, Harlow: Longman.

Dowlatabadi, H. (1994) ' Integrated assessment models of climate change: an incomplete overview', *Energy Policy* 22, 79–91.

Fankhauser, S. (1995) *Valuing Climate Change: The Economics of the Greenhouse Effect*, London: Earthscan.

Grubb, M. (1995) 'European climate change policy in a global context', pp. 41–50 in H. Bergeson (ed.) *Green Globe Yearbook 1995*, Oxford: Oxford University Press.

Haas, P., Keohane, P.O. and Levy, M.A. (eds) (1993) *Institutions for the Earth: Sources of Effective International Environmental Protection*, Cambridge, MA: MIT Press.

Hammond, A.L., Rodenburg, E. and Moonmaw, W. (1991) 'Calculating national sustainability for climate change', *Environment* 33(1), 10–15, 33–35.

Hofrichter, J. and Klein, M. (1991) 'Evolution of environmental attitudes towards environmental issues 1974–1991', Brussels: DG XI Environmental Unit.

Houghton, J.T., Jenkins, G.J. and Ephraums, J.J. (eds) (1990) *Climate Change. The IPCC Scientific Assessment*, Cambridge: Cambridge University Press.

Houghton, J.T., Callander, B.A. and Varney, S.K. (eds) (1992) *Climate Change 1992 The supplementary report to the IPCC scientific assessment*, Cambridge: Cambridge University Press.

IPCC (1994a) 'Radiative Forcing of Climate Change: The 1994 Report of the Scientific Assessment Working Group', Geneva: WMO–UNEP.

IPCC (1994b) Radiative Forcing of Climate Change: Summary for Policymakers, Intergovernmental Panel on Climate Change.

IPCC (1995) Draft summary for policymakers of the synthesis report, Mimeo, Geneva: IPCC.

Jäger, J. (1988) 'Developing policies for responding to climatic change', WMO/TD No. 225, Geneva: World Meteorological Organization.

Jasanoff, S. (1990) *The Fifth Branch: Science Advisers as Policymakers*, Cambridge, MA: Harvard University Press.

Jordan, A. (1994) 'Paying the incremental costs of environmental protection: the evolving role of the GEF', *Environment* 36(6), 12–20, 31–36.

Lanchberry, J. and Victor, D. (1995) 'The role of science in the global climate negotiations', pp. 29–39 in H. Bergeson (ed.) *Green Globe Yearbook 1995*, Oxford: Oxford University Press.

Meyer, A. (1995) 'The results of changing two bases of valuation in the global cost benefit analysis', London: Global Commons Institute.

Rahman, A., Robins, N. and Roncerel, A. (eds) (1993) 'Exploding the Polulation Myth: consumption versus population. Which is the Climate Bomb?' Brussels: Climate Network Europe.

Ramanathan, V. (1975) 'Greenhouse effect due to chlorofluorocarbons: climatic implications'. *Science* 190, 50–52.

Read, P. (1994) *Responding to Global Warming: The Technology, Economics and Politics of Sustainable Energy*, London: Zed Books.

Rudig, W. (1995) 'Public Opinion and Global Warming', Strathclyde Papers on Government and Politics, No. 101, Glasgow: University of Strathclyde.

Subak, S. and Clark, W.C. (1990) 'Accounts for greenhouse gases: towards the design of fair assessments', pp. 78–93 in W. C. Clark (ed.) *Usable Knowledge for Managing Global Climatic Change*, Stockholm: The Stockholm Environment Institute.

Victor, D.G. and Salt, J.E. (1994) 'From Rio to Berlin: managing climate change', *Environment* 36 (10), 6–16, 25–32.

Wang, W.C., Yung, Y.L., Lacis, A.A., Mo, T. and Hansen, J.E. (1976) 'Greenhouse effects due to man-made perturbations of trace gases', *Science* 194, 685–690.

World Climate Programme (1986) 'Report of the International Conference on the Assessment of the Role of Carbon Dioxide and of Other Greenhouse Gases on Climate Variations and Associated Impacts', WMO-No. 661, Geneva: World Meteorological Organization.

Wynne, B. (1994) 'Scientific knowledge and the global environment', pp. 168–189 in M. Redclift and T. Benton (eds) *Social Theory and the Global Environment*, London: Routledge.

2

LEGAL OBLIGATIONS AND UNCERTAINTIES IN THE CLIMATE CHANGE CONVENTION

Elizabeth J. Rowbotham

As was discussed in the first chapter, the *United Nations Framework Convention on Climate Change* (*UN FCCC*)[1] represents a 'delicate balancing' of many different political and economic interests and many complex scientific issues. The political and economic interests reflect, and are highlighted by, the differences amongst the various Parties with respect to levels of development; the relative priority attached to environmental and developmental concerns; the anticipated impacts of climate change; and the equitable apportionment of responsibility for the costs and causes of climate change. The scientific issues include the relationship between greenhouse gas accumulation and the climate system; the inter-relationship between atmospheric gases; and the nature, degree and location of anticipated climate impacts. Appendix 1 provides the full text of the Convention.

Although a general north–south (or developed–developing country) divergence amongst issues and interests can be identified with the north promoting environmental protection priorities and the cost-effectiveness of measures and the south promoting developmental priorities and historical responsibility, alliances have also been formed between countries which do not follow strict north–south divisions but, rather, are based on considerations such as vulnerability to sea-level rise;[2] reliance on fossil fuel production;[3] and the potential for development opportunities.[4] All of these alliances co-exist with countries promoting the diversity of their interests through more than one group. The combination and permutation of interests and issues represented by the alliances underlay the negotiating process and are carefully balanced within the *UN FCCC*. Thus, and for example, the *UN FCCC* recognizes: the need to reduce anthropogenic greenhouse gas emissions;[5] that developed countries should take the lead;[6] that development considerations of developing countries are legitimate and allowances for these considerations need to be made;[7] that certain countries are particularly vulnerable to the impacts of climate change;[8] that some

32

countries rely heavily on the income from fossil fuel production and/or consumption;[9] that the measures to address the causes and effects of climate change should be cost effective;[10] and that the precautionary principle should be applied.[11]

However, owing to the diversity of interests which needed to be balanced, precise wording or expression of concepts was not generally possible in the *UN FCCC* and the text is fraught with ambiguities for which interpretative guidance will eventually be necessary if implementation is to be effective and/or non-acrimonious. As such interpretative guidance will likely emanate from the Conference of Parties (CoP), the interpretation itself will be the result of negotiated agreement and, hence, will continue to reflect the political and economic interests of the Parties as they evolve over time.

The present chapter will discuss, in light of the interests and issues just outlined and with a view to the case studies which follow, those commitments within the *UN FCCC* which are likely to have the greatest influence on domestic policy formation and implementation. These include the ultimate objective contained in Article 2 as well as the commitments concerning: reduction/stabilization; joint implementation; burden sharing; education and awareness; implementation; and compliance.

THE OBJECTIVE

Article 2 defines the 'ultimate' objective of the *UN FCCC* and it is this Article which identifies what the final outcome adherence to the *UN FCCC* should achieve. In its entirety the Article states that

> The ultimate objective of this Convention and any related legal instruments that the Conference of the Parties may adopt is to achieve, in accordance with the relevant provisions of the Convention, stabilization of greenhouse gas concentrations in the atmosphere at a level that would prevent dangerous anthropogenic interference with the climate system. Such a level should be achieved within a time frame sufficient to allow ecosystems to adapt naturally to climate change, to ensure that food production is not threatened and to enable economic development to proceed in a sustainable manner.

There are two main ways in which to approach the interpretation, and subsequent application, of this Article. The first could be called an 'expansive' approach while the second one could be called a 'restrictive' approach.

Under an expansive interpretation, and with application of the precautionary principle referred to in Article 3(3), Article 2 could conceivably operate in a manner which would restrict the activities of the Parties to those which do not increase the concentration of greenhouse gases and which promote sustainable development. Such an interpretation would require the Parties, in particular the developed country Parties, to make significant reduction in their greenhouse gas emissions, especially with regard to carbon dioxide (CO_2). These reductions, possibly in conformity with the 60 per cent (net) emission reduction level suggested by the Intergovernmental Panel on Climate Change (IPCC) in 1990,[12] would serve to reduce the dependency on fossil fuels and would promote energy efficiency. Furthermore, and, *ceteris paribus,* owing to the relationship between energy and consumption, significant reductions of greenhouse gas emissions, particularly in the short term, would likely have a dampening effect on consumption habits, themselves a matter of international environmental concern. In turn, these effects would have a major impact on the economic structure, as well as on social behaviour. In short, an expansive interpretation could likely provide the impetus through which alternative, more socially and environmentally friendly modes of economic and cultural behaviour could be fostered.

The fact that this Article is headed as the 'Objective'; that the Article is referred to throughout the Convention;[13] that the *UN FCCC* took less than 16 months to negotiate; and that it entered into force within 22 months of adoption is indicative of the pressing nature of the issue of climate change and its recognition as a matter of international importance.[14] These factors, together with the political and institutional dynamics discussed in Chapter 1, support an 'expansive' interpretation of this Article. That this approach is favoured by several parties is evidenced by the express and regular reference to the objective in the proposed protocol submitted by Trinidad and Tobago on behalf of the Alliance of Small Island States (AOSIS).[15]

On the other hand, a 'restrictive', or strict, interpretation of Article 2 can also be supported by both the Convention and the behaviour of Parties. Under a restrictive interpretation, state behaviour need not necessarily be significantly altered, particularly in the short term. There are three main arguments supporting a restrictive interpretation. First, Article 2 only requires greenhouse gases to be stabilized at levels that would prevent 'dangerous anthropogenic interference' with the climate system. In order to determine what these levels might be it is necessary to determine what 'dangerous' means. As this is an undefined term in the *UN FCCC*, it can be subject to varied interpretation by the Parties. Given the fact that scientific uncertainty persists with respect to the precise nature and degree of the impacts of climate change, what is considered dangerous by one Party may not be considered as such by another. This problem is compounded by the fact that different states and regions will experience different impacts. While the ongoing work of the IPCC may serve to reduce some of this scientific uncertainty, it is important to note that the IPCC does not have the legal power to compel the Parties to adopt or implement its findings or observations. In the event there is broad political support regarding the

IPCC's findings, it is through the Conference of Parties (CoP) that any new findings would be translated into legal form. Illustrative of this point is the fact that the IPCC's 60 per cent net reduction recommendations[16] were not adopted by the Intergovernmental Negotiating Committee (INC). Rather, the much less onerous commitments contained in Article 4(2)(a) and (b) were adopted (see below).

Furthermore, Article 2 deals with the 'stabilization of greenhouse gas concentrations in the atmosphere' and it is incumbent upon the CoP to achieve that stabilization 'in accordance with the relevant provisions of the Convention'. Given the latitude and flexibility regarding emission reduction built into the Convention on the basis of, *inter alia,* historical responsibility, development status, and cost-effectiveness together with the wording of Article 2, it is apparent that the Article refers to the *global* stabilization of greenhouse gas *concentrations* and not to an individual state, or state by state stabilization of greenhouse gas *emissions*. The former is the objective while the latter is the means to the realization of that objective.

Thus, and with reference to the IPCC, the IPCC's recommendations regarding what is a 'dangerous anthropogenic interference' need to be considered in a global context and it is beyond the scope of the IPCC's legal and political remit to make express comment on individual national contributions to the 'dangerous interference'. This remains a matter for negotiation among the Parties, through the CoP, and as such is subject to the bargaining, protection of vested interests, and horsetrading inherent in all negotiations.

Second, in the event a 'non-dangerous' level can be agreed upon, this level need not be achieved immediately. It need only be achieved before ecosystems or food production capabilities are harmed. Given the complexities of the science involved, at present it is difficult, if not impossible, to ascertain with precision what this time frame may be. Although it could be argued that application of the precautionary principle (Article 3(3)) negates the need for precision and scientific certainty in this regard, a counter-argument also exists which would restrict the application of precautionary measures to those that were cost effective (Article 3(3)). Owing to the fossil fuel dependency of developed country economies, strict application of the precautionary principle and large, short term emission cuts may not be cost effective, particularly if the secondary effects of emission reduction, such as unemployment, were factored into the equation.

The third argument in favour of a restrictive interpretive approach rather than an expansive one is based on the structure of the Convention. In that regard, reference is made to the chapeau attached to Article 1 which states that 'Titles of articles are included solely to assist the reader.' Thus, the identification of an 'objective', 'principles' or 'commitments' section within the Convention is largely superfluous and no section is to be given more interpretive weight or influence than another. However, given the cross-referencing to and between Articles, this argument is relatively weak. More persuasive is the argument that the 'framework' nature of the Convention and the ambiguities contained within was designed not only to balance and accommodate diverse interests, but also to protect them by necessitating the development of greater consensus through continued negotiation. Thus, the ambiguities reflect the limited nature and extent of consensus that existed at the time of the Convention's adoption.

Both the expansive and restrictive interpretative approaches to Article 2 are legally valid interpretations and have been advanced to one degree or another by various delegates to the negotiations. Further, they are illustrative of an interpretive dichotomy which permeates the Convention's provisions. This duality is, in turn, a reflection of the 'delicate balancing' of interests in the international political arena and the text is the legal manifestation of those politics. Thus, although the Convention is a binding legal instrument, and the commitments contained therein are legally binding, considerable scope exists for the Parties to interpret the provisions in a manner which best promotes their political interests. While such 'constructive ambiguity' likely accounts for the willingness of such a large number of diverse states to sign and ratify the Convention,[17] the differences which permeated the negotiations, and which are reflected in the text of the Convention, continue to operate and affect the implementation of the Convention.

REDUCTION/STABILIZATION

Article 4(2)(a) and (b), arguably the most widely recognized provisions of the Convention, is also among the most controversial. It is these provisions which express, *inter alia*, the commitment undertaken by the developed country Parties, including the EC, to reduce emissions of greenhouse gases not covered by the Montreal Protocol and which require developed country Parties to

> adopt national policies and take corresponding measures on the mitigation of climate change, by limiting its anthropogenic emissions of greenhouse gases.[18]

With respect to the commitment to reduce greenhouse gas emissions, the Article does not unequivocally set a strict target, in terms of either dates or emission levels, but rather states that

> these policies and measures will demonstrate that developed countries are taking the lead in modifying longer-term trends in anthropogenic emissions . . . recognizing that the return by the end of the present decade to earlier levels of anthropogenic emissions of carbon dioxide[19]

and

each of these Parties shall communicate . . . detailed information on its policies and measures . . . as well as on its resulting projected anthropogenic emissions . . . of greenhouse gases for the period referred to in subparagraph (a) [above], with the aim of returning individually or jointly to their 1990 levels these anthropogenic emissions of carbon dioxide and other greenhouse gases.[20]

These provisions have generally been interpreted by the respective Parties as requiring them to 'aim to return to 1990 levels' by the year 2000.[21]

Given the convoluted and somewhat confusing language of the Article, this is a valid interpretation. It does not require the developed country Parties actually to achieve this goal nor does it mean that any developed country Party which fails to return successfully to 1990 levels (or which, alternatively, surpasses this objective) will be in breach of its obligations under this provision of the Convention. However, the obligation to 'aim to return to 1990 levels' does, by necessary implication, require that each Party undertake constructive and proactive measures in implementing that 'aim'. The exact nature of these measures will vary greatly among the Parties in both scope and structure but might include:

- Germany's Autobahn user tax and the diagnosis of the energy efficiency of buildings;
- Italy's creation of a National Agency for the Protection of the Environment and gasoline tax proposals;
- Norway's CO_2 tax and Norwegian information campaigns;
- the UK's Energy Saving Trust initiative and the British energy efficiency information campaigns;
- an EC carbon/energy tax and energy efficiency related ecolabelling.[22]

As these examples indicate, the *UN FCCC* affords significant latitude to the Parties with respect to the measures and policies that they need to introduce and implement in order to meet their obligations under this commitment.

As a result of the flexibility in implementation under Article 4(2)(a) and (b) and the fact that there is no obligation on the Parties actually to realize a return to 1990 levels, the provisions have been subject to considerable criticism by NGOs and other interest groups. Critics of the commitment also argue that because 'returning' to 1990 levels does not mean 'stabilizing' at 1990 levels, nothing prevents Parties from increasing their emissions after the year 2000. These are all valid and accurate comments. However, Article 4(2)(d) requires the CoP to review the adequacy of this commitment at its first session and to amend the commitment if necessary. In this regard, the commitment to aim to return to 1990 levels by the year 2000 and all it entails, or fails to entail, should be considered a starting point in addressing the issue of climate change and not an end point.[23]

The review of the adequacy of the commitment is extremely important and, although expressly restricted to Article 4(2)(a) and (b) and joint implementation, it is to be undertaken with reference to the best available scientific information as well as relevant technical, social and economic information.[24] Consequently, the review will not, and cannot, be confined to a straightforward examination of whether the emission level reduction itself is adequate to prevent climate change. Factors which act as barriers to the formulation, adoption and/or implementation of greenhouse gas reduction strategies will, because of their social and economic implications, have a relevant, and likely persuasive, influence on the outcome of the review. Given the difficulties encountered by the Annex 1 Parties in implementing their Article 4(2)(a) and (b) commitments (examined in Chapters 6–10) it is extremely unlikely that reduction levels along the 60 per cent lines initially recommended by the IPCC will be adopted at this stage. Having said this, it should be recognized that the review of the adequacy of the commitment is not an independent one conducted solely by the developed country Parties. It is, rather, a review undertaken by the CoP and as such will also reflect developing country Party concerns that the developed country Parties take the lead in combating climate change, including the reduction of greenhouse gas emissions.[25]

JOINT IMPLEMENTATION

The concept of joint implementation referred to in the *UN FCCC*[26] provides for the collaboration between two (or more) Parties in the reduction of greenhouse gas emissions. Under a joint implementation initiative, country A (the donor) would be able to finance activities which limit net greenhouse gas emissions in country B (the host) and country A would be able to claim credit for the reduction of emissions within its own territory. The objectives underlying this kind of initiative include facilitating the global reduction of greenhouse gas emissions; facilitating technological transfer; and discouraging the movement of pollution intensive industries to developing country Parties. However, because joint implementation is not a defined term under the *UN FCCC*, and is only expressly mentioned once (Article 4(2)(d)), all other references being oblique,[27] there is considerable debate within the INC as to under what condition joint implementation is allowed.[28]

Arguments against the use of joint implementation initiatives to meet national greenhouse gas reduction objectives centre around the fact that each developed country Party is required to adopt national policies and measures to 'limit *its* anthropogenic emissions' (Article 4(2)(a)). This implies that measures within the confines of their own territorial jurisdiction which reduce greenhouse gas emissions in their territory must be undertaken by developed countries. This view is further reinforced by the references within the Article to the historical responsibility of developed country Parties for climate change and the need for developed countries to take the lead internationally in modifying greenhouse gas emission trends. Read in its entirety, it is arguable that policies aimed at returning domestic greenhouse gas emissions to 1990 levels by the year 2000 are a prerequisite under Article 4(2)(a) before joint implementation measures can be undertaken or considered. Such an approach is consistent with the UK's position on the issue.[29] This position is also endorsed by the EC subject to the proviso that joint implementation within the confines of a 'regional economic integration organization' is acceptable. Given the fact that the EC is, to date, the only 'regional economic integration organisation', this consideration is somewhat self-serving. Underlying the exemption, however, one can detect the compromise position it reflects between those EC Member States which do not favour the use of joint implementation in the pursuance of domestic reduction goals (i.e. the UK) and those Member States which do favour its use.

Arguments favouring the use of joint implementation initiatives, on the other hand, highlight the fact that the Convention states that the Parties 'may implement . . . policies and measures jointly' (Article 4(2)(a)); that they may aim to return emissions to 1990 levels 'individually or jointly' (Article 4(2)(b)); and that the CoP may facilitate the coordination of measures adopted by two or more Parties (Article 7(c)). Taken together, these provisions would suggest that joint implementation strategies represent viable and valid measures which the Parties could adopt in 'aiming' to return their emissions to 1990 levels. This approach is favoured by Norway, Germany and the United States and reflects the belief that joint implementation can be taken in conjunction with domestic measures and can form a substantial part of a national policy programme package.

The two positions just advanced, broadly classified as pro-joint implementation and anti-joint implementation, are reflective of the debate which has occurred within the INC and the Berlin CoP. At INC 10, it was generally agreed that a phased approach to joint implementation, beginning with a pilot phase, should be adopted.[30] This recommendation was be accepted at CoP I. Such acceptance does not mean that one of the positions referred to above was more 'legally' right than the other, but rather reflects a split amongst developing countries' opposition to joint implementation. A key factor contributing to the breakdown of a united front involves the uncertainties over the net beneficial economic effect and the opportunities for development and technological transfer in host countries that could result from joint implementation projects. Thus the acceptance by developing countries of joint implementation as a mechanism to mitigate climate change reflects the continued priority of development and poverty eradication concerns by these countries as underlay the initial UN FCCC negotiations and which are contained within the Convention's text.[31]

Notwithstanding this limited conceptual acceptance of joint implementation, however, several additional, controversial issues arise. These issues include the distribution of credits for emission reduction; the type of joint implementation projects which may be undertaken; and the countries which may participate in joint implementation schemes.

The issue of credit is arguably the most contentious at the present time and pertains to who, the donor or the host country of the project, should be credited for the reduction, and if both, how the percentage distribution should be allotted. Although the developing country Parties do not as yet have any specified emission reduction or limitation goals under the Convention, there is considerable discussion in the INC to attempt to set some guidelines, particularly for those with significant greenhouse gas emission potential such as China, Brazil and India. These Parties are anxious to retain the option to utilize the credits within their own accountability. If, however, developed country Parties cannot receive credit, incentive for investment, and the subsequent transfer of technology, an issue of considerable importance for developing countries, diminishes. Hence, there is a need for a further balancing of interests under the *UN FCCC*.

The type of project which would be encompassed in any joint implementation scheme is also the subject of debate and begs the question as to whether or not the enhancement and preservation of sinks would be eligible. Insofar as such sinks absorb greenhouse gases they act to reduce the accumulation of emitted gases and so contribute to the minimization of climate change. Furthermore, because each Party should develop policies that enhance its sinks, it is arguable that sinks are eligible for joint implementation. However, there is significant aversion towards their inclusion by developing countries as such projects are largely viewed as a means of perpetuating the perceived status of the south as the 'sinks' for the north. In addition, Articles 4(2)(a) and 7(2)(c), by stating that developed country Parties should take the lead in modifying emission trends and by recognizing the need for equitable contributions to the reduction objective, are an implicit argument against the development of sink projects through a joint implementation initiative.

With respect to source projects, prerequisite criteria considerations include whether or not reductions in greenhouse gas emissions would have occurred 'but for' the project; whether the reductions gained would have short term and/or long term benefits; whether the project will lead to further greenhouse gas reductions; and what contribution to greenhouse gas emission reduction the project will have (calculation of credit).

The third key area of contention concerns which Parties may engage in joint implementation. In Article 4(2)(a), the provision pertaining to emission limitation and sink enhancement appears to refer only to developed country Parties and Parties in transition. Towards the end of the Article, the *UN FCCC* states that 'these Parties may implement . . . with *other* Parties'.

In this sense, it may be possible that *other* Parties may only refer to developing country Parties, i.e. all Parties which are not developed country Parties or Parties in transition. It therefore follows that joint implementation projects such as those with former eastern bloc countries may not be envisioned or permitted under the Convention. This restriction appears to be carried over into Article 4(2)(b). However, in Article 7(2)(c) no such restriction exists. Any two Parties may request facilitation of the coordination of their policies by the CoP. Notwithstanding the textual wording of the Convention, there appears to be general agreement that joint implementation activities *vis-à-vis* developed county Parties are acceptable.

Although joint implementation will undoubtedly impact on domestic policy formulation and implementation, and in fact countries such as Norway have already entered into some joint implementation-like schemes,[32] it is difficult, given the uncertainties presently surrounding the scope of joint implementation in the Convention, to determine what the nature and degree of that effect will be.

BURDEN SHARING

Burden sharing is a concept similar to that of joint implementation in that it refers to a collaboration of effort between two states in the global effort to reduce greenhouse gas emissions. Further, burden sharing is not a defined term under the *UN FCCC* although it is implicit in joint implementation and is alluded to in Articles 3(1),(3); 4(2)(a),(b),(d),(e); 7(2)(c); and 12(8).

Unlike joint implementation, however, burden sharing looks at the total emissions from two (or more) countries and assesses the emission reduction within this emission bubble. As a result, one country can reduce by a lesser amount than another country, so long as the combined total of their emissions conforms to the Article 4(2)(a) and (b) goal.

As the EC is not a country, but rather a region, the concept of burden sharing is directly applicable to the EC. As a Party to the *UN FCCC*, the EC is required to meet its 4(2)(a) and (b) commitments. Arguably, if the EC is treated as a single unit then it matters not how greenhouse gas emissions are reduced, or within what regions of the EC, so long as the total EC emissions aim at returning to 1990 levels. However, the EC is comprised of separate, sovereign states, each of which is a Party in its own right to the Convention. Further, each Party is a 'developed country' or 'Annex 1' Party under the *UN FCCC*. As such, each of the Member States is required, at international law, to honour its commitments under Article 4(2)(a) and (b). Although Article 4(2)(a) and (b) is ambiguously worded, particularly with respect to the amount and date-lines for reduction, the Article is clear on the requirement that measures need to be pursued which seek to *limit* net emissions.

In view of these independent obligations, the notion of burden sharing within the EC becomes more complex. If burden sharing was to be used by the EC as a means of meeting the Toronto target of a 20 per cent reduction in 1988 CO_2 emissions by the year 2005[33] as was supported in the *Bergen Ministerial Declaration* of 1990,[34] then, because these are not legally binding agreements and thus do not impose legal obligations on the signatories, it would be legally acceptable for certain Member States to increase emissions if other Member States decreased theirs by a sufficient amount. The *UN FCCC* is, however, a legally binding instrument. Thus, if certain Member State Parties plan to increase their emissions (e.g. Ireland) and justify their increase on the basis of historical responsibility and differentiated capabilities, *vis-à-vis* other Member State Parties, such increases will be in contravention of their 4(2)(a) and (b) commitments and they will be in breach of their obligations under the *UN FCCC*. The fact that the EC as a whole will be moving towards realization of the target would be legally irrelevant.

Having made this point, however, it is worth recognizing that because the EC is a distinct anomaly in the world order, special consideration and exception may be made for it. If this were to occur, then the contravening action of certain Member States may not be regarded as a breach of the Convention. Given the fact that the ultimate objective of the *UN FCCC* is essentially a global one, designed to address a global problem, it is entirely possible that latitude would be given by the CoP to individual Member States so long as the EC region as a whole strives to meet its commitment.

On the other hand, given the balancing of interests necessary for the conclusion of the *UN FCCC* negotiations and the reflection of these interests within the text of the Convention, it is equally possible that such latitude may not be forthcoming. Each of the countries within the EC is listed in Annex I and is considered a 'developed' country Party on the international stage – irrespective of its perceptions *vis-à-vis* other countries. One of the underpinning concepts within the *UN FCCC* is that developed country Parties take the lead and, considering the difference in development and standards between the least developed Member States and developing countries, to afford some developed countries an exempt status could prove to be a political minefield with repercussions spilling over into other international fora. At the present time the issue of burden sharing remains unresolved at the EC level and is not a matter of great debate within the INC. However, the issue may gain more prominence once the policies and measures communicated by the Annex I Parties are reviewed, particularly if the EC as a 'bubble' is unable to deliver on its Article 4(2)(a) and (b) commitments.[35]

EDUCATION AND AWARENESS

Article 4(1) of the *UN FCCC* lists a number of commitments, which all Parties, be they developed, developing or Parties in transition, are obligated to perform. With respect to domestic implementation in developed country Parties, those provisions which will likely have the most significant impact can be summarized as:

- the publication of national inventories of anthropogenic emissions of greenhouse gases (Article 4(1)(a));
- the implementation and publication of national programmes to address anthropogenic emissions of greenhouse gases (Article 4(1)(b));
- the promotion and cooperation in the development and diffusion of technologies and practices, in all sectors, which reduce or prevent anthropogenic emissions of greenhouses gases (Article 4(1)(c));
- the consideration of climate change in social, economic and environmental policies (Article 4(1)(f));
- the promotion and cooperation in the exchange of information related to the climate system and to the economic and social consequences of response strategies (Article 4(1)(h));
- the promotion and cooperation in education and training and public awareness related to climate change (Article 4(1)(i)); and
- the promotion and facilitation of public participation in addressing climate change and in developing adequate responses (Articles 4(1)(i) and 6(a)).

Although Article 4(1)(c)–(j) places a mandatory duty to act upon the Parties, the specific provisions within the Article confer a broad discretion upon the Parties with respect to the implementation of those duties. Thus, and for example, while the development and diffusion of technologies is required, the method and process through which the objective is to be achieved are left to the discretion of the Parties. Consequently, it is the sole decision of the Party in question as to whether it uses legislation or economic instruments to force the technological process; or uses hortatory measures to encourage it. This same argument applies for all the commitments contained within Article 4(1)(c)–(j).

While most of the attention to date has focused on the Article 4(2)(a) and (b) commitments; fulfilling that commitment is largely dependent on the effective implementation of Article 4(1). In other words, in order to return greenhouse gas emissions to 1990 levels, national inventories will need to be established for monitoring and evaluation purposes; national programmes and policies will need to be created to guide 'the return of emissions' which will likely impact, and be impacted by, other government programmes and policies. Development of technologies and their diffusion will likely be necessary to implement these programmes; the exchange of information will also be needed to facilitate not only their implementation but also the public understanding of why such implementation is necessary. Further, in order to maximize credibility on the issues and to ensure the participation of all sectors of society, participation of all actors in society is recommended in the development of programmes and policies.

Thus, while Article 4(2)(a) and (b) may prove to be the 'driving' force of the Convention and provide the impetus for cross-sectoral change, it is through Article 4(1) that such change will actually be effected. Consequently, at the domestic level, and from an institutional and implementation perspective, it is Article 4(1) which is likely to be the most influential.

IMPLEMENTATION

Review of Communication

Article 4(1)(a) and (b) and Article 12 require each Party to communicate to the CoP, *inter alia*, national inventories of greenhouse gases not controlled by the Montreal Protocol. Further, developed country Parties are required to communicate a description of the policies and measures designed and/or adopted to implement their Article 4(2)(a) and (b) commitments. These communications will then be reviewed, such process being called the review of communication. Although neither the purpose nor the process of the review of communication is finalized, as at the date of writing, certain key themes can be identified.

For example, it has been agreed that the purpose of the first review of communication is *not*:

- to be a legal assessment of the extent to which the individual Parties are meeting their specific obligations;
- to result in Parties being told to undertake specific actions;
- to be confrontational, judicial or quasi-judicial;
- to be a mechanism for formally assessing compliance with the Convention;
- to be a basis for dispute settlement.[36]

Having eliminated many options, it is likely that the purpose will be restricted to one in which the review becomes a formalized exchange of information on implementation strategies and on the barriers to the implementation of such strategies. Although such a purpose will likely be met with censure from various quarters (i.e. NGOs), it is important to recognize that it is through the exchange of information that issues are identified and without issue identification there can be no constructive negotiation leading to problem resolution. In this way the review of communication is expected to provide the link through which the actual experiences

of the Parties inform the evolution and possible interpretation of the obligations under the *UN FCCC*.

The process for the review, which appears to be favoured by the Parties,
is one which facilitates the exchange of information and experience on
implementation of the Convention; is non-confrontational; is open and is
transparent.[37] It will likely focus on the communications from Annex I
(developed) countries with communications from non-Annex 1 countries as
determined by the Bonn Secretariat and endorsed by the CoP process.
The review of first communications will likely concentrate on six tasks:

1 review key qualitative information and quantitative data points;
2 review policies and measures described;
3 access the information in light of the commitments and to the extent they
   further the realization of the objective of the Convention;
4 describe expected progress in the limitation of the emissions by sources
   and removals of sinks;
5 describe the expected progress in cooperation to prepare for adaptation;
6 aggregate data across national communication with respect to inventories,
   project, effects of measures and financial transfers but without adding up
   the individual national totals for projections and the effects of measures.[38]

Although the review of communication is a distinct and separate undertaking from the review of the adequacy of commitments,[39] the final compilation and synthesis resulting from the review, together with the information
learned in the process, will likely play an important role in determining the
adequacy of commitments and in establishing future commitments.

Review of Implementation:

Pursuant to Article 7(2)(f), the CoP shall adopt regular reports on the
implementation of the Convention. This process, referred to as the review of
implementation (RoI) is a distinct one from the review of communication
discussed above and is not to be duplicative of the review of communication's efforts. As with the review of communication, however, the
process and purpose of the RoI will again likely be designed to be non-
confrontational, consensual and informative. The likely remit of the
RoI would involve the synthesis and compilation of the information provided in the first national communications. Such a synthesis would include
inter alia:

● overall achievements and highlights;
● general considerations including national circumstances;
● inventories of anthropogenic emissions and removal in 1990;
● policies and measures being implemented by Parties to limit anthropogenic emissions and enhancement of sinks and reservoirs;

- projections and effects of policies and measures adopted by Annex I Parties;
- finance, technology and capacity building;
- implementation of other commitments of the Convention such as expected impacts of climate change; adaptation measures; education, training and public participation; integration of climate change concerns in national policies.

Whatever the purpose of process of the reviews of communication and implementation is ultimately decided to be, the fact that they are conceived as being distinct institutional arrangements (as is the review of the adequacy of commitments) will not, however, preclude them from informing each other or from influencing each other at both the functional and conceptual levels. Given this inter-relationship, together with the proposed non-confrontational nature of the process, those factors which influenced the negotiating process in 1991–1992, such as historical responsibility, economic development, environmental protection, competitiveness and sustainability, will remain in play under the CoP and its institutional support structures.

COMPLIANCE

There are three, related, mechanisms under the *UN FCCC* through which implementation and compliance could be assessed. These are (a) review of communication (discussed above); (b) multilateral consultative process; and (c) dispute settlement. The three may be

> seen as a continuum consisting of the communication of information and the review process at the beginning, a multilateral consultative process in the middle and the dispute settlement regime at the end.[40]

Article 13 provides for the establishment, by the CoP, of a multilateral consultative process (MCP) for the resolution of questions regarding the implementation of the *UN FCCC*. Although the establishment of such a process is not mandatory, a working group of technical and legal experts to study the issue was established at CoP I[41] and it is likely that such a process will be put into place.

As the MCP is not yet established it is impossible to state with certainty what will be the scope of its powers in respect to resolution of implementation questions. However, certain emerging themes can be identified. For example, the focus of the MCP could be to resolve questions regarding implementation in a cooperative and consultative manner with a view to assisting, rather than penalizing, Parties which were not fulfilling their Convention commitments.[42] In other words, issues regarding implementation, and hence compliance, would be dealt with in a non-confrontational, non-adversarial manner much like that envisioned for the review of information. The rationales underlying a consultative approach have been identified as:

- the global nature of the climate change issues affect the interests of all states;
- preventative measures have a more benign environmental effect than do remedial measures generally associated with post-dispute settlement;
- a cooperative, consultative process would emphasize cooperation amongst Parties;
- agreement by consensus on issues will increase the likelihood of compliance and thus promote the effectiveness and stability of the *UN FCCC* regime.[43]

While it is clear that the MCP would deal with questions of implementation, it is not clear what powers it would have concerning questions of interpretation. As implementation and interpretation are inextricably linked, it is likely that, in order to be effective, the MCP would need the power to interpret as well. Consequently, additional issues which need to be resolved before the MCP can be established include: which types of questions could be raised in the MCP process (own compliance or the compliance of another Party), who could raise the questions (Parties with reduction commitments (i.e. Annex I Parties), third Parties (i.e. NGOs), the Secretariat); and how could questions be resolved (i.e. recommendation or decision)?

The creation of an MCP would not preclude recourse to traditional dispute settlement mechanisms such as the ICJ, arbitration or conciliation as provided for in Article 14. However, it is anticipated that in the event of the creation of a facilitative, non-confrontational and conciliatory MCP, recourse to traditional dispute settlement procedures would be minimal.

The recurring principle which runs throughout the implementation and compliance mechanisms is the emphasis on non-confrontation and conciliatory dispute resolution. The inclination towards this type of approach is reflective of the 'delicate balancing' of interests upon which the *UN FCCC* is based.

CONCLUSION

This chapter has sought to discuss briefly those commitments within the *UN FCCC* which will likely have the most significant influence on domestic policy formulation and implementation. As a consequence of this process, the ambiguities which permeate the Convention have been illustrated, as have their impact on domestic policy development, i.e. the wide interpretative discretion such ambiguities confer on the Parties. The chapter has also sought to highlight the ongoing negotiations under the INC and latterly the CoP with respect to the resolution of contentious issues and the clarification of the ambiguities.

All three of these factors, commitments, ambiguities, and continual negotiations, are of fundamental importance to the ultimate success of the *UN FCCC*. The commitments are necessary as they legally obligate Parties to undertake certain actions. The ambiguity is necessary given the many scientific, economic and social uncertainties surrounding the issue of climate change. Continued negotiation is necessary to obtain a sufficiently acceptable and precise commitment and to resolve outstanding issues. The three factors are related and dependent and are reflective and representative of the balancing of interests which underpin the Convention. In today's global reality, the *UN FCCC* would not be able to function without them. Over time, as more information and experience is gained by the Parties on the issues and in implementing their commitments, it is likely that less ambiguous obligations will be undertaken and that more onerous commitments, more in keeping with the IPCC estimates with respect to the reduction of greenhouse gas emissions, will be adopted. This evolutionary process is implicit in the 'framework' design and structure of the Convention and is reflective of the changing and fluid nature of international law generally.

NOTES

1 *United Nations Framework Convention on Climate Change* (*UN FCCC*) A/AC.237/18 (Part II)/Add.1; (1992) 31 ILM 851.
2 For example, the Alliance of Small Island States (AOSIS). The Netherlands, although not an island, is one of the most vulnerable developed countries to sea-level rise. It was thus able to benefit from the strong position taken on the issue of climate change by AOSIS.
3 For example, OPEC. Norway, as a major oil and gas producer and with an economy heavily dependent on this industry, was able to benefit from the position taken by the OPEC countries while still maintaining its international 'green' image in the international political arena.
4 The potential for development opportunities is thought to be a significant factor in the breakdown of the G77 plus China coalition on the issue of joint implementation.
5 *UN FCCC, Supra*, note 1, Preamble, para. 2; Article 3(3); Article 4(2)(a), (b).
6 *UN FCCC, Supra*, note 1, Preamble, para. 6, para. 17; Article 3(1); Article 4(1); Article 4(2).
7 *UN FCCC, Supra*, note 1, Preamble, para. 3; para. 21; Article 3(4).
8 *UN FCCC, Supra*, note 1, Preamble, para. 19; Article 3(2); Article 4(8).
9 *UN FCCC, Supra*, note 1, Preamble, para. 20; Article 3(2); Article 4(8).
10 *UN FCCC, Supra*, note 1, Preamble, para. 16; Article 3(3).
11 *UN FCCC, Supra*, note 1, Article 3(3).
12 *Intergovernmental Panel on Climate Change* (*IPCC*) *First Assessment Report: Overview*, 31 August 1990, WMO/UNEP.
13 See *UN FCCC, Supra*, note 1, Article 3; Article 4(2)(a); 4(2)(d); 4(2)(e)(i); Article 7(2)(a); 7(2)(d); 7(2)(e); 7(2)(m).
14 An Intergovernmental Negotiating Committee for a Framework Convention on Climate Change (INC) was established by the UN General Assembly in

December 1990 with a mandate to negotiate a convention. (Protection of Global Climate for Present and Future Generations of Mankind, G.A. Res. 45/212. UN GAOR, 45th Sess., 71st plen. mtg, Suppl. No. 49, UN Doc. A/45/49 (1990).)

The *UN FCCC* was opened for signature at UNCED in June 1992 and required 50 ratifications before it entered into force. The 50th instrument of ratification was deposited by Portugal on the 21 December 1993 and the Convention entered into force on 21 March 1994. As of 31 October 1994, 100 States have ratified the Convention.

15 *Draft Protocol to the UN FCCC on Greenhouse Gas Emission Reductions*, UN Doc. A/AC/237/L./23, 27 September 1994, Preamble, paras 3, 5, 6; Article 1(6); Article 2; Article 4(1); Article 8(1)(a) and *Paper Circulated by Germany at INC 10 on Elements to a Comprehensive Protocol*, UN Doc. A/AC/237/L.23/Add.1, 27 September 1994, p. 1.

16 See Footnote 12 and accompanying text.

17 See Footnote 14.

18 *UN FCCC*, *Supra*, note 1, Article 4(2)(a).

19 *UN FCCC*, *Supra*, note 1, Article 4(2)(a).

20 *UN FCCC*, *Supra*, note 1, Article 4(2)(b).

21 For example, see: *Climate Change: The UK Programme – United Kingdom's Report under the Framework Convention on Climate Change*, Foreword, London, HMSO cm. 2425, 1994.

Some Parties have, unilaterally and outside the scope of the Convention, undertaken to 'aim' for more significant greenhouse gas emission reductions. For example, Germany has pledged to 'aim to reduce CO_2 emissions by 25–30%' of 1987 by the year 2005; Norway has stated that CO_2 emissions should be stabilized by 2000 at 1989 levels; and the EC Council recommends that CO_2 emissions should be stabilized by the year 2000 at 1990 levels (*Climate Change Policy Initiatives, 1994 Update*, Volume 1, OECD/IEA, Paris, 1994). Because these actions have been undertaken outside the scope of the *UN FCCC*, they are largely independent, at international law, of the obligations under the *UN FCCC* to meet the Article 4(2)(a) and (b) commitments.

22 These measures are discussed further in the case studies contained in Chapters 6–10.

23 At the present time there is emerging consensus amongst the Parties that the commitment is not adequate, but debate continues as to how much to tighten the commitment, as to which gases should be targeted, and as to which parties should be bound.

24 *UN FCCC*, *Supra*, note 1, Article 4(2)(d).

25 *UN FCCC*, *Supra*, note 1, Article 4(2)(d) and Article 3(1).

26 *UN FCCC*, *Supra*, note 1, Article 3(3); Article 4(2)(a), (b), (d); Article 7(2)(c); Article 11(5) and Article 12(8).

27 See generally Footnote 265.

28 At INC 10 (August–September 1994) the Parties agreed that joint implementation is a concept embodied in the Convention (*Part II – Final Report INC-X: WG1 Matters: Matters Relating to Commitments*, UN Doc. A/AC/237/76; Part V – para. 59).

29 *Matters Relating to Commitments Criteria for Joint Implementation: Comments from Member States on criteria for joint implementation: Note by the interim secretariat*, 8 December 1993, UN Doc. A/AC.237/Misc.33, pp. 6–9; 97–100.

30 *Final Report*, *Supra*, note 28, Part V – para. 60.

31 *UN FCCC*, *Supra*, note 1, for example Preamble, paras 3; 6; 10; 21.

32 Norway has entered into two joint implementation schemes with Mexico and Poland (*Norwegian Funding of Pilot Demonstration Project for Joint Implementation Arrangements under the Climate Convention: A memorandum of understanding*, Global Environmental Facility, Washington, 1992).

Because of the debate over joint implementation, it is unlikely that Norway will be able to claim any 'credits' through the projects. Rather, the benefits it will accrue will be in the nature of information on implementation; of the costs and benefits; and of the effectiveness of joint implementation generally.

33 *Statement issued by the Participants at the World Conference on 'The Changing Atmosphere: Implications for Global Security' (The Toronto Target)*, Toronto, 30 June 1988, as reproduced in R. Churchill and D. Freestone (eds) *International Law and Global Climate Change*, London: Graham & Trotman/Martinus Nijhoff, 1991.

34 *Ministerial Declaration on Sustainable Development in the ECE Region United Nations Economic Commission for Europe Conference on Action for a Common Future (Bergen Ministerial Declaration)*, para. III 14(b), Bergen, Norway, 15 May 1990, UN Doc. A/CONF.151/PC/10.

35 It is worth noting that the German *Statement on elements for a Comprehensive Protocol to the FCCC* proposes, *inter alia*, that the developed country Parties adopt 'national' policies and measures to improve energy efficiency of small combustion plants, appliances and vehicles; to utilize least-cost planning; and to improve building insulation standards (*Supra*, note 15). The successful introduction of these types of policies and measures at the EC level, and to a degree which would be effective, has been hampered by concerns by certain Member States (i.e. the UK) on the devolution of further competences to the EC, and the requirement of unanimity for measures which affect energy structure and supply. By incorporating relatively detailed 'national' policy and measures guidelines in a protocol, the debate over competences could be defused as Member States could agree, outside the legal framework of the EC, to cooperate with each other on these matters. For a detailed discussion on competences see Chapter 4.

36 Climate Change Secretariat, 'First Review of Information Communicated by Each Party Included in Annex 1 to the Convention', para. 20, Doc. A/AC.237/63, Draft.

37 *Final Report*, *Supra*, note 28, Annex 1: Appendix 1 to Annex II to Decision 10/1.

38 *Supra*, note, Part V, Annex I, Appendix II to Annex II to Decision 10/1.

39 See Footnotes 21–24 and accompanying text.

40 Climate Change Secretariat, 'Consideration of the Establishment of a Multilateral Consultative Process for the Resolution of Question Regarding Implementation (Article 13)', Para. 18, UN Doc. A/AC.237/59.

41 *Supra*, note 28, Part VII, para. 114.

42 *Supra*, note 40.

43 *Supra*, note 40.

3

THE SCIENCE AND POLITICS OF NATIONAL GREENHOUSE GAS INVENTORIES

Susan Subak

INTRODUCTION

Sporting colourful covers with imaginative graphics, the obligatory reports of the Parties to the Convention resemble more the exteriors of popular science paperbacks than the dull manuals of older international environmental agreements, documents that have usually sat forgotten in small UN secretariats. The national communications on greenhouse gas inventories and plans have been the focus of intense involvement by many government agencies and private sector representatives in many countries. Environmental advocacy groups from each communicating party have already issued critiques of each plan. The challenge of devising national plans and baseline emissions has had major implications for the evolution and authority of environmental bureaucracies. In turn, a watchful public are looking for signs that the whole process is accountable and negotiable.

Amongst the Climate Convention's broad objectives and ambiguous commitments, the specificity and immediacy of the Convention's emissions research requirements have stood out. All Parties are committed to publishing estimates of the anthropogenic greenhouse gases they emitted in 1990. The required inventory includes estimates of methane (CH_4) and nitrous oxide (N_2O) emissions, and sources and sinks of carbon dioxide (CO_2), from energy, industrial processes, land use change, and agriculture. Most inventories from developed countries were completed in late 1994. Developing countries can wait up to three years after ratification to turn in their inventories. The inventory requirement was accepted largely without controversy, although it may prove burdensome and expensive for agencies coordinating and sponsoring the research. To complete the inventories, countries are involved in a research effort and an interinstitutional dialogue at the national level that, while inadequate in some respects, still surpasses any equivalent effort for any other international environmental agreement. The inventory requirement, as well as the Convention's 'commitment' to aim to return developed country greenhouse gas emissions by 2000 to 1990 levels, originate in part from the precedent-setting decisions made by European states in 1990 to stabilize CO_2 emissions.

While scientists have investigated greenhouse gas sources and sinks for decades, as was pointed out in the introductory chapter, this work was aimed largely at improving the understanding of biogeochemical processes and their effect on atmospheric concentrations of CO_2 and other greenhouse gases. Beginning in the late 1980s, as national governments began to recognize the importance of global warming as a policy problem, the need for comprehensive inventories with reference to political borders became apparent. The first compendium of countries' greenhouse gas emissions was released by Oak Ridge National Laboratory in the late 1980s, and included a time series of emissions by fuel, starting in 1950 (Marland et al., 1994 and earlier reports). These estimates, which included most countries of the world, were limited to CO_2 from energy consumption and from cement production. In 1990, estimates of national emissions of other gases (Hammond et al., 1990) and with regard to various time frames were published (Subak and Clark, 1990). These raised questions and helped to stimulate a debate about the ability to calculate and monitor the many disparate emissions sources. There was also controversy over the strategic importance of different approaches for comparing the political significance of greenhouse gas assessments (for instance, Smith, 1991; Agarwal and Narain, 1991). Despite the decision to require 1990 inventories for virtually all anthropogenic sources, an outcome accepted without much debate, these contentious measurement and interpretative issues remain. Indeed it is likely that they will move up the negotiating agenda as countries debate additional funding arrangements, extension of targets to all countries, and timetables for deeper cuts.

ARRANGEMENTS FOR ESTABLISHING THE BASELINE

The Framework Convention requires countries to report estimates of all the anthropogenic greenhouse gases that they release, or that they remove through growing trees. This definition excludes non-anthropogenic sources such as CH_4 flux from the ocean and from wetlands. Anthropogenic sources of CO_2 include energy combustion and some industrial processes, the most important of the latter being cement production. Clearing forests and grassland conversion count as emissions of CO_2, whereas abandonment of managed lands and afforestation are treated as anthropogenic removals. The anthropogenic CH_4 sources include fugitive fuel from energy production, human wastes, landfills and energy combustion. Anthropogenic agricultural CH_4 sources are rice cultivation, livestock production and manure, crop residue burning and savannah burning. Nitrous oxide originates from agricultural soils, waste water treatment, energy combustion, particularly mobile sources, and from adipic acid and nylon production.

Fossil fuel burned in international waters and in inner space is currently excluded in national emissions inventories because these emissions are not physically released within any country's boundaries. The matter of where to allocate emissions from these vessels, which are known as 'bunkers', has been the only inventory issue that has received significant attention by the Intergovernmental Negotiating Committee (INC) (INC, 1994a). These emissions only comprise about 1.3 per cent of global CO_2 emissions from energy use, but for some countries, e.g. Panama and Brunei, they represent the majority (Von Hippel *et al.*, 1993). The INC accepted the reporting requirements and format that the IPCC produced, but with the amendment that countries that sell fuel to ships and aircraft for international trips must list this information, although separately from the rest of the fossil fuel emissions.

The decision to centre the baseline on the year 1990 was largely a matter of political convenience reflecting variable reporting histories given that humans have been contributing to the enhanced CO_2 flux for more than a century (Keeling, 1973), and to enhanced CH_4 emissions for much longer (Khalil and Rasmussen, 1994). This is also noteworthy considering that more than a century is required to remove most greenhouse gases from the atmosphere through natural processes. Nonetheless, completing a comprehensive inventory of greenhouse gas flux for one recent year is still a formidable task if countries attempt to use locally derived information rather than 'default' activity and flux estimates based on older research from other countries. This task involves the development and review of new emissions factors in energy, land use, industrial and agricultural sectors. No single bureaucracy, or even country, has had the competence on its own to complete a state-of-the-art assessment without the aid of the international research programme that blossomed in the early 1990s.

The IPCC inventory methodology, as we now know it, reflects a pooling of knowledge that includes Japanese research on N_2O from fuel combustion, Nigerian expertise in biomass burning, US experience in livestock emissions, UK results on CH_4 from coal mining, Dutch research on N_2O soil flux, and Norwegian knowledge of industrial processes affecting trace gas release. Researchers from about 45 countries have taken part in various working groups since 1991 in a programme coordinated internationally by the IPCC/OECD. In the international working groups and in regional workshops beginning in 1993, developing country and Eastern European participants identified additional sources and improved methodologies. The IPCC work programme also included 'transparency' studies in which paired countries, e.g. the United States and China, Canada and Norway, compared results based on their opposite number's methodology and their own. The major products of these various efforts are the IPCC/OECD *Emissions*

Methodology Workbook, a set of calculations software designed by this author, and the Reference Manual released in the spring of 1994. The IPCC/OECD workbook includes more than a hundred tables for entering estimates of emissions-related activities and emissions factors (IPCC/OECD, 1994b). In detail these guidelines provide unprecedented support for complying with the reporting requirement of the Convention, although they were not completed and distributed in time for countries to take into account all of their recommendations before the first communications were due in the autumn of 1994.

The task of completing greenhouse gas inventories has initiated a new level of coordination, or at least dialogue, among different agencies. In the developed regions this has usually meant that the environmental agency commissions a range of research from government laboratories, consulting firms and academics. The inventories are then often compiled by one individual. In developing regions, researchers at the country meteorological offices, which are relatively well organized and funded with help from the World Meteorological Organization, became the initial IPCC technical contacts, although these researchers do not usually have experience in measuring activities that give rise to greenhouse gas emissions. The US country study programme has funded a wide variety of institutions to research emissions and strategies including meteorological, consular and non-profit bodies. Ultimately, the equivalents of the Departments of the Environment or Environmental Protection Agencies have the responsibility for compiling the inventory in most developing countries. The environment ministries, some of them newly created, will in many cases be the agency to review agency policies in sister departments of forestry, agriculture and energy. Funded by the Global Environment Facility (GEF), regional multilateral development banks, and the US and other developed country governments, greenhouse gas emissions research could play a role in helping to strengthen the scientific and public credibility of these organizations. As we shall see later, this would be institutional innovation of a high order.

Compared with past environmental agreements, the demands made by the UN FCCC on scientific and financial resources and the availability of such resources for research on the emissions baseline and plans have been enormous. Shortly before the Climate Convention opened for signature in Rio de Janeiro in June 1992, the United States pledged to provide $25 million for the study of greenhouse gas emissions in developing countries. Pilot studies totalling $2–3 million for national emissions research in 12 countries – developing and in 'economic transition' – were among the first expenditures of the GEF, the international environmental funding arm of the World Bank and the UN Development and Environment Programmes (for an assessment, see Jordan 1994). The Asian Development Bank has also spent considerable resources on emission inventories in South and Southeast Asia.

Different environmental bureaucracies have different views regarding innovation and change; resources expended on inventory and abatement research could enhance the clout of newer agencies that might be less resistant to shifts in policy than the more established agencies. The energy and agricultural agencies are often older and, because of a historic responsibility for energy and food supply, retain greater ties and empathy with the private sector enterprises within their purview. Environmental agencies, on the other hand, founded with the objective of controlling pollution some two decades ago, are likely to adopt more challenging positions on policy *vis-à-vis* the status quo. These agencies often had a key role in international environmental agreements such as those on stratospheric ozone (the Montreal Protocol) and acid deposition agreements (the Long Range Transport of Air Pollution). Their roles in developing measures related to climate change are more *ad hoc* given that, to date, treaty compliance through the publishing of national 'communications' provides broad scope for policy planning as analysed further in Chapter 6. These institutional roles may progress differently in developing countries if a significant proportion of resources for the environmental agencies provided by the GEF or bilateral funding arrangements, or through emissions permits or taxes, is earmarked for greenhouse gas emissions assessment and abatement projects in environmental agencies.

THE EC'S INVENTORY PROGRAMME

In 1993 the European Community agreed, in a decision known as the monitoring mechanism, to evaluate the climate plans of the Member States. This initiative is further discussed in Chapter 6. The monitoring mechanism decision recalls the Long Range Transboundary of Air Pollution Convention of 1979 through which the Community organized a multilateral group of government representatives and academics to evaluate national reports of SO_2 emissions from all the Member States. This expert group helped to harmonize the technical basis of the emissions baselines and of the reduction targets that were eventually agreed upon. In the case of climate change, the monitoring mechanism requires Member States to 'devise, publish and implement national programmes' for limiting their anthropogenic emissions of CO_2. Member States are also required to publish estimates of anthropogenic CO_2 emissions annually. This compares with the FCCC's direction merely to publish inventories 'periodically'. The monitoring mechanism also goes beyond the requirements laid down by the FCCC in calling for trajectories of national CO_2 emissions between 1994 and 2000, and by requiring a description of policies and measures aimed at increasing sequestration of CO_2 emissions.

The EC decision has indirectly improved the reporting and review of national communications. Its annual reporting requirement helps to fill a gap in the Climate Convention, which does not require countries to report annually on plans and emissions after their initial communication for 1990. In addition, the Community announced in 1991 that the CO_2 emissions reports to the EC would be available for public review, a decision that provided momentum towards providing public access to national communications when the issue came up before the INC in 1994. The annual reporting requirement may also help to systematize the inventory research. For example, partly because of this requirement, the UK decided to update its estimates of emissions from all gases annually even though annual reports are not required by the FCCC. The UK believes that this exercise will help its agencies to maintain competence in the emissions estimation procedures in which they have invested heavily (Penman, 1994).

The EC has also prepared its collective emissions communication in its role as a party to the Climate Convention. The communication includes the requisite inventory, EC strategies, and description of circumstances. As the EC ratified the Climate Convention ahead of several of its Member States, the communication was unable to report the sum total of all the member country emissions inventories. Accordingly, the EC derived baseline emission estimates for several Member States. The EC communication makes use of data from the EC statistical bureau, which has the responsibility for reporting information related to the sulphur dioxide and nitrogen oxide agreements. The EC adjusted the information produced by its agencies so that the emissions reports would conform to the IPCC guidelines. The Community also took the further step of changing Member States' emissions estimates if it saw mistakes that it believed were obvious, e.g. the reporting of irrelevant emissions, such as methane from natural, rather than anthropogenic, sources (Lammers, 1994).

This exercise was an important signal in that it revealed the determination of the EC to take a lead in the monitoring process. The steady rise of the European Environment Agency as a collective force in the assimilation and publication of European environmental data and 'State of the Environment' reports is a further indication of the role of data collection on a systematic basis as a political exercise of some significance. Readers wishing to know more about the Agency should consult ENDS (1995, 20–23).

The EC's experience with collaboratively developing emissions inventories precedes the IPCC/OECD work by several years. In the mid-1980s, the Commission of the European Communities and national institutes in the Member States set up a task force at the European Environment Agency to produce a complete and transparent air pollution emissions inventory for EC Member States that would cover SO_2, NO_X and VOCs (including CH_4), the emissions addressed in the Long Range Transboundary Air Pollution Convention. The inventory project on air emissions was called CORINAIR. By 1990 the programme encompassed 29 countries in Europe and included the greenhouse gases CO_2 and N_2O (also CO) in addition to the other compounds. The CORINAIR inventories covered mainly activities related to energy combustion but some fugitive emissions as well. The European emissions inventory provided the IPCC with the challenge of presenting an inventory system compatible with the European inventories, although less detailed. Out of respect for the European emissions research programme, the IPCC/OECD Joint Programme developed documents reporting on how to convert from CORINAIR to IPCC inventories. In part because of the CORINAIR programme, EC members' understanding of trace gas emissions from fossil fuel combustion has been strong compared with some of the other developed regions.

A review of the first group of national emissions communications to be published reveals that compliance with the Convention's reporting requirements by European countries and developed nations in general has been far better compared with other data reporting procedures in other international environmental agreements. For example, fewer than half of the countries required to report halocarbon consumption in line with the Montreal Protocol actually turned in complete reports within a year of their due date (UNEP, 1992). The EC countries failing to produce halocarbon data on time were the EC itself, Greece, Portugal, Italy and the Netherlands. Yet more than 80 per cent of the countries required to turn in their national communications on climate policies did so on time. Eighteen developed countries, including OECD and former Soviet bloc countries, were required to hand in emissions inventories by 21 September 1994. As of 21 October 1994, 15 countries had transmitted their reports to the INC Secretariat (Ellis, 1994). Of the developed country parties required to turn in their inventories in the autumn of 1994, the three that did not do so on time are Portugal, Iceland and the European Community (INC, 1994b).

THE UNCERTAIN BASELINE

Technical support and information exchange may be considered vital components of any international environmental regime (Szasz, 1991). The disclosure of technical information compiled in an approved and systematic fashion capable of being reviewed for accuracy by another party is known as transparency. Often a certain degree of transparency about a country's activities or stockpiles is necessary for other parties to feel reassured that the regime is successful. International agreements that leave the description of relevant activities up to the parties themselves may be setting the conditions for charges of strategic manipulation by member countries. Although there have been few direct charges of cheating concerning the inventories completed by Parties to the Climate Convention, the international team checking the reports admit that they have received insufficient information from the countries to duplicate even the best understood emissions source – energy consumption (INC, 1994b).

As was always intended, the IPCC has not produced a single required methodology and so the inventory requirements are open to individual interpretation to some extent. Countries may use any methodology they like so long as they fill out the mandatory emissions reporting tables. The Convention's language calls for inventories using 'comparable methodologies'. Many countries have in fact used different approaches for estimating emissions, and the workbook sometimes provides as many as three different approaches of varying complexity for the same emissions source. A selection of national inventories published in 1994 indicates that some national communications were based on approaches and default emissions factors that the nascent IPCC/OECD programme published years earlier (OECD, 1991; UK Department of the Environment, 1994a). The emissions factors upon which some countries based their assumptions for the more uncertain sources yield very different estimates than would be calculated in countries making use of the more recent recommendations and default emissions factors. Therefore, the 'comparable' methodologies called for in the Convention are not 'consistent', but are merely available for comparison. While the flexibility allowed in this approach may well contribute to problems in the long run as countries question each others' numbers, in the short run this more adaptable approach has encouraged countries to gain confidence in what is by any standards a complicated enterprise. In any case to require identical methodological approaches would have pulled countries down to a lowest common denominator of relative methodological simplicity because the data available in many countries differ from source to source. Despite the lack of error bars and qualifying statements in the national communications, national governments and their research agencies generally do not believe their own baseline estimates. In the UK, for example, the Department of the Environment convened for three years, through its Watt Committee on Energy, a group of experts on different industrial and agricultural processes related to CH_4 emissions. At the end of the study period, the Committee issued a report providing point estimates for UK CH_4 emissions sources. The researchers attached the data rating 'D' to most of their estimates. Coal mining, which scored a 'B' in data quality, had the highest rating (Watt Committee on Energy, 1994). The uncertainty in the estimates of CH_4 and N_2O emissions is far greater than experienced with SO_2 data regulated under LRTAP, and halocarbon consumption in the Montreal Protocol.

On the other hand, for some countries the accuracy of CO_2 emissions estimates from energy consumption and cement production can be within a few per cent of the measured value for some countries. This is because energy consumption and the carbon content of coal, oil and gas is usually known reasonably well and the possible range of emissions factors from fuel combustion – CO_2/fuel ratios – is narrow for all energy sectors aside from transportation. But emissions estimates for CO_2 from land use changes, as well as all the CH_4 and N_2O sources, are unlikely to approach this level of accuracy within the next few decades, if ever. The cost and hence feasibility of improving estimates of greenhouse gas emissions varies considerably from country to country. Generally, sources characterized by relatively few large sites in a country can be assessed more accurately at given levels of monitoring costs. The higher the proportion of emissions arising from larger sites that are surveyed, e.g. large landfills and coal mines, the greater the potential for more accurate national inventories for these sources. For other sources, such as rice paddies, biomass burning and clearing of primary forests, sufficient improvements in accuracy will be achieved at greater effort and expense through satellite surveillance and, more importantly, ground-truthing.

A selection of the national reports including those of the four countries studied – the UK, Germany, Italy and Norway – reveals some patterns in the reported 1990 baseline emissions that are surprising when compared with estimates completed a few years earlier. The CH_4 emissions estimates published in 1993 and 1994 are higher than estimates released several years before, although the trends vary for the different sources, as shown in Table 3.1. The 'private' sources – energy production and consumption, which the relevant industry usually reports – are much lower than estimated by the IPCC working groups between 1990 and 1992. In contrast, the 'public' sources – namely, landfills – are reported to generate much higher emissions in Europe than estimated in the late 1980s and early 1990s.

Table 3.1: CH_4 emissions (kT CH_4)

Selected European countries	Energy Production		Landfills		Other Emissions[c]
	Country reports[a]	WRI,[b] SEI[d]	Country reports	WRI, SEI	Country reports
Germany	1539	440	2397	1500	2307
		2608		1312	
Italy	–	–	–	–	–
Norway	13	513	167	82	122
		117		78	
UK	1237	2340	1971	1100	1726
		2023		1521	
Total	2789		4535		4165

[a] 1990 base year.
[b] 1987 base year.
[c] biomass burning, livestock, rice cultivation, domestic sewage.
[d] 1988 base year.
Sources: WRI (1990); Subak *et al.* (1993); UK Department of the Environment (1994b); Ministero dell'Ambiente (1994); German Environment Ministry (1994); Norwegian Ministry of Environment (1994); INC (1994b)

One possible explanation is that no sectoral interest has a concern over inflated landfill emissions estimates. Another possibility is that the monitoring technology and professional effort deployed during the last few years have provided a more accurate assessment of these sources. These issues will not be resolved until CH_4 emissions estimates are totalled for many more regions, and the on-site monitoring of these sources is extended and evaluated. The national climate change descriptions that countries released in 1994 do not disclose all of the assumptions necessary to evaluate emissions scenarios. They describe trends and policies that are intended to limit the growth in emissions, but the communications fail to disclose the assumptions that would allow a policy analyst quantitatively to evaluate the credibility of the scenarios. In any case, the reports contain only summary boxes of national inventories whereas the technical description can only be found in half a dozen or more specialized reports. The national climate plans from the major western countries have been written seemingly with the media in mind, not the environmental technician or professional. Each has its own stylistic design and format and some are heavy on political rhetoric. For example, the US Climate Plan contains lists of programmes entitled 'AgStar' and 'GasStar' and so on (Clinton and Gore, 1993).

Irrespective of their non-technical language, most national communications are difficult to evaluate in terms of accuracy or reliability of commitment. NGOs have taken a significant role in reviewing the national plans and policies published to date. Coordinated by Climate Network Europe and Climate Action Network in the United States, a dozen environmental organizations have criticized the climate plans of their respective countries (Climate Action Network, 1994). The NGO studies warn that governments in most countries are not likely to meet their emissions stabilization targets. All NGO evaluations have pointed out the weakness of governmental resolve in alternative planning for high growth sectors such as transport. As a group, these climate networks have called for measures that go beyond emissions stabilization in favour of a 20 per cent reduction target for developed countries by 2005. These groups have been reticent about the emissions baseline, however. This is understandable given that advocacy groups have established positions and experience in such matters as carbon taxes, logging subsidies and halocarbon regulations. The most technically sophisticated environmental organizations have replicated official projections and have also provided their own evidence as to why their country is unlikely to meet a target in view of stated policies and the most recent information on economic growth. A convincing critique of a country's baseline, however, requires access to a large body of 'grey' consultants' reports describing limited site monitoring, unconventional probability distributions, and very detailed activity data. For example, in order to understand the basis for the UK estimate for methane from landfills, the necessary documents include two national laboratory reports, one unpublished, a report from the

Department of the Environment and one from the Department of Energy, as well as two specialized unit reports. Then the baseline is posed in terms of qualitative descriptions of data accuracy and probability distributions based on 'expert' (subjective) opinion. It is not surprising that NGOs have not chosen to invest their time on these matters.

Evaluation of the emissions baseline will therefore depend largely on the skill and thoroughness of the responsible organizations charged with this task by the Conference of Parties. The IPCC guidelines will continue to be developed. Improved default emissions factors should be published based on results from deployment of new monitoring technologies as well as expansion of current research, particularly on agriculture and land-use-related emissions factors. The Subsidiary Body on Science, Technology and Assessment should clarify various aspects of the emissions inventory. The reporting requirements, approved in 1994, include minimum data tables calling for studies of all anthropogenic sources and many sectors, but do not call for thorough documentation or the use of a specific methodology. The first report on national communications which was written by a team of experts from several different countries, seconded to the FCCC Secretariat, revealed a number of lapses in reporting emissions with useful suggestions for improvements. These included questionnaires to be filled in by Parties, and more precise requirements on the background and supporting material to be provided with the data tables (INC, 1994b).

CONCLUSIONS

The international scientific community and national bureaucracies have made dramatic progress during the last several years in improving the under-standing of the likely magnitude of individual sources of greenhouse gas emissions. Nonetheless, we still do not know what is the true baseline for any of the CH_4 and N_2O sources, or for CO_2 emissions and removals from land-use-change. This is to be expected given the site-specific nature of the emissions flux from these sources, and their wide dispersal throughout the world. Additional uncertainties have been introduced into the baseline estimates because many parties have not followed the IPCC guidelines closely enough or have not thoroughly documented alternative approaches. More specific reporting and documentation requirements would improve the ability to review baseline estimates and hence increase international confidence in the climate regime overall.

The INC initial review of national communications has taken care not to charge any individual countries with not complying with the aims of the reporting requirements. It is evident from the text, however, that some countries have not followed the reporting conventions properly. These faults include misallocating emissions, 'adjusting' figures to take into account climatic anomalies, and counting emissions produced in other countries but imported as electricity. Most of the discrepancies noted after review serve to inflate the emissions baseline rather than to diminish it. To maintain the integrity of the reporting regime, future reviews should make explicit that countries must correct accounting errors and thoroughly document all legitimate but alternative methodologies. In addition, the detailed data tables should be made available to all interested parties.

The success of the Convention rests ultimately on timely reductions in greenhouse gas emissions. If the national baseline inventories are distrusted by member parties, countries and advocacy groups are also likely to take less seriously indications of progress towards stabilizing or reducing emissions. If some of the 1990 emissions inventories are based on inflated figures, the value of the stabilization target will be deflated because global emissions will obviously be higher than they are claimed to be.

A dedicated international scientific commitment to standardize this effort, and to assist the developing countries to build their research and reporting capabilities, has emerged over the past few years. This commitment extends to international scientific organizations, national research bodies and many government bureaucracies. Ultimately, the data are only as good as the reported statistics on consumption and production, the assumptions made in the spirit of dealing with uncertainty, and in the quality and quantity of equipment deployed and interpreted. In the past, data anomalies and failures to report have gone unnoticed in many international environmental regimes, as the information systems of environmental treaties too obscure to capture the interest of government parties and a wider public languish unnoticed. This will not be the fate of national climate plans, which are becoming an accessible reference for a range of campaigns on the post-Rio agenda. But as the importance of the climate plans and the commitments that they represent become more firmly established, so too does the sensitivity of the baseline estimates and scenarios. Some countries have and will continue to find pressing reasons to misrepresent the data. It will take time and more institutional commitment to minimize the possibilities for misreporting, although the chief responsibility will in the end rest with national bureaucracies, which have access to the essential information. Increased involvement by subsidiary bodies in CoP, investments in national planning, and critiques by non-governmental groups will all be needed if this promising start is to meet expectations.

Here is a fruitful area for institutional adaptation as discussed in Chapter 4. The key will be the commitment of the reporting states to research, transparent reporting and broad participation in order to provide the necessary information brief. This in turn means that the political credibility of the reporting exercise needs to be sound. The EC effort, outlined in Chapter 6, based as it is on both national reporting and a single data analytical agency, should be most interesting to observe. If both reporting anomalies and interpretation disputes of data reliability can be resolved on a satisfactory basis, both within the European Environmental Agency and through reliable national reporting mechanisms, then the stage is set for an extended variant of this process within the Bonn Climate Change Secretariat. That in turn requires a recognition that national reporting is both a political act and a symbol of good faith in a collective outcome. The fundamental innovation is an attitude of mind that fully accepts the need for independent monitoring of national reporting, and in turn regards the periodic communications as a proper basis for further national action. If the European Union can do this, then the omens are good.

REFERENCES

Agarwal, A. and Narain, S. (1991) *Global Warming in an Unequal World: A Case of Environmental Colonialism,* New Delhi: India Centre for Science and Environment.

Brenton, T. (1994) *The Greening of Machiavelli: The Evolution of International Environmental Politics,* London: Earthscan.

Climate Action Network (1994) Independent NGO Evaluations of National Plans for Climate Change Mitigation, Second Review, Washington, DC, and Brussels.

Clinton, W. and Gore, A. (1993) *The Climate Change Action Plan,* Washington, DC: United States Government.

Ellis, J. (1994) Personal communication, Paris: International Energy Agency.

ENDS (1995) The European Environmental Agency. ENDS 240, 20–23.

GAO (1993) *Environmental Agreements are Not Well Monitored,* Washington, DC: General Accounting Office.

German Environment Ministry (1994) *Environment Policy: Climate Protection in Germany,* National Report of the Federal Government of the Federal Republic of Germany in Anticipation of Article 12 of the United Nations Framework Convention on Climate Change, Bonn: BMU (Federal Ministry of Environmental Protection and Nuclear Safety).

Hammond, A.L., Rodenburg, E. and Moomaw, W. (1990) 'Accountability in the Greenhouse', *Nature* 347, 705–706.

INC (1994a) *Plenary Matters Part V,* Final Report INC-10, Geneva: International Negotiating Committee.

INC (1994b) *First Synthesis of National Communications,* A/AC.237/81, Geneva: Intergovernmental Negotiating Committee.

IPCC/OECD Joint Programme (1991) *Workshop on National Greenhouse Gas Inventories, Estimation, and Reporting Transparency,* Bracknell: UK Meteorological Office.

IPCC/OECD Joint Programme (1994a) *The Greenhouse Gas Inventory Reporting Instructions,* IPCC Draft Guidelines for National Greenhouse Gas Inventories, Geneva: UN Intergovernmental Negotiating Committee.

IPCC/OECD Joint Programme (1994b) *The Greenhouse Gas Inventory Workbook* IPCC Draft Guidelines for National Greenhouse Inventories, Geneva: UN Intergovernmental Negotiating Committee.

IPCC/OECD Joint Programme (1994c) *The Greenhouse Gas Inventory Reference Manual,* IPCC Draft Guidelines for National Greenhouse Inventories, Geneva: UN Intergovernmental Negotiating Committee.

Jordan, A. (1994) 'Paying the Incremental Costs of Global Environmental Protection – The Evolving Role of the GEF', *Environment* 36(6), 12–20, 31–36.

Keeling, C.D. (1973) 'Industrial production of carbon dioxide from fossil fuels and limestone', *Tellus* 25(2), 174–198.

Khalil, M.A.L. and Rasmussen, R.A. (1994) 'Global emissions of methane during the last several centuries', *Chemosphere* 29(5), 833–842.

Lammers, E. (1994) Personal communication, Amsterdam: Institute for Environmental Studies, Vrije Universiteit.

Marland, G., Andres, R.J. and Boden, T.A. (1994) 'Global, Regional, and National CO_2 Emissions', pp. 80–95 in T.A. Boden, D.P. Kaiser, R.J. Sepanski and F.S. Stoss (eds) *Trends '93: A Compendium of Data on Global Change*, ORNL/CDIAC-65, Carbon Dioxide Information Analysis Center, Oak Ridge, TN, Oak Ridge National Laboratory.

Ministero dell'Ambiente, Italia (1994) *Programma Nazionale per il Contenimento Delle Emissioni di Anidride Carbonica entro il 2000 ai Livelli del 1990*, Rome: Italian Environment Ministry.

Norwegian Ministry of Environment (1994) Greenhouse gas emissions in Norway: inventories and estimation methods, Mimeo, Oslo: Ministry of Environment.

OECD (1991) *Estimation of Greenhouse Gas Emissions and Sinks*, Final Report from the OECD Experts Meeting, Geneva: UN Intergovernmental Panel on Climate Change.

Penman, J. (1994) Personal communication, London: UK Department of the Environment.

Smith, K. (1991) *Have You Paid Your Natural Debt?*, Honolulu, HI: Environment and Policy Institute, East–West Center.

Subak, S. and Clark, W.C. (1990) 'Towards the design of fair assessments', pp. 68–100 in W.C. Clark (ed.) *Usable Knowledge for Managing Climate Change*, Stockholm: Stockholm Environment Institute.

Subak, S., Raskin, P. and Von Hippel, D. (1993) 'National Greenhouse Gas Accounts: Current Anthropogenic Sources and Sinks', *Climatic Change* 25, 15–58.

Szasz, P.C. (1991) 'The role of international law in forming international legal instruments and creating international institutions', *Evaluation Review* 15(1), 7–26.

UK Department of the Environment (1994a) *Climate Change: The UK Programme*, The United Kingdom's Report under the Framework Convention on Climate Change, London: HMSO.

UK Department of the Environment (1994b) *National Communication: Greenhouse Gas Emissions*, London: Department of the Environment.

UNEP (1992) *Report of the Ozone Secretariat to the Implementation Committee on the Reporting of Data in Accordance with Articles 4, 7, and 9 of the Montreal Protocol*, Nairobi: UN Environment Programme.

Von Hippel, D., Raskin, P., Subak, S. and Stavisky, D. (1993) 'Estimating greenhouse gas emissions from fossil fuel consumption: two approaches compared', *Energy Policy* 22(4), 691–702.

Watt Committee on Energy (1994) 'Methane Emissions', Report of a Working Group Appointed by the Watt Committee, London: Chameleon Press.

WRI (1990) *World Resources 1990–1991*, Oxford: Oxford University Press.

4

SOCIAL INSTITUTIONS AND CLIMATE CHANGE

Tim O'Riordan and Andrew Jordan

The concept of 'institution' is very broad and diffuse. In many respects, it is another example of what Gallie (1955) terms an 'essentially contested' concept; an idea that can only be clarified through regular argument. The notion of institution applies both to structures of power and relationships as made manifest by organizations with leaders, memberships or clients, resources and knowledge; and also to socialized ways of looking at the world as shaped by communication, information transfer, and the pattern of status and association. This chapter seeks to clarify how the notion of institution extends beyond organizational form, rules and relationships into more fundamental social and political factors that determine how people think, behave and devise rules through which they expect everyone else to play. It then attempts to portray how institutions permeate the politics of climate change, both in its framing as a social and environmental 'problem', and in the devising of solutions, and hence what factors are needed to alter institutions so they can deal more effectively with the delivery of regulation and behaviour designed to reduce global warming.

Creating and managing climate change has to take place through institutional arrangements. Any effort to identify with the causes of that change, or to adapt to its consequences, must also address the medium of institutional behaviour. Institutions permeate all aspects of social organization, both formal and informal. Without institutional arrangements no society could survive as a collective entity. Institutions both create and chase circumstances. As we note in Chapter 1, the science of climate change evolved through an interdisciplinary coupling of formerly separated investigations of climate, oceans, ice and vegetation. In that sense the very concept of 'climate change' was *created* by the institutional alignment of scientific enquiry. But the human response also *chased* the scientific findings by generating an international trans-scientific panel, and by creating an international political and legal agreement to justify and enforce common action.

Here is a good example of the institutional perspective at work. The construct of scientific enquiry dictates the manner of its investigation. Brian Wynne (1994, 171–172) argues that the IPCC process channelled the discussion of climate change into a consideration of what was a tolerable degree of climate alteration, thereby avoiding the moral and political crises of north–south inequality, the debt trap and differential use of the earth's life support functions by rich and poor. Because the north shaped the scientific agenda, so its biases influenced the agreements that were reached as to the severity of the problem, the timing of response, the acceptability of solutions and even the determinants of a cost benefit analysis.

Policy is shaped by existing organizational structures, as well as by informal networks of communication that in turn are the product of values, norms and expectations. The science that provides the perspective of cause and effect is the product of a variety of institutional patterns. These range from the formal deliberating bodies such as the IPCC and research institutes scattered across the globe, to the informal liaisons between policy analysts and policy executives, regulatory agencies and the day to day actions of billions of people. Climate change will be determined by the willingness of those billions of decision makers to change their ways. This could be achieved by persuasion, command, education, taxation, moral arguments, or changes in the notion of property rights. Any combination of these responses involves institutional change of a profoundly social kind.

No matter how chaotic all these relationships may seem, they could not function without some sort of order and guiding principle. This is the essence of social institutions, and this forms the basis of the analysis provided by this chapter. Its purpose is to explore the role of 'institutional' factors in explaining the climate change policies adopted by various European governments. Subsequent chapters reveal enormous geographical and temporal variations in European government responses to the threat of climate change and, more recently, the agreement to a United Nations Framework Convention on Climate Change (UN FCCC). This chapter aims to explain how these cross-national differences are closely linked to the specific institutional settings within which climate policies are made.

WHAT ARE SOCIAL INSTITUTIONS?

There could be a thousand definitions of the notion of institutions and the subject matter would not be exhausted. What tends to complicate matters is the fact that the word appears to have taken on a particular meaning in different disciplines. For sociologists, the study of institutions is central to understanding the organization and functioning of all societies. Indeed, within the functionalist paradigm, sociology *is* 'the science of institutions' and other 'social facts' (Durkheim, 1950, 1). Social theorists such as Eisenstadt (1968, 408–409) regard the concept as a socially organizing set of relationships that govern the basic problems of ordered social life. He visualizes the characteristics of institutions as both a set of normative rules that are shared by a society so as to retain cohesion and control, and patterns of behaviour that operate according to norms and expectations, and in so doing give order and meaning to social life and government. Giddens (1986, 8), meanwhile, refers to institutions as 'commonly adopted practices which persist in recognisably similar form across generations'. 'A society', he explains, 'is a cluster, or system, of institutionalised modes of conduct . . . [which are] . . . modes of belief and behaviour that occur and recur . . . across long spans of time and space'. Smith (1988, 91) visualizes institutions as 'stable, valued, recurring patterns of behaviour', a definition which encompasses 'fairly concrete organisations, such as governmental agencies, but also cognitive structures, such as the patterns of rhetorical legitimation characteristic of certain traditions of political discourse . . . [and] "belief systems"'. Scholars of international relations use the term in a similar fashion: Haas and his colleagues (1993, 5), for example, regard institutions as:

> persistent and connected sets of rules and practices that prescribe behavioural roles, constrain activity and shape expectations. They may take the form of bureaucratic organisations, regimes (rule structures that do not necessarily have organisations attached), or conventions (formal practices).

Finally the *Encyclopaedia of Political Science* defines an institution simply as 'a locus of a regularized or crystallized principle of conduct, action or behaviour that governs a crucial area of social life and that endures over time' (Gould, 1991, 290). Gould (1991, 291) warns against confusing activity with informal patterns which emerge as that activity is repeated over and over again, and more formal rules which are devised to regulate it. An activity may take place through informal arrangements, for example through kinship patterns or the evolution of expected codes of conduct. One example would be neighbourhood home insulation schemes that provide a sense of communal obligation to a shared outcome. Another might be the emergence of a car pooling culture to cut driving needs, or the formation of a cycling fraternity that enjoys the freedom, and discomfort, of avoiding motorized transport. A third case could be the coordination of village communities in Bangladesh to provide single-band radios, platforms of refuge and food redistribution schemes in the event of coastal flooding (see Chapter 11). These activities need neither structure nor regulation, but they are most certainly institutional arrangements of the 'bottom-up' kind discussed by Wynne (1993).

At the very heart of the meaning of institutions, therefore, are a number of highly interrelated concepts, some of which we will subsequently explore in greater detail.

- Institutions embody *rules* that govern values, norms and views of the world. Institutions *regulate* behaviour via socially approved mechanisms such as the rule of law and the accountable exercise of power.
- Institutions have a degree of *permanence* and are relatively stable: They are, for Giddens (1984, 24), 'the more enduring features of social life'. Institutions are *patterns of routinized behaviour*. Nonetheless, institutions are continually being renegotiated, in the ongoing interplay between human agency and wider social structures. There is, in Giddens' (1986, 11) memorable phrase, a 'double involvement of individuals and institutions: we create society at the same time as we are created by it'. Individuals do have an element of free choice in what they do, but they are also institutionally conditioned in what is right and wrong; what is possible and what is not; what is legitimate and what is plainly unacceptable in a modern society.
- Institutions are *cognitive and normative structures* which stabilize perceptions, interpretations and justifications (Olsen, 1992, 12). Institutions are attitudes of mind; they are not directly observable. Organizations, as we have seen, have a physical presence.
- Institutions determine what is *appropriate, legitimate and proper*; they define obligations, self-restraints, rights and immunities as well as the sanctions for unacceptable behaviour. What is or is not 'appropriate', 'proper' or 'right' is normally not an individual matter. Individuals continuously relate their own needs and preferences to socially determined structures such as institutions via a process of 'reflexivity' (Giddens, 1984, 1991).
- Institutions help to *structure* the channels through which new ideas are translated into policy and new challenges receive a government response. Institutions, Hall (1993b, 109) posits, are 'critical mediating variables, constructed by conscious endeavour but usually more consequential than their creators intended. They are not a substitute for interests and ideas as the ultimate motors of political action, but they have a powerful effect on which interests prevail'.

Institutions are vital in the processes of identifying and responding to threats to survival or to conflict and social order. The mechanisms of predicting outcomes, or of organizing response, of preparing for possible danger and of accommodating to stress or hardship are constitutive of the political debate about global environmental change. This is why institutions have to involve rules, regulation and legitimacy – or social justification of the exercise of authority – for the social good. It is also why institutions are constantly changing, adjusting subtly and sometimes turbulently to the needs of the times as interpreted by society in a myriad of ways. The direction of change is therefore not random or unpredictable. Change can be analysed in terms of the existing relationship of values and norms, of social and political organization, of the jockeying for power and of the constant interchange between internal views of the world and external forces which are not always controllable by those in power.

INSTITUTIONS AND THE 'NEW INSTITUTIONALISM'

As may be expected with such an amorphous concept, theorists have tried to slice the institutional cake in a variety of ways. In the analysis which follows we describe and explain those strands of contemporary political and socio-logical theory which have some affinity with 'new institutionalist' theory. Our analysis is in no way comprehensive or exhaustive; the literature is simply too diffuse. Rather our hope is that the broad framework we develop will be useful to those who wish to make sense of the complicated process by which different European states are attempting to respond to a common problem, namely climate change. Each strand of theory illuminates a differ-ent role for institutions in the patterning and dynamics of climate politics. Their compatibility stems from the fact that they all see individual prefer-ences and perceptions as being crucially shaped by the institutional context within which they operate, be that a bureaucracy or an organization, a grouping of like-minded people (such as a policy community), or a pre-determined set of social commitments. In the main, they tend to down-play more 'agency'-centred theories, which portray the individual, rational decision maker as the primary unit of analysis.

The study of institutions is undoubtedly experiencing a renaissance throughout the social sciences (Di Maggio and Powell, 1991, 2). In eco-nomics, politics, public choice, sociology and organizational analysis the study of institutions has emerged in the last 15 years as a 'new' focus for analysis (Blumstein, 1981; Kiser and Ostrom, 1982; Ostrom; 1986; Hodgson, 1993). For Powell and Di Maggio (1991), the renewed interest in institutional phenomena stems from a growing dissatisfaction with the atomistic, agent-centred, 'behavioural' accounts of social processes that characterized the social sciences in the 1950s, 1960s and 1970s.

In political science, the leading proponents of the 'new institutionalist' perspective are James March and Johan Olsen (1984, 1989). They emphasize the important role institutions play in structuring politics: 'social, political and economic institutions have become larger, considerably more complex and resourceful and *prima facie* more important to collective life' (March and Olsen, 1984, 734). March and Olsen also suggest that 'most of the major actors in modern economic and political systems are formal organizations, and the institutions of law and bureaucracy occupy a dominant role in con-temporary life' (March and Olsen, 1984, 747). In attempting to define the new institutionalist perspective, they aver that:

the new institutionalism insists upon a more autonomous role for political institutions. The state is not only affected by society but also affects it. . . . The bureaucratic agency, the legislative committee, and the appellate court are arenas for contending social forces, but they are also collections of standard operating procedures and structures that define and defend interests. They are political actors in their own right.

(March and Olsen, 1984, 738)

At its simplest the new institutionalism should be viewed as a heroic attempt to join two very different kinds of analysis, each of which refers to institutions in a quite specific manner. The first is the *behaviouralist, rational actor view*. This has dominated mainstream political science, economics and public choice theory building for the last three decades. In this view, human behaviour is portrayed as the result of rational calculations designed to further self-interests. It thus has more in common with agent-centred theories of sociology, which lay stress on the fact that individuals are thinking, feeling and acting subjects who create the world around them, rather than being the products of that social world (as presented by Weber, Mead, Simnel). The values as preferences of individuals, the power and resources they hold, are treated as exogenous 'givens' by agent-centred theories. Rational actions ultimately result in rational, efficient and functional outcomes. Behaviouralists tend to see institutions as the conscious products of human design; as epiphenomena or 'congealed tastes' (Riker, 1980). Game theorists tend also to view institutions as dependent variables, specifically as the 'equilibrium outcome[s]' (Sened, 1991, 381) of successive rounds of 'games' (Schotter, 1981; Sugden, 1986). But rarely do they ask where individual preferences originate, or consider the inter-relationships between interests and institutions. Institutions are largely viewed as the conscious and deliberate outcome of human action.

On the other hand, there are *functionalist* accounts of human behaviour which are more structurally centred and 'sociological' in their perspective. From a functionalist perspective, the structural properties of society form a constraining influence over individual actions. Institutions regulate the use of authority and power and provide actors with resources, legitimacy, standards of evaluation, perceptions, identities and a sense of meaning. Much Marxian theory also falls into this category since it downplays the importance of individuals and groups in favour of explanations of human behaviour that emphasize broader social class conflicts, processes of capital accumulation and longer term shifts in the economic 'base' of society and the 'impersonal logic' (Dunleavy and O'Leary, 1987, 211) of historical materialism (see also Dunleavy, 1985).

New institutional accounts of politics attempt to find a balance between these two traditions by emphasizing the relative *autonomy* and importance of social institutions while giving attention to the attempts made by individuals and groups to shift them to their advantage. It is worth reflecting on what this means. First, institutions are viewed as political actors in their own right; they constitute the 'rules of the game'. They help to determine who interacts in political conflicts, how these 'battles' over issues are in turn biased in favour of some participants and against others, and how the outcomes are affected by what is seen to be legitimate, 'acceptable' or 'proper' in society. In terms of climate politics, existing institutionalized patterns of behaviour would rule certain policy responses in but rule others out. All the national case study chapters which follow, for example, indicate how difficult it is for countries to grapple with the transport sector in formulating CO_2 reduction plans. The freedom to move about and to drive or fly without restriction is, it seems, regarded as right and proper. These widely held social beliefs are rarely if ever seriously challenged when it comes to making decisions about climate policies. More broadly, we see how each country's response to the climate change issue is severely conditioned by the pre-existing configuration of ministries and regulatory offices, economic incentives, societal expectations, and political opportunities. For instance, the Germans have capitalized on the economic restructuring of the former East German *Länder* to reduce CO_2 emissions reduction, put into place with relatively little public dissent. And because institutions in the broad sense accepted here can reinforce greenhouse gas emitting behaviour, so any institutional reform to redirect emissions and land use practices will have to evolve from within an alert and accommodating society. This is why institutional adjustment for climate change is proving so profoundly problematic. Put simply it has to emerge 'inside out' in a modern democracy when all the 'outside-in' signals are propelling society in the other direction.

Second, institutions have 'two faces' (Windhoff-Heritier, 1991, 41): they *enable as well as constrain choices*. This is a key tenet of Giddens' theory of structuration (Giddens, 1984, 169–226). Institutional routines and repertoires provide stability in a complex and changing world, and make it easier for decision makers to come to decisions. A fundamental aspect of structuration theory is the concept of routinization – of doing things habitually and without conscious choices – what Giddens refers to as the 'recursive' nature of social life (1984, xxiii). It is evidently easier to carry on and do what has always been done than rapidly change direction. In the arena of climate politics, for example, we see how the need to reduce greenhouse gas emissions has been incorporated into a pre-existing pattern of policies and practices. Again, the norms of public acceptability and tolerance rule certain options 'in' and others 'out'. The Norwegian government found out this aspect too late. Its early enthusiasm to impose a carbon tax ran into a huge resentment over the effect on the cost of ferry transport, so vital to the economy of coastal communities whose votes were so vital to the coalition government. The resulting mix of a reduced carbon tax and other regulations was a necessary political accommodation. But as a result, Norway may not meet its targets. Similarly, pre-existing institutional arrangements – the composition of elite policy communities, constitutional arrangements, social norms and 'styles' of decision making – predispose certain countries to particular forms of policy intervention. Weale (1992) applies a broadly 'institutional' model of politics to inform an analysis of what he terms the 'new' politics of pollution control.

At a much broader level, the new institutionalism marks another stage in the ongoing dispute within social science over the extent to which individual conscious 'agents' are constrained to think and act in the ways that they do by broader social structures (Lukes, 1977, 3). Behaviouralists (instrumentalists) tend to place the human conscious agent at the centre of their analysis, giving much less attention to the structural constraints on free choice and action. Behaviouralists generally study observable behaviour, noting how individuals choose voluntarily between a variety of possible alternatives. Functionalists (structuralists), on the other hand, seek to show how overarching structures – be they 'class structures, kinship structures, occupational structures, opportunity structures, age structures' (Lukes, 1977, 8) – limit the free choice of individuals to a range of predetermined choices. On this view, political outcomes are not determined by the actions of individual agents, but by more fundamental aspects of social change such as class relations and economic restructuring. Benton and Redclift (1994, 8), for example, note how individuals are '"locked into" patterns of daily activity . . . which they know to be environmentally destructive' even though they may want – and even try – to act in a more environmentally sustainable manner. Clearly, it would be misleading to analyse an individual's or a pressure group's behaviour without also taking cognizance of the complex of social, historical and economic context within which they make choices, or to take an excessively 'structuralist' view and assume that individual choices are simply determined by their class position, bureaucratic role or whatever. Rather, we need to recognize that

> social life can only properly be understood as a dialectic of power and structure, a web of possibilities for agents, whose nature is both active and structured, to make choices and pursue strategies within given limits, which in consequence expand and contract over time.
>
> (Lukes, 1977, 29)

ELEMENTS OF AN INSTITUTIONAL EXPLANATION

Thus, the so-called 'new institutionalism' – like post-modernism – is an attempt to resurrect past theories in contemporary costumes (see March and Olsen, 1984). It is also a brave effort to cross the social science disciplines of political science, sociology, economics, law, anthropology and philosophy. But the lax and inconsistent manner in which different writers have sought to interpret the new institutionalist literature has hitherto militated against a steady, cumulative development in theoretical thinking. One sceptic, for example, claims that:

> The ambiguities in [March and Olsen's article] have allowed very different academic tribes to see it as a source of encouragement. . . . New institutionalism is a label indicating a disposition to oppose the political mainstream rather than agreement on the content of a new approach.
>
> (Jordan, 1990, 482–483)

Nonetheless, we remain convinced of the general efficacy of a broadly 'institutional' approach to the analysis of climate change politics in and across different nation states. In the remaining part of this chapter we identify those aspects of contemporary social theory which we feel could be, and in some cases have been, integrated into a much broader 'new institutionalist' paradigm. These are:

- *Institutions as types of policy networks.* Policy networks of like-minded people cluster around agents of action such as government departments or industrial trade associations to promote certain policies and edge out others.
- *Institutions as structures of political power and legitimacy.* Established procedures, routines, codes of conduct and *modi operandi* are the means by which political power can be applied and protected in liberal democratic societies (Saunders, 1976).
- *Institutions as 'standard operating procedures' and barriers to 'rational' decision making.* Individuals within organizations are motivated partly by their personal ambitions but also the 'role-interests' given by their agency position.
- *Institutions as national policy 'styles'.* National political systems tend to have a particular 'style' of processing policy problems, irrespective of their origins and precise characteristics. The national 'style' is said to be intimately related with the institutional configuration in a given country.
- *Institutions as international regimes.* Independent sovereign states co-operate through the negotiation of international regimes to resolve conflicts, mediate tension and pursue peaceful cooperation.

To guide the reader, we show how these different theories – and in some case fragments of theories – can be used to shed light upon the direction and pace of climate politics in Western Europe, and especially the EC.

Institutions as policy networks

According to Jordan and, more recently, Smith (1993, 70), one of the most important institutions within which policy making and implementation takes place is the policy network or community (Smith, 1993, 70). A policy network is a generic, meso-level concept, designed to shed light upon the relationship between state and non-state actors such as pressure groups (Rhodes, 1990). The policy network idea is based on the observation that policy making tends to be fragmented into discrete 'issue areas', and that most non-controversial issues are dealt with in an incremental manner by a limited number of actors within small groups of participants from governmental and non-governmental agencies. For Jordan and Richardson (1987, ix):

> political activity [is] largely conducted in policy communities of interest groups and government departments and agencies. In these communities . . . party politics has only limited impact . . . most political activity is bargained in private worlds by special interests and interested specialists.

Network theory posits that within each specialized 'issue area', interest groups cluster around one or a number of government departments and other key decision fields in the hope of influencing policy. But the flow of resources is not merely one way; the state also needs the support of certain groups to make, legitimize and eventually implement policy. At one end of a continuum of policy network types, Marsh and Rhodes (1992) identify the development of tightly formed, fairly autonomous and highly stable *policy communities*, normally formed around one particular government department. Policy communities are stable networks, with a limited number of participants. According to Jordan (1990, 327):

> A policy community exists when there are shared 'community' views on the problem. Where there are no such shared attitudes, no policy community exists.

The link to institutional theory lies in the notions of a shared ideology or world view which participants in the community share. To all intents and purposes, a policy community is a strongly institutionalized form of policy making arrangement. Players within the community share an appreciation of what is important in political life; there are shared values and a specific 'world view' within the group. Members of the group interact frequently, there is a constant and two-way flow of resources between actors, and members tend to agree upon what specific problems justify a policy response and how, in turn, that response should be structured. Within the community there is a shared ideology or an 'acceptance or dominance of an effectively unified view of the world' (Dunleavy, 1981). In the case study chapters that follow we see policy communities operating in the areas of fiscal (or tax) matters, transport and notably the car and the road lobbies, and industrial competitiveness. These form powerful client groupings around finance, transport and industry–employment ministries respectively. Though environmental groups have succeeded in entering some of these communities, more by force than by invitation, they are not yet in a position to bring about a realignment or to undermine the extant policy paradigm (see below).

At the other end of the continuum is the much more weakly institutionalized *issue network*, characterised by, *inter alia*, a greater number of participants, a lack of consensus on the problems at issue and the means by which they should be tackled, more open access to different groups and, consequently, less permanence and stability. Issue networks tend to encompass two or more government departments and there is plenty of opportunity for other groups to become involved. The type of policy network which develops is critically determined by the characteristics of the issue area concerned. Smith (1993, 234) suggests that policy communities tend to develop in situations where the state needs the assistance of nongovernmental groups to implement policy, while issue networks tend to develop in more politicized issue areas or where resource dependencies are not as pronounced.

In general, climate politics, unlike say agricultural or defence politics, are *pervasive* in the sense that they impinge upon a number of different policy communities and government departments. What emerges, then, is a situation where different policy communities manoeuvre within a much wider issue network and the government is forced to coordinate policy across a range of departments and interest groups. Where policy problems – and climate change is a paradigm example – overarch several policy sectors we are more likely to see what Rhodes (1985) terms a 'policy mess', characterized by policy coordination problems, conflicts between government departments backed by their respective policy communities and legitimation crises (created as central government attempts to parry the onset of political or economic embarrassment by driving inconsistencies into other realms of policy). The following chapters are replete with examples of 'policy messes' that have been formed as states have started to address the issue of climate change in their policy and programme making. The evidence will form the basis of the concluding comments in Chapter 12.

If policy networks are partly the structuring institutions how do they influence the development and direction of policies? First and foremost, policy networks provide an enabling and constraining filter to the development of policy. Within a community, consensus and ideology restrict the discussion of future policies to certain alternatives judged to be permissible and legitimate within the group. In the chapters that follow we will see the influence of stable policy communities operating to thwart the carbon energy tax proposed by the European Commission, and to limit technological innovation in the efficiency of electrical appliances and thermal insulation. But we shall also spot dynamic coalitions around anti-road protests, integrated transport options and job creation around household energy efficiency. For the purpose of this analysis, the significance of policy networks lies in this relative freedom of formation, their consensus creating role around zones of conflict (or the opposite, namely their power to block change), and their evaluating function both within their own networks and amongst society at large. In the study of European climate politics, the policy communities that appear to matter revolve around science, energy supplying, generating and using bodies, regulatory agencies, and bureaucracies of governmental machinery controlling finance, trade, industry, transport, the environment and foreign affairs. A critical factor in the shifting coalitions of these various communities lies in the need for cooperative and coordinated international action backed by legal convention and moral responsibility for the tolerance of ecosystems and the well-being of peoples yet unborn. But as noted earlier, we show that climate politics only evolve within policy networks with their own policy biases: they are not policies in themselves.

Second, the network model shows how institutions determine the pattern of interest group participation in decision making. The relative 'openness' of the network is not only determined by organizational factors (for example, there are always important committees to attend, august bodies to be members of and advisory panels to sit on): less perceptible factors such as ideology, 'world view' and legitimacy also play a role. Participants in stable communities are bound to abide by certain rules, which dictate standards of reasonableness, legitimacy and acceptability. To enter a community, pressure groups must show that they will abide by the prevailing ideological consensus within the group. Groups which do not − the following chapters give myriad examples − are routinely excluded from the locus of power and decision making. In some cases, a conscious decision is made to exclude other groups from important meetings, but in others groups are excluded not because of the conscious decisions of individuals but by the very process of decision making (Smith, 1993, 73). This is because, over time, patterns of consultation and formal and informal interaction (i.e. exclusion and integration) become habituated and codified within the policy making system. Certain groups hold power because they have always held power; they are organized 'into' politics because they can be trusted to observe the norms of the community. Meanwhile, decision making within the group almost becomes depoliticized; that is, unless 'outsiders' can mobilize sufficient support to challenge the status quo.

The chapters which follow indicate the emergence of a similar pattern across a variety of policy sectors in every country studied. It is important, therefore, to view policy networks as institutionally structured and therefore semi-permanent, but not immutable: it is the 'opening' and 'closing' of networks which gives the policy community a dynamic quality which is sometimes missing from more deterministic accounts of politics. Policy communities are never completely closed; even the most cohesive of communities can undergo changes over time as groups make the transition from 'outsider' to 'insider' status, or as structural and otherwise 'external' factors undermine the status quo in the policy domain (Maloney *et al.*, 1994; Richardson *et al.*, 1992; Smith, 1991). Smith (1993, 91–97), for example, shows how policy networks shift according to changes in political party, problem analysis, technology, external relationships, key personalities entering and leaving, and internal restructuring. We shall see this in the chapters to come most especially in the realm of transport. The anti-road protest manifest in the rise of eco-vandalism, the formation of alliances between middle class property owners and the rootless 'traveller' culture is the very stuff of incipient institutional realignment. Couple that protest to significant policy moves on gasoline taxes and congestion levies and one can see signs of gradual institutional reform, especially if all this relates to local community involvement in new mixes of mobility and shifts in land use decision making.

Third, policy networks are an important means to envision and explain continuity and change within political life. As far as continuity is concerned, stable policy communities – for example, those which cluster around the transport and electricity generation sectors – become an effective constraint to radical policy change; they tend to change slowly if at all. A largely non-executant government faced with a new policy problem such as climate change is forced to negotiate with and gain the support of relevant and important communities if it wants its policies implemented. In fact, we may go further still and hypothesize that the precise mix of policies chosen by central government may be a function of what can reasonably be achieved rather than what is best. In terms of change, other analysts have shown how policy networks *open and close* in relation to the stages through which a policy issue passes in its lifetime (i.e. policy communities crystallizing out of networks and then, with time, breaking down again) (Richardson *et al.*, 1992). Sabatier's model of policy change over longer time periods would also be of great relevance in tracking longer term shifts in policy (Sabatier, 1987, 1988). We return to the subject of institutional change below.

Despite some progress on the transport front, we will discover from the case studies that policy realignment, so critical for the successful reduction of greenhouse gases, is proving to be more difficult than had been hoped. In Italy there is virtually no movement in a country beset by political scandal and warring governments. In Germany, the industrial interests remain to be persuaded to pass beyond an efficiency motive. In the UK deep political controversy surrounds tax and regulatory policy at a time when cohesion is regarded as vital. Norway simply cannot do the job without a carbon tax, because of the high carbon element in transport fuels, but cannot also generate the new alliances between environmentalists, community activists, regional economic development strategists and macroeconomists to make headway. Climate politics is a mighty tough nut to crack.

To conclude, network models of policy and politics give a more concrete example of how the 'organization [of politics] makes a difference' (March and Olsen, 1984, 747). It is difficult to explain political outcomes by focusing purely on the structure and resource capabilities of individual actors such as departments of state and pressure groups. Rather, outcomes are determined by the interactions between groups, the underlying rules of the game, common values and shared assumptions which underpin the procedural values which regulate interaction in a policy community, and the perceptions which one group has of another (which are engendered by past activities).

Institutions as structures of political power and legitimacy

To understand the ebb and flow of the climate change issue in national and international contexts requires an appreciation of the way in which political power is exercised by different groups in pursuit of their aims and objectives. Analyses of who has power, how that power is applied and to what end are one of the central elements of all political analyses of decision taking and policy making. Power, or the lack of it, lies at the heart of the processes through which policy networks open and close, and different groups are either pulled into or excluded from the loci of decision making.

In the case studies that follow power is not always revealed in observable conflict between individuals, pressure groups and government departments. The simple definition of power is the ability to get what one wants, usually at the expense of the interests of others, though that is not a requirement (see Lukes, 1974, 26–27). Power is exercised by a variety of means, including the rule of law (which awards certain groups rights over others); coercion by virtue of military muscle; political position; superior knowledge or economic sanction; paying off political debts; or through example, inspiration or sheer persuasive leadership. We shall see that traditional pluralistic accounts of policy making emphasize the importance of observable conflict over 'key' issues, such as open debates conducted through the media, in local governments and parliaments (Ham and Hill, 1993, 65). Clearly, conscious, observable actions and decisions do have an influence on policy outcomes, but there is also what Lukes refers to as a 'second' dimension of power which operates more covertly to keep 'key' issues off the political agenda and away from public scrutiny. Second-dimension power is organizationally and institutionally framed so that it tilts the political playing field to the advantage of certain groups. Second-dimension power is exercised when issues are removed from the agenda for action or the political arena, by preemption and exclusion (Saunders, 1980, 28; Clegg, 1989, 75–82). These advantages are not created by the deliberate observable acts of individuals or groups but develop over time as the result of a whole range of acts, intended and unintended consequences, and the repetition of routines (Giddens, 1984). In other words, there is an endemic bias in favour of the powerful and the privileged, contained within the very structure and institutions of society. These structural, historical and institutional factors tend to be missed or only partially observed by pluralist analyses of power.

Where does all this fit into climate change politics? In the UK and in the United States, in particular, business interests are colluding to finance scientific interpretations that are contrary to established IPCC viewpoints. The American Petroleum Institute, for example, is cooperating with the coal industry to review and critique global circulation models. In the UK, the Confederation of British Industry (CBI) is actively assisting the Conservative administration in blocking any introduction of a carbon tax. Both groups are manoeuvring behind the scenes to ensure that industry is not unduly disadvantaged by any proposals to meet the 1990/2000 CO_2 stabilization programmes. Hence the focus is on domestic insulation and 'win–win' industrial energy conservation schemes. The role of trade unions is also supportive: in a period of prolonged recession and uncertain recovery, organized labour is not anxious to raise the costs to industry in case it imperils the prospects for employment.

Non-decision making is by no means a simple concept to understand. It is, for example, not merely a decision not to act – it is also a situation where a need for a decision does not even arise. Here we see the influence of routines, established procedures, traditional and close ties between certain groups and reputation. In total they help to keep many unarticulated grievances and what Lukes terms 'potential issues' (Lukes, 1974, 19) from ever reaching the political arena. The notion of policy networks being 'closed' or 'open' is a useful heuristic. Certain interest groups ('outsiders') may find they are deflected from the locus of decision making by their inability to penetrate 'closed' (institutionalized) patterns of consultation and representation, whereas others (insiders) pass through the gate of influence more easily when the network is more 'open' (less institutionalized). But there is also an ideological aspect to non-decision making which provides a link to more structuralist explanations of environmental politics (Sandbach, 1980). Other demands – those of the 'deep greens', for example – may simply be deemed illegitimate or inappropriate by the majority of the public. On the other hand, institutionalized patterns of behaviour tend to give the business sector a more privileged position in policy making:

> Capital is different from other interests because it exercises power and influence in two ways – directly through interest groups and structurally because of the crucial role boards and managers exercise over the production, investment and employment decisions which shape the economic and political climate within which [all] Governments make policy.
>
> (Marsh and Locksley, 1983, 59)

As Lindblom (1977) explains, the 'Grand Majority' issues in political life – such as private property, enterprise autonomy, and economic growth – are kept off the political agenda by processes of social conditioning and indoctrination. Political conflict is thus relegated to 'an ever-shifting category of secondary issues – such as tax rates and particulars of regulation and promotion of business'. Because of this, materialism, greater prosperity, and 'indiscriminate growth remain the unofficial [and legitimate] policy in every nation' (Walker, 1989). In terms of the climate change issue, differential patterns of interest group mobilization and representation help to sustain the growth of road traffic and the provision of extra energy generation facilities, rather than sustained attempts to manage the demand for travel or energy.

In the studies that follow we argue that scientific communities in terms of both formal review structures and informal peer review networks are less influential in terms of second-degree power than they were. We argued in Chapter 1 that the science of climate change is now in the 'counting the costs' phase of the issue attention cycle, where no significantly new evidence is being created, where the first period of public alarm has been accommodated, and where the costs and benefits of early versus delayed action are beginning to be more critically examined. We can also see that concern over global warming has stabilized in those countries where the scientific debate has been longest drawn out. It is somewhat of a paradox that as the scientific consensus grows so public interest wanes. This does not imply there is no willingness to act; only that the willingness to take bold initiatives is less evident and that the issue may have become too familiar.

What these interests are and on what terms they pass through the 'gate of influence' depends in part on the changing dynamics of second-dimension power. An organization does not need to be physically present to alter an outcome. It merely needs to create a viewpoint that cannot be ignored if political legitimacy is to be retained. Second-dimension power is the avenue to institutional adaptation. The shifting of public demands and political priorities are key ingredients for these give legitimacy to interests at the edges of policy communities. Groups fade and glow in political legitimacy according to shifts in the public mood, the articulation of argument that both captures and reinforces that mood, and the aggregation of incipient or floating opinion that seeks an organizational anchor to gain political recognition.

The last sentence needs amplification, otherwise it remains an academic generalization. Take the theme of bicycles as an alternative to car transport in a modern city. Environmental groups such as Friends of the Earth and Greenpeace International have members who are genuinely anxious to increase the use of the bicycle as a commuting vehicle. Their profile depends largely on the degree of public interest and in their systematic evidence that the road and planning lobbies at local level still collude to exclude safe bicycle lanes from major vehicle routeways. A cycle accident affecting a young child glows briefly in the public imagination. A prestigious report by a learned body such as the Royal Commission on Environmental Pollution (1994) sparks off short-lived public debate on safer cycle routes. Until highway authorities are allowed to allocate more funds to cycle lanes, or place priorities on scarce funds to do so, action will be slow and patchy. Some companies shift policy and provide showers for sweaty executives, as well as a cycle allowance in lieu of a (lower) car allowance. Activist members within the companies campaign for more privileges. Unless the chief executive is similarly minded, or the image of the company executive as a 'climate friendly' organization catches on, the culture of cycling to work will not take on institutional support. This is why such action is best found at the level of the community, the household, the canteen or cafeteria.

Institutions as standard operating procedures and barriers to 'rational' decision making

Any attempt to understand how institutions structure the pace and pattern of policy making must take into account the important role of organizations and bureaucracies. Most policies are developed in and then implemented by organizations, be they government departments, organs of the European Community, the United Nations, or small pressure groups. In an influential book published in 1971, Graham Allison set out to explain how decision making is crucially determined by the *organizational context* in which it takes place. Although Allison is not without his critics (Smith, 1981), the models he presented have been used by a number of scholars to appreciate the mechanics of decision making in large organizations. On this analysis, organizations are not monolithic units with a set of clearly defined goals – they are internally divided into components and subcomponents, each of which follows its own set of repertoires and *standard operating procedures* (SOPs) (defined by Allison (1971, 68) as 'the rules by which things are done' within an organization or government department). All organizations develop SOPs as a means to deal with the uncertainty, conflict and ambiguity in policy making. Clearly, it is easier for organizations to fit problems to an existing template of 'solutions' than to work out continually new solutions to each new problem – in other words these institutionalized routines and repertoires pre-exist any rational assessment of the problem at issue. They mean that all problems tend to be processed in a regular fashion, so existing practices tend to become important determinants of the way in which new problems are treated.

An appreciation of the role of SOPs has important implications for the study of climate change. First, SOPs and other routines (Saunders, 1976) can have an important effect upon the substantive outcomes of decision making by regulating the access of participants, the patterns of negotiation and consultation. By affecting the participants' allocation of attention, their standards of evaluation, priorities and perceptions, identities, and resources

81

are very much influenced. On a day to day basis, SOPs can have an important influence on who is consulted and in what form that consultation takes place, in other words what kinds of evidence and information are passed to ministers and what is routinely screened out as being 'irregular', infeasible or dysfunctional. Similarly, an individual bureaucrat's motivations and perceptions are determined by his or her own preferences, but also the role-interests given by his or her agency positions. That participants tend to opt for courses of action that reflect their own institutionalized position within the bureaucracy is a theme running through many accounts of politics and is summed up by Allison's pithy aphorism 'where you stand depends upon where you sit' (Allison, 1971, 176). (Allison suggested that 'the stance of a particular player can be predicted with a reliability from information about his seat' (1971, 176).)

Third, Allison went on to explain how SOPs become converted into specific 'ideologies' or world views within entire departments, a phenomenon sometimes known as 'departmentalism'. For example, in all the countries studied here, departments taking responsibility for energy-related issues tend to favour gas over coal, and supply addition over demand reduction. The fact that departments are physically orientated in a certain way — the allocation of staff, resources and time, for example, tends to predetermine particular policy responses to up and coming problems, as does the configuration and ideological basis of the relevant policy network (Jordan and O'Riordan, 1993). Departmentalism is not hard to find in political memoirs: politicians, for example, have been known to display a fundamental ideological 'conversion' when they move onto new portfolios. Ministries tend to be associated with particular forms of policy involvement and not others. This aspect of policy making is captured by another well-used aphorism: the departmental 'view' (e.g. the 'Treasury view', the 'European Commission's view'). Thus it would be expected for an environment ministry to adopt a position on CO_2 that would, amongst others, focus on the car as an emitter, while industry and finance ministries would see the car as a manufacturing issue or a catalyst for economic growth. The way that departments 'view' existing problems in a way pre-determines how they approach new policy problems. As one British Treasury official once explained:

> We . . . had a framework for the economy [which was] basically neo-Keynesian. We set the questions which we asked Ministers to decide arising from that framework so to that extent we had great power. . . . We were very ready to explain it to anybody who was interested, but most Ministers were not interested, were just prepared to take the questions we offered them, which came out of that framework without going back into the preconceptions of them.
>
> (Lord William Armstrong, 1976 (quoted in Greenaway, et al., 1992, 30))

That policies have 'taken for granted' premises which are rarely opened to critical public scrutiny is another, inherently 'institutional', aspect of politics in liberal democracies (Majone, 1989; Sabatier, 1987). There are clearly linkages here with the stabilizing institutional structures underpinning the existence of policy communities, and the whole literature on belief systems (defined by Weale (1993, 197) as 'systems of ideas'). Various writers have sought to show that policies have more stable 'cores' ('fundamental normative and ontological axioms' (Sabatier, 1988, 145)) and 'peripheries' (the type of policy instruments to be used to effect a series of less stable 'peripheral' elements (determining, for example, the specific policy instruments to be used to obtain those underlying goals)). There are also links here with Lindblom's (1977) 'Grand Majority' and secondary issues; the former insulated by second- and third-dimension power, the latter the focus of observable conflict between groups. One way of looking at this is the bureaucracy and policy community interpretation of institutions as to visualize both a 'core' of the status quo in policy formation, as well as a 'periphery' of tension creating adaptation. This is not a locational distinction: more one of tradition and dominant viewpoint at the core and unconventional thinking and alignments at the periphery.

In the case studies that follow a pervading example is the struggle between the carbon tax and monetarist macroeconomic interests favouring low costs and deregulation. In Norway the tax was modified three times to accommodate wider political and economic requirements, despite extensive psychological preparation. In Italy, because the fuel tax is a routine way of raising any government revenue, the public suspects the motive despite a popular concern over climate change. In Germany, a coal tax remains an unresolved dispute between environmental and economic ministries. In the UK, the VAT on fuel policy badly misfired, seriously affecting the credibility of the whole CO_2 strategy. In every case the 'non-carbon tax core' was not shifted by activity on the 'carbon tax periphery'.

In another of Allison's (1971, 162–181) models of decision making (the 'bureaucratic politics model'), policy outputs emerge out of a process of bargaining and conflict ('bargaining games') between key personnel within an organization, or between different parts of an organization. These 'bargaining games', as Allison terms them, are not haphazard, but follow specific 'action channels'. In other words:

> The individuals whose stands and moves count are the players whose positions hook them on to the action channels. An action channel is a regularised means of taking government action on a specific kind of issue.
> (Allison, 1971, 169)

Action channels are relatively long standing arrangements governing the procedures to be used for tackling an issue. Actions channels, like SOPs, are institutions: they are routinized patterns of behaviour.

Together, action channels and SOPs help to explain why policy integration is so hard to achieve without formal guidelines and clearly supported priorities for the top of the political pile. If there is no backing from the office of the head of state, then policy integration becomes a game played by departmental bureaucrats according to their political purchase power, cabinet pecking order, patronage and the relative strength of their supporting policy communities. Policy integration is also difficult to implement when operating procedures are so closely connected to bureaucratic structures and purpose, yet so varied between departments and regulatory bodies. The main point of interest lies not in the established culture, but in the scope for change. In the chapters that follow two innovations in bureaucratic behaviour reflect institutional adaptation to climate change in Europe. One is the formation of a host of interdepartmental committees at both ministerial and official level. The other is the creation of environmental policy units within government departments, bold enough to change viewpoints within established organizations, yet not too obvious to create opposition and dismissal. This is why, for example, connections between response to climate change and economic, industrial and employment policy are becoming more acceptable and innovative.

Institutions as national policy styles

A large body of political theory is geared towards identifying and explaining the existence of specific national policy styles. The underlying assumption of this work is that individual countries tend to process problems in a specific manner regardless of the distinctiveness of the problem at issue. In other words, transport problems are dealt with in the same pre-programmed manner as economic, employment or environmental issues: there is, *au fond*, a routinized (institutionalized) method of dealing with issues; a style or a legitimate 'way of doing things' (Richardson *et al.*, 1982, ix). Governments would deal with climate change in much the same way in which they deal with other social, economic or foreign affairs.

According to those who have studied policy styles, in the *UK* there appears to be a predilection for 'consultation, avoidance of radical policy change and a strong desire to avoid actions which might well challenge well entrenched interests' (Richardson *et al.*, 1982, 3). This tends to generate a very reactive approach to problem solving. *Germany*, on the other hand, is believed to have a more anticipatory, consensus-orientated style of policy making, dominated by a strong and interventionist state (Richardson and Watts, 1985, 21). Scandinavian countries such as *Norway* tend to place greater emphasis on consultation, yet also have a tendency for policy innovation and a more anticipatory style of policy making in some sectors (Norway's introduction of a carbon tax is a good example). The whole idea of a national policy 'style' is, of course, at odds with Lowi's celebrated thesis which holds that it is the nature of the problem itself (e.g. climate change, ozone depletion) that shapes the manner in which it is processed by the political system (Lowi, 1964). In other words, distinct types of politics flow from different types of policy issue; there can thus be no national style. Rather, the same issue – in our case climate change – is dealt with in a similar manner by all countries.

With this in mind, two questions seem particularly pertinent to a study of national government responses to climate change. The first is the extent to which climate change has been processed in the same manner by different states, and whether national responses are consonant with a pre-determined national style. According to Lowi (1964), we should expect some convergence in style, whereas Richardson's work portends a more variegated series of responses, moulded and structured by the precise configuration of institutions within each state. The second is whether the Europeanization of environmental policy making – catalysed especially by the intervention of the European Community in the domestic affairs of Member States – is eroding (or at least deflecting) established policy styles. The case studies which follow indicate an intriguing blend of convergence and divergence. The European Commission itself is struggling with its own policy style, being uncertain of how far to integrate social and environmental policy. Similarly the legal aspects of external community competence, discussed in Chapter 5, suggest enormous ambivalence over any imposition of a 'European' policy style on suspicious and inherently independent Member States. Nevertheless, the formation of national monitoring and reporting mechanisms will tend to create a more uniform approach to the compliance mode of the UN FCCC within all Member States (see Chapter 6). This could well be the focal point for long term convergence of policy style.

Institutions as international regimes

Here there is an enormous literature. Scholars of international relations have rejected the Hobbesian (essentially anarchic) state of nature, and have set about exploring the conditions under which cooperation is engendered. In order to collaborate, individual sovereign states negotiate multilateral agreements and create international organizations. Both of these are encompassed by the generic term 'regime'. Institutional regimes can be defined as

> sets of implicit or explicit principles, norms, rules, and decision making procedures around which actor's expectations converge in a given area of international relations.
>
> (Krasner, 1983, 2)

The UN FCCC, the World Bank, the United Nations are all relatively formal institutions 'in that they build upon, homogenize, and reproduce standard expectations, and in so doing, stabilize the international order' (Di Maggio and Powell, 1991, 7). Conventions and protocols on the other hand are less formal in the sense that they have no physical form. The European Community meanwhile is a rather unique example of a supra-national institution that actually possesses sovereignty in the sense that it can adopt legislation that is binding on its Member States.

In general, the debate about the importance of international laws in alter-ing state behaviour is split between the 'institutionalists' on the one hand and the so-called 'realists' on the other. In terms of the distinction we made earlier, the institutionalists see regimes as independent variables in inter-national politics whereas the realists view them as dependent variables — the simple outcome of states acting in a rational and self-interested manner. The former approach places its emphasis on models of collective action set in a relationship that recognizes mutual gain from mutual action. This is typical of managing common property resources such as whales or migratory birds, or inland seas. Here the line is not unlike that followed with the theory of the policy community, namely quasi-formal clusters of interests that redirect national biases from the core of regime formation and action. On the basis of a cross-national analysis, Haas and his partners (1993, x) conclude that international environmental institutions have contributed to more effective national efforts to protect the environment. Institutionalists tend to take a much broader view of politics in that they analyse the behaviour of state and non-state actors (international networks of scientists, international pressure groups). Once states are 'locked into' an international regime, institutionalists point out that the force of habit, bureaucratic inertia and routine make it difficult for states to simply 'opt out' once they are parties to an international agreement. In the main, realists give little atten-tion to the role of international regimes; they contend that international law has little effect on state behaviour. When 'compliance' does occur it is a spurious correlation because states always act in a self-interested manner, paying little attention to the obligations and expectations embodied in inter-national regimes (Mitchell, 1994).

In the main, regime theorists have tended to concentrate on how inter-national agreements are negotiated, rather than the manner and extent to which they are actually implemented, although welcome efforts are now being made to address this lacuna (Haas, et al., 1993). The case studies which follow indicate that the implementation phase is every bit as political as the negotiation phase. Individual states must find ways of reconciling their inter-national environmental commitments with their obligations and commit-ments in other areas. Domestic policies can change, sometimes quite dramatically, as international agreements are implemented at the national level for reasons that have nothing to do with international environmental issues, yet are germane to implementation. The case study chapters are replete with examples within the European sphere.

Another problem with the models propounded by international relations scholars lies in the nature of climate change science and politics. When the science is unclear, reference has to be made either to the significance of peer review or to the role of so-called expert networks or 'epistemic communities' (see Haas (1990) for a review). These serve as informal linkages of knowledge that collectively influence policy actors, but in a micropolitical manner. Possibly of greater interest would be the scope for studying the inter-relationships between expertise groupings and established stakeholder policy communities. Patterns of ministerial advice would coalesce around both. Equally of significance would be the scope for studying the relationship between the semi-formal structure of policy review — commissions of investigation and legislative committees, for example — and this inter-relationship. For here might be revealed the shifting links between morality, international obligation, the rule of international law, and the opportunity to apply precaution as part of a shifting interpretation of international regimes in an ecologically interconnected but ethically fragmented world.

We shall see in the chapters that follow that an emerging version of this relationship is becoming a little clearer. The German Enquete Commission has become embedded in the Bundestag and in local Agenda 21 politics. The Italian Climate Commission has access to a broader economic forum. The relevant UK parliamentary committees did not give way in the face of government dismissal, but take heart from international research findings and the conclusions of the standing Royal Commission on Environmental Pollution. In Norway, as indeed in the UK and Germany, the Foreign Office played a vital role in shaping national negotiations. No move by an environmental agency could be made without the international dimension being considered. This criss-crossing of institutional innovation is beginning to be replicated in many nation states.

CLIMATE POLITICS AND INSTITUTIONAL CHANGE

It should now be clear to the reader that there is no one single 'institutional' theory of politics or policy. The foregoing discussion simply illustrates a number of theoretical works which are broadly institutional in their orientation and, wherever possible, we have sought to show why this is the case and how and why the different theories interconnect in a way that can be used to illuminate the wider political process. In total, we argue that one of the best ways to apprehend institutions is as systems of prior 'social commitments' (Redclift, 1992), which serve to embody the core values and established patterns of behaviour in all (and especially western) societies. As such, unsustainable development becomes pre-ordained rather than a conscious choice for individuals (Jordan and O'Riordan, 1993). Redclift (1992, 40), for example, suggests that:

> Environmental problems . . . are the outcome of a series of choices, many of which we make collectively as a society. The epicentre of these choices is the developed world, and most of these choices are so culturally grounded that few people recognise them as choices at all; they are routinely depicted as 'needs' rather than 'wants'.

The foregoing illustrates how the deeper institutional structure, which governs and constrains society's interaction with the environment, locks the human race into an unsustainable trajectory of development, the effects of which are sometimes cumulative and not always well understood. Strong vested interests in the form of pressure groups and established procedures serve to protect and reinforce the status quo (Jordan and O'Riordan, 1993). Much the same can be seen in European government responses to climate change. For Redclift (1992, 38), then:

> [t]he underlying assumptions about our relationship with the environment, that support the idea of 'progress' in advanced industrial societies, have tended to become normative impositions, and environmental policy is increasingly the battleground on which conflicting views of human possibilities are fought out. The implications of examining the underlying social commitments of our society, rather than alternative refinements in environmental policy options, may be very radical indeed.

The text up till here reviews the various conceptualizations of institutions and draws out a selection of analytical arenas in which any study of institution can be developed. Recall that we regard institutions more as social and political stabilizing mechanisms that create and maintain order and a sense of shared commitment to a society and a policy rather than organizational structures with mandates, budgets and consultative arrangements both internal and external. Remember too that we see institutions as multidimensional phenomena, embracing social, political, economic, cultural and international relations dimensions, yet coalescing around new theories of social learning, precaution, sustainability and empathy to the natural world. Now we look at a typology of how institutions change, but with special reference to climate change issues. In every sense, all institutions are in ferment, so 'change' in this sense relates to particular driving forces as selected in the introductory series of chapters to this volume. In this section and the final one, we briefly examine some of the forces which may cause institutions to change. In general, institutionalist accounts of politics give a good account of political continuity but they are relatively quiet on the subject of change. According to Thelen and Steinmo (1993, 14):

institutional approaches have often been better at explaining what is not possible in a given institutional context than what is. What has been missing is more explicit theorising on the reciprocal influence of institutional constraints and political strategies and, more broadly, on the interaction of ideas, interests, and institutions.

This is probably because institutions are more commonly seen as relatively constant and predictable features of political life, rather than as targets for political pressure and change (but see Olsen, 1992). In some respects, Giddens' theory of structuration is a useful antidote to the more deterministic accounts offered by some of those working in the institutionalist canon, but it is very general and needs to be fleshed out. Krasner has offered a model of 'punctuated equilibrium' (Krasner, 1984), which posits that institutions are characterized by longer periods of stability, and much shorter periods of crisis that bring institutional change. Krasner suggests that institutional change is driven by periodic shifts in the external environment, but his model is still lacking in explanatory power. Smith (1993, 93–97), in a helpful but brief contribution, conceives of institutions at the national level adapting to eight stimuli including, *inter alia*, *changes in external relations* (including international interventions and ongoing developments within the EC), *economic and social changes* (the rise of so-called 'new social movements' (Rucht, 1991) and the development of post-material values within an ongoing 'culture shift' (see Inglehart, 1989)), *the emergence of new problems* (such as climate change), *internal divisions within established policy communities*, and *new technologies* (the discovery of a proven technological 'fix' to climate change would be a good example). Smith's contribution is a useful one because it indicates how the institutions which filter and mediate politics at the national level are themselves mediated by the broader political and economic context. It may be the coming together of various elements in Smith's list that provides a context that is conducive to institutional change. Sabatier (1988, 136), for example, opines that policies shift slightly in response to relatively *stable* factors (the nature of the problem; the distribution of resources), but more fundamentally according to *dynamic* (systemic) events such as changes in the world economy, the imposition of policies from international organizations such as the EC and the unforeseen impacts from other policy domains. The case study chapters which follow contain numerous examples of both.

Some of the most interesting models of institutional change focus on the interaction of ideas, arguments and debates within institutionally structured contexts. According to Hall (1993a, 290) policies change over time as a result of learning, political discourse and an iterative process of trial and error:

> Organised interests, political parties, and policy experts do not simply 'exert power'; they acquire power in part by trying to influence the political discourse of their day. To the degree they are able to do so, they

may have a major impact on policy without acquiring the formal trappings of influence.

In a similar vein, Dunleavy (1981) is one of a number of theoreticians who have sought to emphasize the importance of ideological structures in society. He notes, for example, that:

> [t]o win a policy decision it may be necessary to mobilise power, to organise politically, to invoke sanctions or rewards in an effort to structure the terms of a decision, to deploy 'power resources', to seek allies, to isolate opponents, etc. But it may equally be important to win a rational argument, to undermine a policy 'paradigm' intellectually, to solve specific 'technical' problems to demonstrate a shift in the 'intellectual technology' of the policy area.

The same point also forms a springboard for a series of studies by Majone (1986; 1989), who has argued that:

> public policy is made of language. Whether in written or oral form, argument is central in all stages of the policy process. . . . We miss a great deal if we try to understand policy making solely in terms of power, influence, and bargaining, to the exclusion of debate and argument.
>
> (Majone, 1989, 2–3)

By emphasizing the importance of rational debate between individuals and groups, Majone appears to take a rather narrow 'agency-centred' view of the policy process. Gone are the 'invisible' structures of influence and power that are so difficult to analyse in any concrete manner, leaving us with the much more recognizable face of politics we witness in the media each day as politicians spar over the relative merits and demerits of a particular policy. 'Winning' a rational argument is, of course, one of the primary strategies adopted by pressure groups in their battle to win access to the gate of influence. In the transport sector, for example, climate change is but one of a number of arguments used by the environmental movement to lobby for more sustainable forms of development. But structural and institutional factors are also at play when actors seek to undermine another's policy paradigm (Smith (1989, 162), quoting Gramsci, refers to this as a 'war of position').

Rational argument, the presentation of evidence and the process of persuasion are, to put it simply, bounded by institutional limits. May (1992, 351), for example, hypothesizes that 'learning' takes place within advocacy coalitions of like-minded individuals and groups: like many aspects of institutional change it cannot be observed directly, but is manifested in the adoption of new policy tools or implementation designs. Certain solutions immediately win favour because they contribute to or are in accordance with the 'ideological cohesion which may exist on substantive policy questions'

(Dunleavy, 1981). Novel solutions or policy prescriptions which challenge the status quo or the power and influence of those in power may, on the other hand, be discounted at a relatively early stage in the process. Indeed, the more radical ideas may not even be articulated for fear of rejection or failure. At this point, Lukes' (1974) second and more 'radical' third dimensions of power become more manifest. There are also strong links with the policy network literature, for the policy paradigm will be stronger and more resistant to attack when there is a cohesive policy community in place. Where there is a more open issue network, the policy agenda will be more 'open', and policies will change noticeably over time as different groups win and then lose access to the ear of government. But this is not to say that 'outsiders' can never become 'insiders': ideas that may once have seemed outlandish and radical may become accepted over time as they are discussed and debated. We see this as much in the social construction of global warming as a 'problem' in need of a policy solution as in the subsequent debate about the form those solutions should take (Buttell *et al.*, 1990). Meanwhile, scientists continue to argue about the precise rate, scale and effects of global warming – a debate in which interest groups have been actively involved.

TRIGGERS FOR INSTITUTIONAL CHANGE

What follows is a possible list of important triggers, within the environmental realm, that might allow the opportunity to be created for a shift from the institutional periphery to the core, and hence from the more ephemeral to the more central groupings and knowledge networks of influence. Hall (1993a, 278–279) refers to three levels at which policy change may occur: changes to the setting of policy instruments (first order); changes to the policy instruments and their settings (second order); wholesale changes to the policy itself (third order). Very similar typologies have been offered by Sabatier (1988, 145) and Majone (1989). In the realm of climate politics, at least three triggers of institutional change can be identified: *scientific knowledge, circumstances* or *events*, and the *repercussive effects of small but purposeful steps*. The possible relationships involved are presented in Figure 4.1.

In institutional terms, *science* is an articulation of knowledge, a structure of self-examination, and a product of dominant political and social interests. Science proceeds by testing evidence, by erecting plausible interpretations of assumed reality, and by subjecting itself to self-criticism. But as Wynne (1992), Jasanoff (1990) and others have noted, science is a socially constructed phenomenon which reflects the dominant beliefs and which is designed to convey a pre-determined sense of order and understanding about an otherwise chaotic world.

```
┌─────────────────────────┐
│    Levels of Change     │
└─────────────────────────┘
```

Policy integration through formal and informal structures

Development of a mix of policy instruments including taxes, permits, voluntary agreements and negotiated regulatory packages

Policy settings with organizational structures and interdepartmental coordinators

```
┌──────────────┐                              ┌──────────────────┐
│   Triggers   │                              │ Transformations  │
└──────────────┘                              └──────────────────┘
```

Changes in scientific knowledge

Circumstances or events

Progressive shifts in small steps

```
┌─────────────────────────┐
│   Global commitment;    │
│                         │
│   National cohesion,    │
│                         │
│     Local action        │
└─────────────────────────┘
```

Connections to other policy arenas mostly by happenstance

Stricter application of the precautionary principle

Recognition of the moral advantage of collective gain by mutual action

Realignment of power by shifting responsibility globally and locally

Social learning by incorporating moral and legal awareness

Cosmetic compliance by lip service response

```
┌─────────────────────────┐
│    Degree of Change     │
└─────────────────────────┘
```

Figure 4.1 Institutional adaptation and climate change politics: a summary. This is a diagrammatic representation of the points made towards the end of Chapter 4, and which form the principal theoretical basis for the institutional dynamics that form the content of the case study chapters. Readers will judge for themselves how representative is this depiction.

Note: The box in the centre covers the three levels at which any change politics must work, namely a national framework that extends globally through moral and legal agreements, and offers opportunities locally for people-centred action through formal rules and informal networks.

The triggers may be fresh perspectives on the political significance of scientific knowledge, or the opportunistic advantage taking of circumstances or other policy initiatives, usually revealed through small changes. The transformations in outlook begin through these policy connections, extend through a greater political willingness to adapt the precautionary principle with its overtones of fairness of treatment and allowing the planet room to breathe, and the grafting on of a moral perspective. Usually these responses should occur sequentially.

The policy scene is initially structural, involving organizational change, better coordination and more accountable audits. But this is assisted by the shifts in regularity procedures as well as a change in moral and legal outlook.

The degrees of change usually begin with fairly unsurprising and low key initiatives, especially in an era when the precautionary principle is not involved. But that change moves outwards in the policy sphere through a process of social learning and ultimately realignments of power form the nation state to global political and logical mechanisms on the one hand, and to local formal and informal initiatives on the other. The process requires the adoption of a strong morality and potentially supportive policy integration procedures.

As outlined in Chapter 1 the science of global climate change has gathered sufficient credibility – by networking, by peer review, by high profile spokespeople, and by sympathetic media coverage – to play this role. The credibility of climate change science rests particularly with the IPCC because of its pre-eminent role in influencing the UN FCCC, and because of its adaptation to include more policy-relevant analysis and social scientists in its second round of operation. Science has also influenced national governments through formal commissions of inquiry into climate change issues. For the purposes of this study, the role of the German Enquete Commission and the UK Royal Commission on Environmental Pollution will be compared in terms of the way they have used science. These bodies articulate the established traditions of scientific consensus, create plausible evidence of possible futures, indicate the consequences of doing nothing as opposed to doing something, and convey a sense of urgency that captures an inexpressible but deeply felt public concern. By so doing, they bestow a credibility on science that generates its legitimacy in influencing difficult and controversial policy shifts. But it is the institutionalized frame that such mechanisms bestow on science that awards science its status. There never can be a free-standing climate science.

For our purposes, the interesting point is how the broad scientific consensus over climate change is holding up in the face of public apathy, diverted political attention and sniping attacks by penetrating critiques of the models and the scenarios. While we argue that lack of new evidence is not pushing the debate forward, nevertheless the buttress of strong science provides the basis for the CoP process, a post-Rio protocol and a tougher political commitment to go beyond the 1990/2000 stabilization agreements. In this case, the relative stability of science creates the dynamic for evolving organizational change.

Circumstances or events act as metaphors or symbols of the dilemma. A bad harvest, unusual rainfall or temperatures, high profile scientific discoveries of say species loss or gain, policy measures that appear to ignore the pleas of scientifically founded recommendations – all or more of such things spur anxiety and reinforce the demand for a focus for social articulation, strong enough to propel a formerly marginal pressure focus through the gate of influence. Thus environmental groups such as Greenpeace and Friends of the Earth adopt science of a sort to gain leverage in the debate. Jeremy Leggett (1993), scientific adviser to Greenpeace International, for example, constantly emphasizes the 'doom' scenarios – the worst case possibilities that tend to be dropped by mainstream scientific judgement. This tactic, which is quite reasonable in a battle for attention, tends to attract the anger of sceptics, or inflame latent anxieties. But it also serves to remind participants that precaution can be given a higher place in practical judgement, if the stakes are set high enough.

Small steps of innovative responsiveness charge the radicals with greater impetus. If energy efficiency begins to get a fair wind, so the lobbies that long have championed its cause gain strength. As carbon taxation becomes an essential means of raising public revenue, so supporters of directed tax revenue gain sympathy. As a new interministerial or intersectoral consultative mechanism is created, irrespective of political cosmetics or media toothlessness, so adherents of more formal and legally binding policy coordination advance their cause with more vigour and stridency. The UK government contemplates market-based economic incentives without actually implementing many. But this is sufficient to legitimize the advocacy of environmental economics in a host of organizations.

These observations serve to show that climate change politics explicitly require a very broad social and economic framework in which to evolve. Technological breakthroughs in CFC substitutes that are less globally warming, or alterations to educational curricula to imprint the social, cultural and moral elements of the sustainability debate into all subject matter, or school budget responsibilities that enable schools to gain financial advantage over energy efficiency – all of these are examples of small steps that signal institutional change via opportunism and realignment. The socializing interconnection of a host of supportive processes, usually via informal networks, helps mould modern climate change politics.

Position is all the more important if the policy process itself is accommodating. The complement to structure and positioning is that of procedure or rules of engagement. Shared rules abound in the institutions covered in this study. Thus regulatory bodies build a collective viewpoint around custodianship, the defence of the public interest, rule formation and target setting and enforcement via negotiation or programmed compliance. In a similar way, industrial lobbies share views about competitive position, minimization of government interference, relaxation of financial impediments to innovation and market success, and social responsibility in their corporate performance. Environmental NGOs rely on contested science, support by maverick but authoritative figures, non-conformist thinking on industry and government statistics, high campaigning profiles, and publicly embarrassing government or industry by pointing out how rhetoric or public relations fails to be matched in practice. Threats to their credibility by, say, scientific criticism or by adverse publicity caused by the misfire of a campaign publicly hurt NGOs a lot. They have high internal demands on credibly influencing the rules of engagement. But in climate politics, they have a critical amount of influence, whether in or out of ministerial chambers.

Such rules cannot remain static. Institutions evolve as actors learn from the tensions and loss of respect that continuance of the status quo bequeaths. Procedural innovation, like the dynamics of positional influence, does not take place rapidly. It cannot do so if institutional order is to be maintained. But it does happen, and for the purposes of this investigation we need to know why. So what are the powerful transformational forces shaping institutions with regard to contemporary environmental politics? We recognize three: *environmental empathy*, *precaution*, and the *failure to manage the commons*.

Environmental empathy is an elusive notion. It applies to the incorporation of environmental concerns into institutional procedures across a broad range of activities. This is part of the notion of 'ecological modernization' advanced by Weale (1992, 75–76). Resource conservation, protection of life support systems, granting rights of existence to ecosystems, recognizing that natural functions service society and should be appropriately paid for and incorporated into regulatory behaviour and innovative resource allocation – all those forces that promote the acceptance of sustainability are becoming part of the modern psyche. No institution can afford to ignore this transformation of societal values. Goodin (1993) provides a useful perspective.

Precaution is both a philosophy and a practice (O'Riordan and Jordan, 1995). The precautionary principle reflects the concern that margins of ecological and biogeochemical tolerance are possibly being threatened with dire consequences for habitability and freedom of manoeuvre for future populations. Precaution also accepts that social action must precede full scientific understanding for the simple reason that to delay where there is reasonable and sufficient evidence of calamity would be counterproductive and probably far more expensive than anticipatory action. Precaution also changes the bias of proof: in effect it bestows rights on possible victims, including future generations, not on developers and innovators who do incomplete environmental homework. The governments studied in this book admit that their climate policy is based on the precautionary principle, though they are ambivalent over how far to include this in justification of early action.

The perspective of the commons means shared responsibility and shared disaster if the common good is undermined. The global 'commons' has no property right except the well-being of all who are alive today and born tomorrow. The life support systems of the planet, the stabilizing gyroscopes of the great biogeochemical cycles and the symbiotic mutual reliance of all ecosystems suggest that the commons is more ubiquitous than we have ever imagined. Just as the commons means collective action built upon by collective enforcement, and hence collective shifts in attitudes of mind, so the commons also allows for shared agreements such as carbon fixing exchanges or bilateral studies of mutual reduction of CO_2 emissions. The commons is a force for highly innovative cooperative behaviour in a world where most institutional procedures are based on self-serving mindsets. The commons provides a fresh perspective on what truly is self-interest – namely, mutual cooperation in the name of a habitable planet. This is the focus of attention of the study by Haas *et al.* (1993) who see international environmental institutions as magnifiers of concern, facilitators of agreement and builders of response capability across national borders.

THE EVALUATION OF INSTITUTIONAL CHANGE

Institutionalized world views and internal behavioural codes are responding to the mood of self-preservation and planetary responsibility inculcated by the 'new environmentalism' (McGrew, 1993). For the most part these changes will be marginal, grudging, and at times tempestuous. The full transition can often take four stages, each one of which will be analysed in the study that follows, each one steadily more transformational. We argue that the influences of science, event triggers and incrementalism begin this process, while the notions of environmental empathy, precaution and common representations influence subsequent response.

Acknowledgement by cosmetic compliance. The first response is to pay lip service to external change, the more so if positional influence is high, and procedural change takes place on hierarchical lines. This is typical of established lobbies and senior governmental departments located at the core of the 'old' policy machinery. Cosmetic compliance is a good example of the use of second-dimension power (non-decision making) to keep highly contentious issues off the political agenda. It tends to be typical of established science and, intriguingly, NGOs who do not see an advantage (or find difficulty) in collaborating across the environment–consumer–development divides. We shall examine how far governmental, industrial, trade union, environmental–consumer–developmental NGOs, and science institutions perform on this initial response criterion.

As noted in Chapter 1 and as we shall conclude in Chapter 12, lip service is still the prevailing response in climate change politics. Climate change policies acquire status and respectability through issue connections, but while innovative in principle, the actual degree of policy change is pretty small. Paradoxically it is necessary for climate politics to be buttressed by other, more robust policy arenas, but in being so dependent, the degree of change is modest and generally uncontroversial.

Adoption of new perspectives by social learning. This does not necessarily mean behavioural changes but it certainly applies to a shift in priorities and perceptions, and thus to political bias and policy goals. Social learning is a codified series of accommodating procedures through which new perspectives are first tolerated and subsequently embraced, fresh alliances are formed, and innovative lobbying positions are established (Parson and Clarke, 1991). These tend to be more likely in less centrally positioned organizations, as well as those with a more open learning structure where democracy by internal debate and consent changes the policy positioning at the top. More often than not social learning is encouraged by sympathetic senior officials who provide the intellectual space for new ideas to be incorporated. We shall see in the concluding chapter that social learning plays a vital part in institutional adaptation. This is especially the case with local community activities to reduce energy wastage and restrict the need for private vehicle mobility. But it is also a function of mobilizing public opinion, admittedly slowly, via new policy measures and educational techniques.

Formal realignment of position and influence. Here the innovation is regarded as sufficiently important as to shift political power and strategic alliances. This means the formal accommodation of views usually involving one or more organizations, so that their collective authority is enhanced. It is particularly attractive to environmental NGOs, to innovative environmental science, and to international organizations jockeying for power over national bodies. Normally such realignment comes through a series of stages in innovation. First, acceptance of principle and perspective; second, shared analyses and prediction; third, agreed joint lobbying action; fourth, positional success to achieve policy change.

Second-order stability around new norms. Institutions succeed more in their stable form than in their dynamic manifestation. Convulsions are necessary and indeed may be essential for institutional change, but a substantial period of consolidation is also required of both first-degree power and political influence, retained or acquired over a long period if it is to have any meaning. Consolidation does not mean that the internal ferment is ended, or that interorganizational relationships are all sweetness and light. Consolidation is simply the solidification of change by both attitudinal and behavioural responsiveness resulting in a new order and outlook. Arguably this is the manifestation of comprehensive institutional adaptation – so long as it is always realized that no period of stability can and should last for long. In this day and age the length of such consolidation periods may be becoming uncomfortably foreshortened, destabilizing the very essence of institutional effectiveness.

We see this in the very process of responding to the 'soft law' of the UN FCCC. This is described in Chapter 2 and exemplified in the EC case study in Chapter 5. The conflict over the carbon energy tax and the burden-sharing agreements are examples of foreshortened consolidation, and hence 'implementation failure' (Sabatier, 1986).

EVALUATING INSTITUTIONAL CHANGE

This study concerns itself with two aspects of institutional adaptation: one is the processes of *learning, adjustment and change* – of spotting tension, of identifying how to cope, and of actually altering form, structure and procedure. This has been the purpose of the analysis so far. The second part is the examination of the *effectiveness* of that adaptation process. We now turn towards a theory of institutional effectiveness through change.

The proof of effectiveness is altogether more problematic. Institutional change takes place through a myriad of influences, only some of which are to do with overt environmental matters. The chance conjunction of two policy shifts – say carbon or energy taxation and a forcibly driven need to raise public revenue – may make an environmentally driven institutional adaptation look thoroughly successful. In another vein a dominant political personality who adopts environmental values for public image purposes – for example, the current Clinton administration in the United States – may energize the regulatory arena purely by indicating a strong political preference for strong regulatory action.

Accordingly any study of institutional effectiveness through change has to be fairly circumspect. We can identify four possible criteria, as follows.

Transferring credibility to previously marginalized positions or influences. Credibility is the amalgam of public empathy plus authoritative support. One way of achieving this is to adopt a form of analysis which, while novel, actually carries conviction. Thus we focus on the *changing role of environmental science* as a basis for assessing transferred credibility. This obviously reflects the authority of science combined with the acceptance of innovation in proposals, in economic assessments, in ethical positioning and in international commitment also outlined in the discussion of the dynamics of perceptual change and lobbying influence. In particular we examine the significance of science-based assessments of climate futures and societal impacts by international and national commissions or working groups as stimulating fresh perspectives in interdisciplinary science and credibility of analysis and recommendation.

Influence on structural realignments. Putting it simplistically we can envisage an institutional 'map' both before and after the Climate Change Convention was signed as formal UN law. This 'map' would chart the structure, position and world view of various organizational agencies – government at national, local and international level, industry at international and national level, trade unions, NGOs of various stripes and science bodies. These realignments can be examined on the basis of the four stages outlined in the earlier analysis. Their significance can be measured by:

- internal cohesion of new perspective;
- degree of vigour in presentation and lobbying;
- persistence in the face of obstruction;
- willingness to reconsider and reconstruct in the light of modest success or continued rejection;
- new interorganizational contacts, networks and relationships that in themselves confer fresh legitimacy.

Influence on policy shifts. This is the most problematic area as well as the most central for the study. It is problematic because policy shifts are always happening. They cannot always be predicted, they are sometimes event driven, catapulted by some surprise or release of pent-up tension, and they can be shaped by dominant political personalities. All this we have already mentioned. But it is worth stressing because it is probably better to envisage policy making as a pool of actors and positions swimming about in a kind of nebula waiting for a trigger to be activated. This view is advocated by Kingdon (1984), who visualizes policy formulation as a series of opportunistic processes through which interests capture attention when the circumstances are appropriate. Much of the text in this part of this chapter discusses the conditions of 'appropriateness' of circumstantial opportunity. Klandermans (1989, 388) outlines these factors more explicitly, but covers similar ground as explored here.

The manifestation of this influence would partly be unobservable, namely a phase change in the institutional mindset that might only be recorded via interview. But there are other ways and we use them in our analysis:

- legislation which sets targets and devises accountable mechanisms reviewing arrangements that are open and subject to democratic review;
- ministerial statements of intent, a performance as enshrined in declarations or policy documents, policy measures of a specific kind – the focus of this report – such as target setting, fiscal measures, regulatory activity, education, grant aid, international agreements via joint implementation, etc.;
- the influence of campaigns of more peripheral groups in policy communities especially as those campaigns exploit the uncertainties of scientific prognoses;
- shifts in the public mood as evidenced by media reporting, poll data and documentaries. This is somewhat ephemeral material, but it can be deployed as supporting evidence.

Influence in cognate policy areas. Again this will be very difficult to prove. But the link between environmentally enshrined policy change, international obligation and a wider arena of policy is an inexorable and necessary one. Indeed we argue that without such links to 'non-environmental' policy arenas then full effectiveness cannot be said to have been achieved. This would refer to the following:

- fiscal matters – taxation, directional revenue, public expenditure considerations, industrial matters – trading arrangements, relative competitiveness, joint action over targets and regulation, common position in national lobbying, technological innovation, adjustments to whole management procedures;
- transport policy and especially the relationship between private and public transport, land use planning and mobility generally, and the scope for information technology to reduce mobility;

• international measures, including joint implementation schemes, international aid (that may operate through such schemes) and commitments to international agreements. Such commitments require both a political initiative and a moral force, so the analysis will examine both of these as they apply to all the points above.

CONCLUSION

We have seen that institutions can be as broad as they are long. They may be very vague notions almost akin to cultural norms, they may be sharply focused concepts such as bureaucracies and pressure groups, they may apply to structures or to ways of thinking and approving action. They may be the product of conscious design or the outcome of local and political change. They may be formal or informal, they may block or innovate, they may guide or impede. There is no single concept or single theory of institutions and institutional change, nor is there ever likely to be one. Politics is, to quote Dunleavy (1990, 56), 'a multi-theoretical discipline in which issues of interpretation are of central intellectual interest'. Using different theories is the way to find out what works and what doesn't.

What we have sought to represent here is an approach that tends to regard institutions as quasi-formal structures or organizations and ways of looking at the world that link the individual actor to the social realities of politics, peer groups and customary norms. Above all, institutions are the vehicle through which any social change is mediated, and in the process of identifying and responding to threats or requirements for unexpected and unaccustomed cooperation, institutions themselves change. Sometimes institutional change is slow and deliberate; at other times it is rapid and unpredictable. One way we might apprehend institutional change is in terms of different 'levels':

> social order and individual [action] arise in and through a process of ongoing negotiation about who shall be whom and what order shall pertain. These negotiations may take place on relatively small scales ... or on large scales. ... In fact, ... [the] ... smaller scale negotiations are continuously taking place in very large numbers within the context of the larger scale arrangements which are changing more slowly and less visibly to participants. The larger scale arrangements appear to individuals at particular times and places as 'givens'; 'the system'; 'the natural order of things', even though a larger scale (i.e. macrosociological and historical) perspective shows them as changing, often 'rapidly'.
>
> (Gerson, 1976, 796)

By using some of the institutional theories identified in this chapter, it is possible to elucidate the multilevel and inter-related social changes that are currently taking place within the states of Western Europe in response to the threat of climate change. At the international level, the debate about the likely scale and pace of global warming continues; at the European level, the Community struggles to coordinate the policies of its Member States; at the national level, independent states are now beginning the slow process of fulfilling their obligations under the UN FCCC; and at the local level, a variety of informal initiatives are being tried and tested to provide a more 'bottom-up' response to the problem. The problem for the analysts is that these levels are interacting, with actors influencing the activities at a variety of different levels simultaneously.

We remain cautious, however, about a complete analysis of the effectiveness of institutional adaptation. On our part, the theory of transition is more important than final proof of the manifestation of its effectiveness. This is not because the latter is less significant. It is largely because, ironically, the most startling indication of effectiveness is policy change that is so ubiquitous and stimulated by so many influences that it is extremely difficult and possibly inconsequential to disentangle the environmental component from the general ferment.

And we have only treated this via an interpretation of institutions as formal and observable processes. There is a whole theory of institutions as informal, quasi-anarchic, networks through which individuals and groups come to terms with the incompatibilities of their private and public lives. We have put this to one side for the present, but will return to it in the final chapter. Arguably it will be through this dimension that climate change politics will effectively be converted into transformational social behaviour and world views, not through the observable but messy *realpolitik* of the governing fray.

ACKNOWLEDGEMENTS

We would like to thank Neil Adger, Geoff Hodgson, Landis MacKellar, Peter Rawcliffe, Steve Rayner, Michael Thompson, Heather Voisey and Brian Wynne for their invaluable comments on earlier drafts of this chapter. Any remaining errors and omissions are the sole responsibility of the authors.

REFERENCES

Allison, G. (1971) *Essence of Decision*, Boston: Little, Brown.

Benton, T. and Redclift, M. (1994) Introduction, in: Redclift, M. and Benton, T. (eds) *Social Theory and the Global Environment*, Routledge: London.

Blumstein, J. (1981) Resurgence of institutionalism, *Journal of Policy Analysis Management* 1, 1, 129–132.

Bryant, C. and Jary, D. (eds) (1991) *Giddens' Theory of Structuration*, London: Routledge.

Buttell, F., Hawkins, A. and Power, A. (1990) From limits to growth to global change, *Global Environmental Change* 1, 1, 57–66.

Clegg, S.R. (1989) *Frameworks of Power*, London: Sage.

Di Maggio, P. and Powell, W. (1991) Introduction, in: Powell, W. and Di Maggio, P. (eds) (1991) *The New Institutionalism In Organizational Analysis*, Chicago: University of Chicago Press.

Dunleavy, P. (1981) Professions and policy change: towards a model of ideological corporatism, *Public Administration Bulletin* 36, 3–16.

Dunleavy, P. (1985) Political theory, in: Short, J. and Bryzinski, Z. (eds) *Developing Contemporary Marxism*, London: Macmillan.

Dunleavy, P. (1990) Reinterpreting the Westland affair, *Public Administration* 68, 29–60.

Dunleavy, P. and O'Leary, B. (1987) *Theories of the State*, Basingstoke: Macmillan.

Durkheim, E. (1950) *The Rules of Sociological Method*, Glencoe, IL: Free Press.

Eisenstadt, S.N. (1968) Social institutions, in: D. Sills (ed.) *International Encyclopaedia of the Social Sciences*, London: Macmillan.

Gallie, W.B. (1955) Essentially contested concepts, *Proceedings of the Aristotelian Society* 56, 167–198.

Gerson, E. (1976) On 'Quality of Life', *American Sociological Review* 41, 793–806.

Giddens, A. (1984) *The Constitution of Society*, Cambridge: Polity Press.

Giddens, A. (1986) *Sociology: A Brief But Critical Introduction*, London: Macmillan.

Giddens, A. (1991) Structuration theory: past, present and future, in: Bryant, C. and Jary, D. (eds) *Giddens' Theory of Structuration*, London: Routledge.

Goodin, R. (1993) *Green Political Theory,* Cambridge: Polity Press.

Gould, S.J. (1991) Institution, in: Bogdanor, V. (ed.) *The Blackwell Encyclopaedia of Political Science*, Oxford: Blackwell.

Greenaway, J., Smith, S. and Street, J. (1992) *Deciding Factors in British Politics: A Case Study Approach*, London: Routledge.

Haas, P. (1990) *Saving the Mediterranean*, New York: Columbia University Press.

Haas, P.H., Keohane, R.O. and Levy, M.A. (1993) *Institutions for the Earth: Sources of Effective International Environmental Protection*, Cambridge, MA: MIT Press.

Hall, P.A. (1993a) Policy paradigms, social learning and the state, *Comparative Politics* 25, 3, 275–294.

Hall, P.A. (1993b) The movement from Keynesianism and Monetarism, in: Steinmo, S., Thelen, K. and Longstreth, F. (eds) *Structuring Politics*, Cambridge: Cambridge University Press.

Ham, C. and Hill, M. (1993) *The Policy Process in the Modern Capitalist State*, London: Harvester Wheatsheaf.

Heclo, H. and Wildavsky, A. (1974) *The Private Government of Public Money*, London: Macmillan.

Hodgson, G.M. (1993) Institutional economics: surveying the old and the new, *Metroeconomica* 44, 1, 1–28.

Inglehart, R. (1989) *Culture Shift*, Princeton, NJ: Princeton University Press.

Jasanoff, S. (1990) *The Fifth Branch: Science Advisers As Policy Makers*, Cambridge, MA: Harvard University Press.

Jordan, A. and Richardson, J. (1987) *British Politics and the Policy Process: An Arena Approach*, London: George Allen and Unwin.

Jordan, A.G. (1990) Policy community realism vs 'new' institutionalist ambiguity. *Policy Studies* 38, 470–484.

Jordan, A.J. and O'Riordan, T. (1993) Implementing sustainable development, in: Pearce, D. *et al.*, *Blueprint 3*, London: Earthscan.

Kingdon, J. (1984) *Agendas, Alternatives and Public Policies*, Boston: Little Brown.

Kiser, L. and Ostrom, E. (1982) The three worlds of action: a metatheoretical synthesis of institutional approaches, in: Ostrom, E. (ed.) *Strategies of Political Inquiry*, Beverly Hills, CA: Sage Publications.

Klandermans, B. (1989) *Organising for Change: Social Movement Organisations in Europe and the United States*, London: Jai Press.

Krasner, S. (ed.) (1983) *International Regimes*, Ithaca, NY: Cornell University Press.

Krasner, S. (1984) Approaches to the state: alternative conceptions and historical dynamics, *Comparative Politics* 16, 2, 223–246.

Leggett, J. (1993) Who will underwrite the hurricane?, *New Scientist*, 139, 1885, 29–33.

Lindblom, C. (1977) *Politics and Markets*, New York: Basic Books.

Lowi, T. (1964) American business, public policy, case studies, and political theory, *World Politics* 16, 4, 677–715.

Lukes, S. (1974) *Power: A Radical View*, London: Macmillan.

Lukes, S. (1977) Power and structure, in: Lukes, S. (ed.) *Essays in Social Theory*, London: Macmillan.

McGrew, A. (1993) The political dynamics of the new environmentalism, in: Smith, D. (ed.) *Business and the Environment*, London: Paul Chapman.

Majone, G. (1986) Mutual adjustment by debate and persuasion, in: Caveman, F., Ostrom, V. and Majone, G. (eds) *Guidance, Control and Performance Evaluation*, Berlin: DeGruyter.

Majone, G. (1989) *Evidence, Argument and Persuasion in the Policy Process*, New Haven, CT: Yale University Press.

Maloney, W., Richardson, J. and McLaughlin, A. (1994) Interest groups and public policy: the insider-outsider group model revisited, *Journal of Public Policy* 14, 1, 17–38.

March, J. and Olsen, J.P. (1984) The new institutionalism, *American Political Science Review* 78, 734–749.

March, J. and Olsen, J. (1989) *Rediscovering Institutions*, New York: Free Press.

Marsh, D. and Locksley, G. (1983) Capital in Britain, *West European Politics* 6, 36–60.

Marsh, D. and Rhodes, R. (1992) Policy communities and issue networks, in: Marsh, D. and Rhodes, R.A.W. (eds) *Policy Networks in British Government*, Oxford: Clarendon Press.

May, P.J. (1992) Policy learning and failure, *Journal of Public Policy* 12, 331–354.

Mitchell, R. (1994) Compliance theory: a synthesis, *Review of European Community and International Environmental Law* 2, 4, 327–334.

Olsen, J. (1992) *Analysing Institutional Dynamics*, Working Paper 92-14, LOS Senteret (Norwegian Research Centre in Organisation and Management), Bergen, Norway.

O'Riordan, T. and Jordan, A.J. (1995) The precautionary principle in contemporary environmental politics, *Environmental Values* 4(3), 191–212.

Ostrom, E. (1986) An agenda for the study of institutions, *Public Choice* 48, 3–25

Parson, E. and Clarke, W. (1991) *Learning to Manage Global Environmental Change*. Working Paper 91-13. Centre for Science and International Affairs, John F. Kennedy School of Government, Harvard University, Cambridge MA.

Powell, W.W. and Di Maggio, P.J. (eds) (1991) *The New Institutionalism in Organizational Analysis*, Chicago: University of Chicago Press.

Redclift, M. (1992) Sustainable development and global environmental change, *Global Environmental Change* 2(1), 32–42.

Rhodes, R.A.W. (1985) Power-dependence, policy communities and inter-governmental networks, *Public Administration Bulletin* 49, 4–31.

Rhodes, R.A.W. (1990) Policy networks, *Journal of Theoretical Politics* 2, 3, 293–317.

Richardson, J. and Watts, N. (1985) National policy styles and the environment: Britain and West Germany compared, WZB IIUG DP 85-16, Berlin: IIUG.

Richardson, J., Gustaffson, G. and Jordan, G. (1982) The concept of policy style, in: Richardson, J. (ed.) *The Concept of Policy Style*, London: George Allen and Unwin.

Richardson, J., Maloney, W. and Rudig, W. (1992) The dynamics of policy change: lobbying and water privatisation, *Public Administration* 70, 157–175.

Riker, W. (1980) Implications from the disequilibrium of majority rule for the study of institutions, *American Political Science Review* 74, 432–446.

Royal Commission on Environmental Pollution (1994) *Eighteenth Report: Transport and the Environment*, Cmnd 2674, London: HMSO.

Rucht, D. (ed.) (1991) *Research on Social Movements*, Frankfurt: Campus Verlag.

Sabatier, P.A. (1986) What can we learn from implementation research?, in: Kaufmann, F., Ostrom, V. and Majone, G., (eds) *Guidance, Control and Performance Evaluation*, Berlin: DeGruyter.

Sabatier, P.A. (1987) Knowledge, policy-orientated learning and policy change: an advocacy coalition framework, *Knowledge: Creation, Diffusion, Utilisation* 8, 4, 64–92.

Sabatier, P.A. (1988) An advocacy coalition framework of policy change and the role of policy-orientated learning therein, *Policy Studies* 21, 129–168.

Sandbach, F. (1980) *Environment, Ideology and Policy*, Oxford: Blackwell.

Saunders, P. (1976) They make the rules: political routines and the generation of political bias, *Policy and Politics* 4, 1, 31–57.

Saunders, P. (1980) *Urban Politics*, London: Macmillan.

Schotter, A. (1981) *The Economic Theory of Social Institutions*, Cambridge: Cambridge University Press.

Sened, I. (1991) Contemporary theory of institutions in perspective, *Journal of Theoretical Politics* 3, 4, 379–402.

Smith, M.J. (1989) Changing agendas and policy communities, *Public Administration* 67, 149–165.

Smith, M.J. (1991) From policy community to issue network, *Public Administration* 69, 2, 235–255.

Smith, M.J. (1993) *Pressure, Power and Policy: State Autonomy and Policy Networks in Britain and the United States*, London: Harvester Wheatsheaf.

Smith, R.M. (1988) Political jurisprudence, the 'new institutionalism', and the future of public law, *American Political Science Review* 82, 1, March, 89–108.

Smith, S. (1981) Allison and the missile crisis, *Millennium: Journal of International Studies* 9, 1, 21–40.

Sugden, R. (1986) *The Economics of Rights, Co-operation and Welfare*, London: Basil Blackwell.

Thelen, K. and Steinmo, S. (1993) Historical institutionalism in comparative politics, in: Steinmo, S., Thelen, K. and Longstreth, F. (eds) *Structuring Politics*, Cambridge: Cambridge University Press.

Walker, K.J. (1989) The state in environmental management, *Political Studies* 37, 25–38.

Weale, A. (1993) Ecological modernisation and the integration of European environmental policy, in: Liefferink, J.D., Lowe, P.D. and Mol, A.P.J. (eds) *European Integration and Environmental Policy*, London: Belhave Press.

Windhoff-Heritier, A. (1991) Institutions, interests and political choice, in: Czada, R. and Windhoff-Heritier, A. (eds) *Political Choice: Institutions, Rules and the Limits of Rationality*, Boulder, CO: Westview Press.

Wynne, B. (1992) Reconceiving science in the preventative paradigm, *Global Environmental Change* 2, 2, 111–127.

Wynne, B. (1993) Implementation of greenhouse gas reductions in the European Community, *Global Environmental Change* 3, 1, 101–128.

Wynne, B. (1994) Scientific knowledge and the global environment, in: Redclift, M. and Benton, T. (eds) *Social Theory and the Global Environment*, London: Routledge.

Zucker, L. (1983) Organizations as Institutions, in: Bachrach, S.B. (ed.) (1983) *Research in the Sociology of Organizations*, Greenwich, CN: JAI Press.

5

THE EUROPEAN COMMUNITY AND CLIMATE CHANGE

The role of law and legal competence

Richard Macrory and Martin Hession[1]

THE COMMUNITY LEGAL ORDER AND INTERNATIONAL RELATIONS

Any evaluation of European Community[2] participation and implementation of the UN Framework Convention on Climate Change (UN FCCC) must be understood in the context of the distinctive institutional structure which binds its Member States together in common action. The legal framework of the Community has a fundamental effect on the freedom of the Member States to pursue independent action with other countries and has created a new international personality through which the traditional functions of the sovereign state are exercised.[3]

Yet the dividing line between Member State and Community competence is an ambiguous one.[4] It rests on both political compromise and the operation of general legal principles derived from a treaty. The Treaty of Rome with its amending documents[5] itself forms a compromise. It has not as yet created a state to supersede the Member States nor has it left the Member States with the full attributes of statehood. Quite unlike other such systems 'the new legal order'[6] of Community law results in a quasi-federal system of government in which the Member States maintain a foreign relations capacity.[7]

The purpose of this chapter is to consider the effect of Community and international law and, in particular, the Community law concept of competence on the participation and implementation of the Convention by the Member States. The framework of principles concerning competence,[8] and the process and formulae by which the broader international community recognizes their legal effect, do not always provide clear and coherent answers to specific questions of responsibility and implementation.[9]

In a truly federal state the room for dispute, inherent in a federal division of powers, is an internal matter impacting on external relations only insofar as it impedes the implementation of those agreements concluded by central

106

government.[10] For the Community, the division of competences between it and its Member States in the external sphere impact at all stages, including the negotiation and adoption of international agreements. Because the legal principles peculiar to Community law cut across so much of the Member States' activities, they have become the common currency not only of internal debate amongst the Member States but also of international relations – in some cases so much so that the Community and its Member States can be subject to the criticism that substantive issues become lost in internal procedural wrangling.[11]

The increasing identification of environmental interests with a broader range of economic and social concerns brings into sharp relief the underlying structure of Community rules on competence. Comprehensive international agreements with an environmental motivation are not easily related to the system of segregated legal bases the Treaty prescribes. As Community powers are divided according to a classical framework of policies, competence is divided, not only between the Community and its Member States, but also according to distinct legal bases. As the legislative procedure for the adoption of measures and international agreements may differ according to policy area, choosing a particular basis is of some political importance to the institutions.[12] Some attempt to reconcile the broad range of environmental concern and the environmental impact of many policies has been attempted in recent amendments to the Treaty. This involves the recognition that individual environmental measures may be adopted in the context of other policies and that all policies must as a requirement of Community law integrate environmental protection requirements into their definition and implementation. Nonetheless, balancing what are sometimes conflicting interests, while at the same time ascertaining the appropriate legal basis for a particular measure, has remained a matter of controversy.[13]

Given the rather open-ended nature of the language involved in the UN FCCC the basis for Community participation to date has been a political rather than legal problem.[14] Nonetheless even the current relatively soft commitment has raised the issue of competence and the allocation of responsibilities between the Community and the Member States.[15] As further commitments and specific programmes for action are adopted under the Convention and as these commitments are implemented (whether by the Community or the Member States), the impact of the UN FCCC on the Community structure and the substantive requirements of Community law increases and therefore the rules of competence designed to preserve and protect them become central issues.[16] Furthermore the increasing availability of qualified majority voting by the Council of Ministers is likely to intensify the significance of the legal dimension.[17]

Table 5.1 Some legal bases relevant to the environment

Procedure	Legal basis	Objectives
Co-decision	Article 130s(3) ENVIRONMENT	General Action Programmes setting out priority objectives to be attained
	Article 100a Approximation of Laws (by way of derogation from Article 100)	Harmonization with the aim of establishing the internal market
Cooperation	Article 130s(1) ENVIRONMENT as a contribution to preserving, protecting and improving the quality of the environment, protecting human health, the prudent and rational-utilization of natural resources and promoting measures at an international level to deal with worldwide environmental problems	Action to achieve the objectives referred to in Article 130[1]
	Article 75 TRANSPORT the objectives of the Treaty in matters governed by the transport title	Common rules on international transport across the territory of one or more Member States etc.
Qualified majority voting	Article 130s(2) ENVIRONMENT	
	Article 113 COMMON COMMERCIAL POLICY to contribute in the common interest to the harmonious development of world trade, the progressive abolition of restrictions on international trade and the lowering of customs barriers	Measures to implement the Common Commercial Policy
	Article 43 AGRICULTURE increased agricultural productivity, a fair standard of living for the agricultural community, the stabilization of markets, the availability of suppliers and that supplies reach consumers at reasonable prices	Measures for the common organization of the market in agriculture

Table 5.1 (continued)

Procedure	Legal basis	Objectives
Unanimous voting	Article 130s(2) ENVIRONMENT	Provisions primarily of a fiscal nature Measures concerning town and country planning, land use with the exception of waste management and measures of a general nature, and management of water resources Measures significantly affecting a Member State's choice between different energy sources and the general structure of its energy supply
	Article 100 APPROXIMATION OF LAWS	Directives for the approximation of laws which directly affect the establishment of functioning of the common market
	Article 75(3) TRANSPORT (by way of derogation from Article 75(2))	Transport policy: where would have a serious effect on the standard of living and on employment in certain areas
	Article 99 HARMONIZATION OF INDIRECT TAXATION As Article 100a	Provisions for the harmonization of legislation concerning turnover taxes, excise duties and other forms of indirect taxation necessary to ensure the establishment and functioning of the internal market
Budgetary procedure	Article 130d STRUCTURAL FUNDS to promote overall harmonious development ... leading to the strengthening of ... economic and social cohesion	The tasks, priority objectives and organization and general rules to ensure the effectiveness of the Structural and Cohesion Funds

[1] Detailed to the left of this box.

INSTITUTIONAL DYNAMICS AND THE SIGNIFICANCE OF LEGAL PRINCIPLE

It would be too simplistic to characterize the Community process as reflecting simply a tension between pragmatism without principle and mechanistic legalism. The structure of the founding treaties, based as it is on the indeterminate objective of 'even closer union among the people of Europe',[18] and, more pragmatically, a framework for economic cooperation in a common market, provide the states with a deepening mutual self-interest and range of policy alternatives around which constructive political compromises must ultimately be made through the common institutions.

The role of law in international negotiation should not be underestimated. The constitutional traditions of the Member States give primacy to the rule of law and the notion of limited powers. International treaties are themselves legal expressions of political compromise and the legal formalism of debates surrounding their negotiation and ratification is an essential part of the process.[19]

The Community especially is not merely a political entity governed by the mutual self-interest of its Member States. The role of law is fundamental and it exhibits a growing constitutionalism in its procedures.[20] The legal framework has given each Community institution a place in the process, different interests in the outcome, and the right to resort to law in the European Court of Justice (the European Court) to secure its position.[21] As a result, Community law has force in a way many other international arrangements do not.[22]

The European Court of Justice and the application of law

The Court of Justice[23] has the function of seeing that the law is observed and the onerous task of interpreting the sometimes opaque treaty framework.[24] As the final arbiter of inter-institutional disputes concerning external competence, the Court has the potential to act as a key influence on the external relations of the Member States and the Community as it has done in many other areas of policy. The Court has adopted the philosophy of integration as its own and developed general principles of Community law with this end in sight.

In doing so, it has promoted an intensification of Community law and given flesh to the framework of the treaties even in the absence of a clearly defined consensus among the Member States,[25] to such an extent that in several areas it has forced the pace for other institutions. In the international sphere in particular the Court has gone a long way to extend the breadth of Community competences at the expense of the Member States. As a result it has been open to the criticism often levelled at constitutional courts: that of introducing government by judges.[26]

110

Others would argue that the Court's promotion of integration and intensi-fication is based on a legitimate interpretation of the Treaty freely entered into by the Member States. In recent years, though, some commentators have detected a more conservative and consolidatory approach and a firmer line on the extent of Community powers.[27] This is reinforced by the introduction of a new Article 3b to the Treaty which emphasizes the limits to Community competence, including the much discussed subsidiarity principle which has been seen as a pointer for the Court in the future.[28]

A further effective limitation on the power of the Court is its reliance on the other institutions to refer questions to it for decision.[29] The number of disputes coming before the Court are a tiny proportion of the whole and one might argue that in practice the process is dominated as much by internal political negotiation as by the application of hard legal principle. Where the Court has supplied a rule its application requires the cooperation of the other institutions, and, for better or worse, is usually sufficiently flexible to allow the Community parties to settle disputes in a non-doctrinaire manner.[30]

The Council and intergovernmentalism

The Council, which is made up of representatives of the governments, has preserved a central and often final role for the Member States in the adoption of Community measures.[31] The increased use of qualified majority voting[32] in the decision making of the Council and the final demise of the 'Luxembourg Compromise'[33] points to a transformation in the nature of this institution. The gradual abandonment of the consensual approach which underpins intergovernmentalism within the Community structure suggests that the ability of an individual state to enforce its position by political means will gradually be reduced. Inevitably the legal process will be strength-ened as Member States outvoted in the Council will increasingly rely on the Court to protect their interests.[34]

In terms of intergovernmental cooperation, national governments are unused to the types of legal restraint implied by restrictions on competence and judicial review. Nowhere is this resentment better illustrated than by the framework of the new foreign and security policies of the EUT.[35] Under these policies the Council still operates as a body of Member State's repre-sentatives but the European Court as supervisor has been expressly excluded. The Commission's exclusive power of proposal is also removed.

The Commission and the *esprit communautaire*

The Commission is made up of persons appointed by common accord of the Member States and is guaranteed legal independence from government interference.[36] With an express mandate to act in the Community interest and the sole right to propose Community legislation, the Commission has historically provided the Community with a *communautaire* conscience.[37] The Commission, along with the European Parliament, presents an integrationist view sometimes at odds with the interests of some or all of the states represented in the Council and might therefore be termed a more 'native' institution. Certainly most disputes on the competence question and over the related question of the appropriate procedures for the adoption of particular measures have been between the Council and the Commission. In the international context, the Commission has the role of sole negotiator of international commitments on behalf of the Community which it has vigorously defended before the Court over many years.

The European Parliament and the democratic deficit

The European Parliament[38] has been given an increasing role in the Single European Act and Maastricht Treaties and the express right to defend its prerogatives before the Court.[39] Directly elected and claiming a mandate to fill the democratic deficit so often discussed, it has, by effective utilization of limited prerogatives, extended its influence. Parliament has the greatest claim to represent Europe independent of the mediation of the Member States and so is the most federal of the major institutions. While its interests in environmental matters have often coincided with the Commission, the increasingly diverse and delicately balanced legislative procedures suggest that Parliament will have an interest in undermining compromises between the Commission and the Council which operate to exclude it from decision making. In particular Article 228 of the Treaty has been amended by Maastricht to give Parliament the power to reject certain categories of international agreement.[40]

THE CLIMATE CHANGE CONVENTION AND SEPARATE COMPETENCES

The Community acceded to the UN Framework Convention on Climate Change with the Member States and has therefore arrived at a full participatory role in future developments of the Convention, including the Conferences of Parties.[41] Article 22 of the Climate Change Convention,[42] following the formula first adopted in the Negotiations on the Vienna Convention on the Ozone Layer[43] provides for ratification of the Convention by states and 'Regional Economic Integration Organizations'. By late 1995 14 of the 15 Member States and the Community had done so.[44]

From the Community perspective the underlying reasons for joint participation of the Community and the Member States are relatively simple:[45]

- only where the entire subject matter of an agreement falls within the scope of exclusive competence[46] has the Community the sole right to participate as a matter of Community law;
- in the rare areas where the Community has no competence at all only the Member States may participate;
- in areas of potential Community competence (also termed concurrent competence[47]) the Member States remain free to participate subject to some rather general limitations.[48] But, should the Community decide, it may exercise competence, and participate to the exclusion of the Member States.[49] In practice it appears the Community rarely exercises its external potential competence.[50]

Determining the nature of the competence involved in each of the categories is not straightforward, and can involve difficult legal principles which are considered in more detail later. But it is clear that:

the subject matter of many international conventions will not fall neatly into one or other of these three categories and designation is made all the more difficult by the very complexity of the concepts involved. As a result, in most cases the Community and the Member States are mutually dependent on each other to allow for full participation in the agreement. In effect, the competences are 'mixed' giving rise to what is called a 'mixed agreement'.[51]

The Framework Convention is therefore just the latest of such agreements and an inevitable product of the division of competence between the Community and the Member States. However, joint participation by the Community and its Member States raises certain difficulties for other parties and has been made subject to certain conditions which third states impose for their own legal protection.[52] Third states are concerned to ensure that the integrity of an international agreement, and an equality of rights and obligations of all parties, are maintained where there is a division of competence between certain parties.

The basic issues are to ensure:

- all the obligations of a given convention are to be undertaken by one or other of the joint parties;
- the rights of joint parties must not be exercisable concurrently;
- more controversially third parties must be able to attribute particular obligations to one or other joint party at any given time.

The Framework Convention therefore contains what is a now familiar clause in respect to participation by the Community intended to meet these requirements.[53] Article 22(2) requires that, where a regional economic integration organization (effectively the Community) accedes alone, it is bound by all the obligations of the Convention. As is the position for the Community, where it accedes with its Member States, the organization and its Member States 'shall decide on their respective responsibilities' and 'shall not be entitled to exercise rights under the Convention concurrently'.[54]

The third concern, allowing third parties to attribute individual obligations, is more controversial. The Court of Justice has stated that third states need not concern themselves with the precise division of functions between the Community and the Member States in a mixed agreement.[55] Nevertheless Article 22(3) provides that 'in their instruments of approval such organizations shall declare the extent of their competence with respect to matters governed by the Convention'.[56]

At present the bland declaration of Community competence presented to third parties perhaps fails to reflect what is a much more fluid and ambiguous position. The Community has provided the Convention with a declaration drafted in very general terms.[57] Given the generality of the current commitments under the Convention and the current state of Community law with respect to climate change the statement is probably the best that can be made. In the absence of a clearly defined area of exclusive Community competence for climate change and in the absence of a clear obligation detailing specific action it is extremely difficult to isolate Community and Member State obligations.[58]

In particular it remains unclear whether the central commitment of the Member States and the Community under the Convention to aim at stabilization of greenhouse gas emissions at 1990 levels for the year 2000[59] applies to each Member State individually or to the Community jointly. The former interpretation is one the poorer Member States will find difficult to accept.[60]

In the following sections we attempt to describe the basis and operation of rules on competence in Community law. The rules can be criticized as obscure and therefore unhelpful in the sense that they fail to allocate responsibility in a clearly defined manner. Even so these rules form the basis of the Community and Member States interaction in the external field and the basic parameters of Community participation.

We argue that in respect of the Climate Change Convention legal principles developed in Community law are likely to have growing significance for both Member States and the international community in the future. As the obligations under the Convention become more crystalline and detailed, and as the Community adopts further measures affecting the implementation of additional international commitments, further refinement in the application of competence rules becomes possible. Indeed, further definition becomes politically and legally *inevitable*.

A NEW LEGAL ORDER FOR COMMUNITY COMPETENCE[61]

The style of the European Court of Justice is much affected by the fact that its judgments are prepared by consensus of the members of the Court. The emphasis on broad principle and general statements makes the Court's exposition of the law often difficult to interpret.[62] Nevertheless, it is possible to identify a number of general legal principles which the Court has developed to guide its analysis of the relationship between Member States and the Community in international affairs. In doing so, we can identify two distinct and key questions: (i) When does the Community have the power or *competence* to act? (ii) What is the effect of the Community possessing and/ or exercising that power? In particular, is action by Member States prohibited in these cases? The **legal basis** of Community action is of critical relevance to the first question, while the concepts of **superiority, pre-emption, exclusivity and exhaustion** shape the response to the second. Underlying both issues is the general approach of the Court in interpreting the Treaty and its concern to establish an integrated and consistent system of Community law.

Principles of interpretation, teleology, *effet utile*, and implied powers

The first principle is one of interpretation.[63] Under Article 164 the Court has a duty to ensure the law is observed in the interpretation and application of the Treaty. One function of law is to provide a degree of certainty, and courts consequently often adopt a conservative approach to interpretation, particularly of constitutional documents, to preserve this characteristic. Faced with the Treaty as it is, the Court is not permitted this luxury; the style and structure of the Treaty has forced it to adopt a distinctive approach. The founding treaties are programmatic in nature in that there is no final list of Community powers but a list of functions which may be dedicated to achieve broadly defined ends. Where the Treaty grants express power to the institutions to adopt legislative acts the only express limitation on the exercise of these powers lies in the definition of the objectives the Treaty lays down for that particular power. To interpret the limits of Community power the Court must look at the objectives adopting what is termed a teleological approach. The Court is also concerned to preserve the overall integrity of the Treaty text, Community law and Community processes, and to achieve this it adopts an interpretation of maximum effectiveness determined to give each provision a useful effect (*effet utile*).[64] Figure 5.1 portrays this relationship diagrammatically.

The combination of the teleological method and the concept of *effet utile* has led the Court to adopt an expansive approach to the question of powers. The Court assumes that the Community must have all those powers necessary to give full effect to the tasks it has been established to pursue. Even

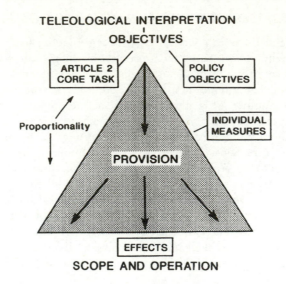

Figure 5.1 Teleological interpretation of individual provisions is based on determining what is necessary or effective to achieve the objectives of that provision. The means adopted should be proportionate to these objectives. (See Article 3b of the Treaty of Rome)

where the Treaty provides no express power the Court finds an implied power if this is necessary to achieve Community objects.[65]

Principles of competence and separate legal bases

All authority or competence to adopt measures on the part of the Community derives ultimately from the Treaty. The Treaty lays down a formal hierarchy of objectives, with a core task described in Article 2 which includes the promotion of 'sustainable and non-inflationary growth respecting the environment'. Article 3 then lists a number of policies through which the objectives in Article 2 are to be achieved, and these include 'a policy in the sphere of the environment' (Article 3(k)). Subsequent Articles, including Article 130r–t (Environment), may expand on the objectives in particular spheres of activity.

Despite the broad nature of these stated tasks and policies, it is well-established law that the adoption of any Community measure must itself have a particular legal basis. While the Court has never yet found that a legal basis does not exist for any proposed Community action, the significance of determining the correct basis lies in the different procedural rules for the adoption of measures. In practice, therefore, the question of legal basis does not constrain the Community from taking action *per se* but establishes the ground rules for how it takes decisions to act.

There are different procedures for adopting measures which are determined by the legal bases upon which they are made. Each procedure provides a distinct role for the various Community institutions and will affect their power and involvement in the process, and as the procedures are usually incompatible with each other, a particular measure must be adopted on the appropriate legal basis.[66] These differences give rise to fertile ground for dispute which, though not concerned with competence *per se*, demonstrate the Court's approach to the limits of Community power as defined by objectives.

Given that an individual measure may contribute to several objectives, the teleological method has serious limitations in this context. Ascertaining the primary purpose of a measure where there are perhaps several different purposes to which a provision contributes is a process which is legally difficult.[67] The Court has stated that the legal basis should be chosen by reference to objective factors (as opposed to purely political factors), factors it has had some difficulty in defining. (Figure 5.2 sets out these relationships in a simplifying manner.)

This is particularly important as in the international context ascertaining the appropriate legal basis for an international treaty involves an assessment

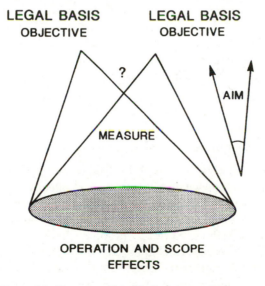

Figure 5.2 Choosing a legal basis for a particular measure must be based on objective factors, which involves an assessment of the measures, aim and effects

of the legal basis which would be required for the adoption of internal rules. In particular, where international negotiations are taking place the appropriate legal basis for a particular matter may shift as several separate issues are discussed, but the decision authorizing negotiations and concluding the agreement must, as with the principle for internal measures, have a single legal basis.[68]

The legal basis for environmental measures

A specific and express treaty basis for environmental measures may be found in Article 130r–t of the Treaty, which was introduced in 1987 by the Single European Act and revised in the Treaty of Maastricht. Article 130r–t recognizes that Community policy must 'contribute to the pursuit' of four environmental objectives, expressed in broad terms:

1 preserving, protecting and improving the quality of the environment;
2 protecting human health;
3 prudent and rational utilization of natural resources;
4 promoting measures at an international level to deal with regional or worldwide environmental problems.[69]

The Treaty also provides that environmental protection requirements be integrated in other areas of Community action.[70] Building on this integration principle, case-law has confirmed that measures with an environmental orientation can be adopted on legal bases other than Article 130r–t. Nonetheless, the tension between the institutions over the procedures to be adopted for particular measures remains, and disputes over the legally appropriate basis for such measures are likely to continue.[71]

Article 5 and the superiority of Community law: introducing pre-emption, exclusivity and exhaustion

So far we have essentially concerned ourselves with the question of when the Community has the power to act but not the effect of such action on the Member States' own freedom of action. This second question is clearly politically the more sensitive, and in the international context the issue will affect the freedom of Member States to negotiate individually and as they wish. The delicacy of this area of Community law is reflected in the nature of the case-law of the European Court on the subject. The Court has elaborated general principles but often appears reluctant to pursue these to their logical conclusion when it comes to individual agreements.[72] Not unnaturally, the possible implications of the case-law have given rise to the closest scrutiny from Member States.

Article 5 of the Treaty[73] provides a starting point for analysis by creating a general obligation for Member States in respect of Community law:

> *Member States shall take all appropriate measures, whether general or particular, to ensure fulfilment of the obligations arising out of this Treaty or resulting from action taken by the institutions of the Community. They shall facilitate the achievement of the Community's tasks.*
> *They shall abstain from any measure which could jeopardize the attainment of the objectives of this Treaty.*

In essence this is merely a restatement of the general duty of states to act in good faith in respect of obligations in respect of any treaty. The terms of Article 5 create both a positive duty to act and a negative duty to refrain from acting.[74] Here we are concerned primarily with the negative duty in the second paragraph which forms the basis for the principles examined below.[75]

It is clear in the context of measures adopted to attain a particular Community objective that Member States have a duty not to undermine the operation of such measures. National measures which have this effect cannot be applied nor may they be adopted. As a result Community law may be said to be superior to national law and to have what might be termed 'pre-emptive effect'.

Nonetheless, determining which Community provisions have pre-emptive effect and which national measures are thereby pre-empted is a matter of some complexity and the terminology adopted by the Court and commentators alike can be deceptive.[76] We can summarize three areas of difficulty.

First, it is clear that while all Community provisions are formally superior not all provisions are pre-emptive. If all provisions are considered to have a discretionary and a non-discretionary element,[77] at one end of the scale formal rules dictate the means by which objectives *must* be achieved and therefore raise the possibility of conflict,[78] whereas at the other end of the scale, objectives merely establish ground rules for implementing measures which cannot in themselves override national action.[79] Figure 5.3 describes these relationships.

Second, the concept of conflict between Community and national measures is variable and deciding whether or not there is a conflict is not merely a question of simple textual analysis.[80] The scope and purpose of the measures are also examined to established whether one measure adversely affects the intended operation of the other[81] (see Figure 5.4).

Lastly, in an apparent extension of the second rule, complementary measures are permissible in some instances, but not in others where the Community measure is exhaustive. A measure may be termed exhaustive where it is interpreted to form a complete scheme and even measures which neither directly conflict with nor affect the operation of the scheme are prohibited[82] (see Figure 5.5).

LEVELS OF CONFLICT

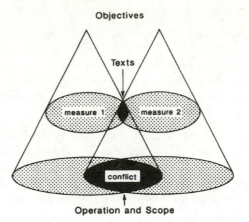

Figure 5.3 Conflict may arise at several levels and can be established through a comparison of texts or analysis of the effects of one provision on the operation or scope of another

MANDATORY AND DISCRETIONARY
ELEMENTS OF A PROVISION

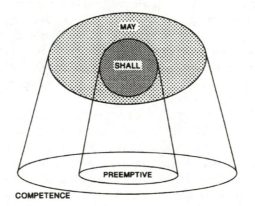

Figure 5.4 All treaty or legislative provisions might be said to include mandatory and discretionary elements, the former, statements of obligation, are pre-emptive in nature; the latter, directory statements, descriptive of power or competence to act

120

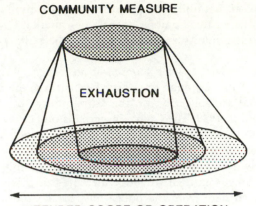

Figure 5.5 Exhaustion is a concept intimately related with that of scope. Determining whether a measure is exhaustive involves determining the scope it intends to cover

With all their inherent limitations, these concepts – pre-emptive effect,[83] exhaustive provisions,[84] and the scope and operation[85] of pre-emptive measures – form the grammar of the Court's decisions on competence. No one principle can be said to have an existence independent of the other.

The application of these principles to the Treaty and to Community measures forms the foundation of exclusive competence which has its origins in the pre-emptive effect of particular treaty provisions, and the so-called ERTA effects of individual pieces of legislation – legislative pre-emption.

Exclusivity Policies and Treaty-derived pre-emption

Exclusivity is clearly one of the key concepts which determines the nature of Community involvement in the international arena. There are two areas where the Court has found that provisions of the Treaty itself require pre-emption of Member States' action in the external sphere.

The first is the Common Commercial Policy where the external competence involved is explicit. In several cases[86] the Court justified its conclusion that Member States could not participate in agreements falling within the scope of this policy on the grounds that the Treaty itself lays down a requirement of strict uniformity in the conditions of external trade. The effect of various provisions was dealt with compositely to achieve this result. The Court reasoned that any unilateral action in this sphere by the Member States would lead to disparities in these conditions and a distortion of competition between undertakings in external markets.[87] The Court, faced with the conclusion that unilateral measures must be disallowed to preserve the integrity of the customs union and common market – which forms the bedrock of the Community system – declared the policy exclusive to the Community.[88]

The second involves marine conservation aspects of the Common Fisheries Policy. This was found to be exclusive because the Act of Accession of 1972 contained an obligation on the Community to adopt a common system by the end of 1979. The Court, in perhaps one of its more politically controversial set of decisions, first warned,[89] and then confirmed,[90] that even though the Community had failed to adopt the necessary rules, in the absence of a fully developed set of conditions for uniformity,[91] the Member States were no longer entitled to adopt measures in this area unilaterally. Exclusivity derived here from the obligation to take action in Article 102 of the Act of Accession which, with Article 7 of the Treaty, required an equality of legal conditions in the area of fisheries. This decision was particularly inconvenient but the finding of exclusivity created a sanction for deadlock in the Council. The Council finally adopted the necessary measures in 1983.[92]

In both these cases the Member States have been very reluctant to concede that the scope of particular international agreements fall entirely within these policies.[93] They have preserved their participation in agreements concerning trade and fisheries policy by claiming that various individual provisions fall outside the proper scope of these policies. More ingeniously, Member States sometimes justify continued participation on the grounds that they must represent overseas territories and protectorates which are not part of the European Community.[94]

Non-pre-emptive treaty provisions: concurrent or potential competence and legislative programmes

Outside these spheres in areas of the Treaty, where action by the Community is permitted but not required, uniformity in the legal system is the product of positive acts of the Council. The Treaty articles creating competence do not in themselves require an absolute unity of conditions but provide guidelines for action. In these circumstances, in the absence of implementing measures, Member States retain the freedom to adopt national measures.

As both the Community and the Member States may act the competence may be termed concurrent.[95] As the Community has not yet acted the competence is also termed potential competence. Following from this, the competence derived from Article 130r of the Treaty is concurrent or potential competence.

Although there is no case-law on the subject, it is clear that the environmental policy is not exclusive in the same sense as fisheries or external trade.[96] Article 130r of the Treaty dealing with Community environmental policy states that Community policy is a 'contribution' to the stated objectives. Furthermore, Article 130t expressly allows Member States the freedom to adopt more stringent measures than those adopted at Community level.

Given this framework, Member States may suspect that the Commission will present what are in reality environmental measures as falling within the scope of trade or fisheries policies in order to secure exclusivity. In the case of specific environmental agreements which utilize trade instruments to achieve their objectives there has been some argument by the Commission that these are, in effect, trade agreements. To date, however, the Council has succeeded in adopting such agreements for the Community on other bases.[97]

Legislative pre-emption and the shifting boundaries of exclusive competence

When the Community does act within its potential competence, it creates a requirement of uniformity by virtue of individual pre-emptive measures, and the Member State is prevented from derogating from the effect of the particular Community rule adopted. In this sense the formally potential competence has become exclusive as the Member State is pre-empted from adopting a conflicting measure.[98]

Two situations, which have only become clear over the years, must be distinguished. In the first, by adopting internal measures Member States lose external competence and the power to conclude international agreements in the area of that internal measure. In the second, by adopting international commitments the same principle ought to apply, but the practice of adopting mixed agreements, in which the Community exercises competence which is already 'exclusive', appears to obviate the possible pre-emptive effect of such agreements.[99] (See Figure 5.6.)

The ERTA Case: the necessity of implied powers and pre-emption

Until 1970 it was assumed that the Community's ability to act in the international sphere was confined to the Common Commercial Policy and Association Agreements which provided expressly for the conclusion of international agreements. The ERTA case[100] of 1970 extended external competence in one vast stroke. In an area of transport policy the Court expounded both the doctrine of implied powers defining areas of potential external competence and laid down the principles upon which exclusivity of these implied competences are based.

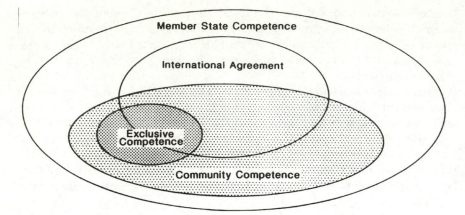

Figure 5.6 Mixed agreements occur because neither the Community nor the Member States have the competence to negotiate, conclude or implement a convention alone. Different areas of a possible agreement will regularly fall within different areas of competence

The Community had adopted a regulation governing aspects of road transport and Member States now proposed to conclude an agreement which would interfere with its operation.

On the application of the Commission, the Court ruled that the Community did have external power even in the absence of an express provision on the matter. In essence the pre-emptive effect of rules created the necessity for a Community power to undertake international agreements which might affect those rules. The argument went as follows.

In particular, where the Community had laid down common rules, Member States no longer had the right, acting individually, or even collectively, to contract obligations towards non-Member States affecting these rules. To the extent these rules had come into being, the Community alone was in a position to assume and carry out contractual obligations towards non-Member States affecting the whole sphere of application of the Community legal system. The Community authority excluded the possibility of concurrent authority on the part of the Member States since any initiative taken outside the framework of the common institutions would be incompatible with the unity of the Common Market and the uniform application of Community law.

The ERTA judgment is not without its ambiguities and is subject to at least three interpretations.[101] Later case-law has confirmed the broader interpretation, namely that the adoption of internal rules is not the only basis for the implied external competence, but that such authority could 'equally flow from other provisions of the Treaty'.[102] Community power exists externally insofar as it is necessary to achieve the Treaties' objectives on the same basis and according to the same rules and considerations which apply to the adoption of internal measures. The powers are said to be *in foro interno, in foro externo.*

The existence of concurrent external powers brings into question how it may be determined that such powers are exercised, the basis upon which they are exercised, and the effect of their exercise on the internal legal order and the division of competences.

Before turning to the exercise of such potential external competence the nature of the pre-emptive effect should be examined in more detail. Both the form of the measure adopted and the substance of its provisions and their interpretation lead to different results.[103]

PRE-EMPTIVE EFFECT

Some general principles might be identified. Non-pre-emptive provisions which guide the exercise of power, such as a statement of objectives or principles upon which action may be taken, define the extent of a particular power – its scope. In certain circumstances the exercise of such power may fall outside the scope of the authority laid down in these provisions.[104] Other provisions contain a substantive element which may either ban independent action by the Member States[105] or, in rare cases, both Member State and Community action.[106] An individual provision may contain elements which amount to a substantive rule which overrides action by the Community and the Member States, and a pre-emptive rule which overrides Member State action alone. Nowhere in the Treaty is there a provision which directly prohibits action in a particular field. Pre-emption derives rather from an implied prohibition based on substantive rules in the treaty system. Individual provisions may contain both pre-emptive and non-pre-emptive elements which both define the limits of Community power and the limits of Member State action.

The effect of particular provisions cannot be divorced from their invocability. In fact the pre-emptive effect of particular provisions is established and enforced in several ways as follows.

The institutions and the Member States as subjects: pre-emptive effect

First, Community law is invocable before the Court of Justice, not only by the Member States, but also by the Council, the Commission, and the Parliament. In particular the Commission has a special responsibility to enforce Community law under Article 169 and may take enforcement action against states for breach of Community law.

The enforcement of Community obligations by the Commission is a regular occurrence. Though international agreements adopted by the Community are expressly stated to be binding on the Member States, in practice the Commission does not enforce international agreements to which the Community is a party. In respect of other provisions of Community law a finding that a Member State is in breach of its obligations is only rarely expressed in terms of pre-emption. The matter is dealt with more simply with a finding that a Member State is in breach of its obligations.

Individuals as subjects: direct effect

Second, Community law is invocable in the national courts by individuals pursuant to the principle of direct effect.[107] Directly effective principles of Community law are necessarily pre-emptive but it appears that not all pre-emptive measures are directly effective.[108]

Direct effect gives pre-emptive principles a particularly extended grip. This is a unique facet of Community law and has particular consequences in an international context. Under international law, the effect of international treaties in national law is a matter for the states themselves. Most states require some express incorporation of international law according to national procedures before an international obligation may be invoked in the national courts.

Community law is unique amongst international treaties as, under Community law, provisions which have direct effect are automatically invocable in national courts without the need for incorporation. By virtue of Article 5 of the Treaty national courts are bound to take into account all elements of Community law and to disapply national measures which are inconsistent irrespective of implementation by the Member State.

The availability of the doctrine of direct effect to individuals presents the possibility that international agreements adopted by the Community might give rise to direct effect. This is a controversial proposition as it runs contrary to the practice of many states. It allows for indirect incorporation of international agreements through Community law in the absence of any implementing measures.[109]

The form of the measure: degrees of being bound

Article 189 of the Treaty of Rome lays down three legally binding measures by which the Community may act: the regulation, the directive and the decision.

The regulation operates much like national legislation and is stated in Article 189 to be generally and directly applicable. This means that the measure has full legal force in the Member States without the need for any implementing measures and, being superior to national law, it renders contrary provisions of national law inapplicable as between the individual and the Member State, and as between individuals.[110]

The directive is binding but requires implementation. The Member States are bound to adopt national measures to achieve the result a directive specifies but are free to choose the form and method of implementation. The amount of discretion involved in implementation has lead to doubts over whether a directive, as opposed to the common rules in the form of regulations (as in ERTA), can have pre-emptive effect.[111] The ILO case has confirmed that directives do have pre-emptive effect and logically this must be the case.

The directive which is the preferred instrument of environmental policy can be very detailed as to the result to be achieved. As a result the level of discretion afforded to Member States in implementation is practically quite limited. It is clear, in any event, that provisions of unimplemented directives, which are precise and unconditional in their terms, give rise to direct effects, and it would be inconsistent to argue that such provisions are not in any sense pre-emptive.[112]

The decision is binding on those to whom it is addressed. Decisions are therefore not generally applicable law and have not as yet been ruled to have either direct or pre-emptive effect. As international agreements are most often adopted by decisions addressed to the Member States they have some importance in arguments over the direct or pre-emptive effects of Community agreements.

International agreements. The Court has ruled that the category of Community acts is not closed and is willing to interpret other decisions and measures intended to have legal effects. International agreements themselves appear to be capable of falling within the Court's definition of such acts.[113]

Interpreting measures: the relationship between exhaustion, scope and conflict

Quite apart from the form in which a measure is adopted, the actual substance of the provisions is important and may leave room for Member State action. In essence these are questions of scope and the interpretation of individual provisions of Community law.

The first question of interpretation is that of exhaustion. Whether or not a particular rule is exhaustive or not is of particular importance to the pre-emptive effect of Community rules in the internal sphere but also helps demonstrate the problems of applying ERTA to the external sphere.

In the area of agriculture policy the Court has interpreted measures to give rise to different effects based on whether they are intended to form a complete scheme which permits no further national regulation or a non-exhaustive set of rules which permit complementary Member State rules to supplement the Community scheme.[114]

The question can be categorized as one of scope – whether the Community scheme covers the area of proposed regulation – or one of conflict – whether particular measures conflict with the objectives of the Community scheme.[115]

The ERTA case reflects this analysis in the principle that international agreements which are likely to affect Community rules or alter their scope are pre-empted.[116] Conflict defined by such vague concepts as 'likely to affect' or 'alter' and 'the scope' leaves a broad margin for interpretation as to whether particular measures are in fact pre-empted by Community rules[117] (Figure 5.4).

The freedom to adopt more stringent measures and minimum standards

In our description of the legal basis for environmental measures we mentioned that the Treaty expressly provides that Member States may adopt more stringent measures than those the Community has adopted. Individual measures may also preserve this Member State freedom. Thus, the express provisions of the Treaty or of a particular measure may prevent the general pre-emptive effect of Community standards over more stringent requirements.[118]

In the ILO Convention case[119] the Court of Justice examined the consequences of such a provision in the context of the social policy which contains a similar provision to that contained in Article 130t.[120] The directives in question included measures adopted under the social policy (Article 118a) and measures adopted under Article 100 with an express provision in the measures text incorporating the freedom to adopt more stringent measures.[121] The ILO Convention itself laid down minimum standards in respect of safety of chemicals in the workplace.

The Court clearly stated that insofar as the Member States are free to adopt more stringent measures internally they may do so by external agreement at least provided the agreement itself contains minimum stringency measures. In this sort of case there can be no conflict with existing Community law and, as importantly, no prejudice to Community action in the future. Both the Member States and the Community remain free to adopt or maintain more stringent standards, as a matter of Community and international law respectively.

Only in the case of directives which laid down absolute standards could the Community be said to have exclusive competence by reason of the ERTA rule as here the Member States were free neither to reduce nor to tighten standards but were obliged to conform to the absolute Community norm.[122]

In particular the Court rejected the Commission's argument that the difficulty in ascertaining whether a given rule was or was not in fact a more stringent rule could not of itself pre-empt the Member State from adopting the agreement in question.[123]

The Rhine Case: pre-emption or non-delegation: prejudice to institutional balance

To return to the possibility raised in 5.4 the Rhine case[124] confirms that the Community may exercise a potential competence in an international agreement and therefore that internal and external competences are parallel. The adoption of an agreement does not require the prior adoption of internal and pre-emptive measures.[125]

The Rhine case is unusual in that the Court in the absence of an adopted and pre-emptive measure found that the Member States were precluded from entering the Agreement on a Laying Up Fund for Rhine Tugs.[126] A proposal for a measure did exist but had not yet been adopted by the Council and the Court confirmed that Member States could participate but only insofar as this was necessary so as to rescind earlier agreements (and indeed Member States were obliged to do so to facilitate the Community entering the new arrangement). In the absence of a pre-emptive measure which has the effects described in ERTA the reasons for effective exclusivity here must lie elsewhere than in the model of legislative pre-emption.[127]

Expressly, the Court ruled that institutional arrangements in respect of institutions which would exercise functions reserved to the Community called 'into question the power of the Community' and altered 'in a manner inconsistent with the treaty the relationships between Member States within the context of the Community'.[128] In particular, the agreement proposed by the Member States contained a variety of rights and obligations, some applicable to all the Member States, some to the Member States with one exception and some to five states. Furthermore the participation of the Community institutions was extremely limited.

Deriving from this there appears to be a tentative rule against agreements by Member States in areas of potential competence which are incompatible with the Community institutional balance. The consequences of such a rule are wide ranging. It prohibits the adoption of an agreement whereby Community potential competence is exercised by institutions which might be said to substitute the Community. It might be characterized therefore as a rule against unconstitutional delegation of functions applicable as much to the Community as the Member States.[129] The Rhine case has not been extensively examined but recent opinions of the Court in the European Economic Area References echo the language of the Rhine case and confirm that agreements which prejudice the autonomy of the legal order are contrary to the Treaty.[130]

In the EEA[131] opinions the Court has spoken of the need to preserve the autonomy of the Community legal order[132] – which involves at least the power of the Court to determine a mix of competence and responsibility between the Community and the Member States[133] – but authorizes a sharing of competence provided this does not change the nature of the powers of the Community and of its institutions as conceived by the Treaty.[134]

The prejudice to the legal order prohibited by such a rule is far from easy to establish or define. In some sense all agreements which provide for a machinery in which further binding commitments might be agreed outside the Community context are possibly prejudicial to the Community's institutional balance. If substantive commitments may be agreed in an area of potential competence, the future exercise by the Community of that competence in a manner which conflicts with those commitments becomes impossible. To preserve potential competence for the Community all competences should be exclusive. If, as appears to be the case, not all competences are exclusive the factors which prevent Member States adopting commitments within the Community's potential competence ought to be more closely defined.[135] The duty of solidarity or unity expressed most recently in the ILO case may form the basis for such a definition in the future.[136] (See Figure 5.7.)

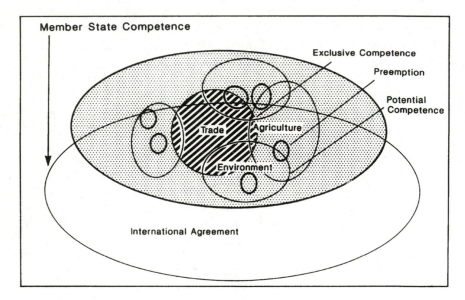

Figure 5.7 The overlap of competences in a particular agreement may be so extensive and various as to make the exercise of an independent Member State competence very difficult

Beyond competence: unity and solidarity: specific duties in the international context

Irrespective of the rules on competence and pre-emption, the first paragraph of Article 5 containing a more general duty to take action in support of Community objectives gives rise to specific and supplemental duties to promote Community competences where these are established. As a consequence of the positive duty contained in Article 5:

- where the Community has exclusive competence Member States are required to make all reasonable efforts to secure Community participation in a relevant agreement and, failing this, to act as trustees of the general interest;[137]
- where the Council has failed to act in an area of exclusive competence Member States must act in consultation with the Commission to secure an interim solution;[138]
- where it is possible to extrapolate a general duty of cooperation. Where an agreement is a mixed agreement as the exercise of Community exclusive competence in such agreements is predicated upon the assistance and agreement of Member States in the exercise of their retained competence.[139]

The general duty of solidarity to be found in the ILO case may also derive from Article 5. Its exact consequences must be speculated upon but may include a duty to abstain from action in the absence of an agreed common position where there is a mix of competences.[140]

Logically, the rule that the Member States must make every effort to secure Community participation in an international agreement pertaining to its exclusive competence could extend to those agreements within its potential competence. Nonetheless, Member States have acceded without Community participation for various reasons.

It appears that the Council at least is prepared to accept the practice of Member State sole participation in agreements within the Community's potential competence and exclusive competence while choosing to regulate the consequences of an agreement internally. In this way the Community has regulated Member State accession to the UN Liner Conference code subject to conditions (which the Court has tacitly approved), as well as the implementation of the Washington CITES Convention though supporting Community accession.[141]

SCOPE AND PRE-EMPTION

As discussed, questions of scope and exclusivity leave room for considerable ambiguity. Teleological interpretation of Community law means that the boundaries between areas of competence and the limits of pre-emptive effect are dependent on a clear determination of the objectives and effects of both the Community instruments and national measures concerned.

Fish or fowl: teleology and scope in Antarctic fisheries

The Convention for the Conservation of Antarctic Marine Living Resources (CAMLR) may be given as an example. The Community has adopted the Agreement on the basis of the exclusive Common Fisheries Policy.[142] The Member States retain an unspecified competence.

Article II of the Convention provides that harvesting should be conducted in accordance with listed principles of conservation which include the maintenance of the relationship between harvested, dependent and related populations. The fisheries policy objectives are the conservation of the resources of the sea, a less inclusive objective than the more holistic formula adopted by the Convention. Under CAMLR the effect on other species must be considered before a total allowable catch (TAC) for fish can be set.

Member States have argued that certain measures, in particular a TAC which takes into account pressure on penguin populations, fall outside the Community's exclusive competence as its objective is not the conservation of marine resources but the conservation of land-based penguins.[143] This is debatable in itself but, assuming Member States' arguments are correct, the Community has a potential competence in respect of environmental protection in general even though the competence is not exclusive. The Community's Bird Directive is limited in its territorial application to the European Territories of the Member States and, in any event, for perhaps obvious reason, does not explicitly apply to penguins.[144]

Lists and more stringent standards: exclusivity, exhaustion and the duty of solidarity in CITES

By mid-1995 the Community had not yet acceded to the International Convention on Trade in Endangered Species (CITES) but had implemented the Convention by regulation.[145] The regulation explicitly preserves the freedom of Member States to impose more stringent measures to protect such species.[146] Even though the regulation effected a trade ban and trade restrictions it was not adopted on the basis of the Common Commercial Policy as it was considered that its objective, that of nature protection, fell outside the ambit of the trade policy.[147]

CITES bans trade in certain listed species and on occasion these lists are amended. On amendment the Community regulation is updated. In international negotiations the Member States must act as trustees of the Community interest where the competence is shown to be exclusive. Prior to the ILO case, arguments arose as to whether or not the issue of adding new species to the banned list is a matter of exclusive competence. The Commission argued that as such additions fell within the scope of the agreement, the matter was therefore one of exclusive competence.[148] Though the regulation could not be considered to be exhaustive it should be remembered that residual arguments over the trade effects of CITES remained a concern and formed an alternative basis for argument.

Eventually, in the absence of agreement among the Member States, several states abstained on a vote to amend the lists owing to concerns as to the legal position.[149] Ultimately this was resolved by an agreement between the Community and the Member States whereby all CITES issues were to be dealt with according to the same procedure. The ILO case confirms that Member States are subject to a general duty of unity in respect of CITES but that the regulation as a minimum standards measure does not create exclusive competence for the Community in respect of measures to protect listed species, nor can there be exclusive competence in respect of unlisted species.

It is important to note that the freedom to adopt more stringent measures may be subject to several limitations. First, these measures must be consistent with the general framework of the Community law, as it is possible that stricter standards in one area may undermine the minimum standards in another. Under Article 130t Member States can adopt more stringent measures compatible with the Treaty but only if these are reported to the Commission. Second, the duty of solidarity mentioned in ILO support is of uncertain effect but it appears even if it cannot be said that competence has formally, or absolutely, passed to the Community; in these circumstances Member States must abide by elements of the Community framework.[150]

THE PROCESS OF COMMUNITY NEGOTIATION

At the beginning of this chapter we saw how agreements such as the Climate Change Convention explicitly allow for participation by regional economic bodies such as the Community. The principles which determine how individual states participate on the international stage are drawn from national constitutional provisions which are recognized as part of the recognition of Member State sovereignty. Similarly, the procedures concerning Community participation are determined by internal Community rules whose basic framework is found in the Treaty. The recognition of the Community's international capacity is a matter of some greater difficulty, however, and it is to this question we now turn.

Gaining recognition: the status of the Community in international law

Status in international law is dependent on recognition. The traditional position has been that only states are recognized as having legal personality in international law and therefore only states are capable of maintaining rights and contracting responsibilities. The rigidity of the traditional position has always been subject to exceptions but has certainly become more relaxed in the latter half of this century with the recognition of the functional capacity of international organizations and even in certain circumstances individuals. The right to participate in the many international and regional fora is governed by the constitutional instruments of these organizations and, while states maintain their primacy at the United Nations where only states are entitled to membership,[151] other international organizations may have different rules.

The European Community, initially through the mediation of its Member States, has had some success in achieving participation in regional agreements.[152] Amendments have been made to the governing instruments of the FAO and ILO to allow full participation.[153] Given the split in competences it is usual that the Member States maintain a presence along with the Community and where this is the case the Community has the same number of votes as there are Member States even in respect of matters falling within its exclusive competence. To a certain extent therefore, there is some benefit to the Community in denying its exclusive competence at least in fora which operate by less than a consensus vote.[154] Nonetheless, the Community has acceded alone to various fisheries conventions and to trade agreements.[155]

In the environmental sphere exclusive competence has never been claimed and the Community block vote has at times become a matter of controversy. This is particularly the case over questions involving so-called mixed competence, where the exclusive and the potential competences involved cannot sensibly be separated. Resulting disagreement between the Member States as to their obligations may lead to abstention or delays in voting on particular issues.[156] In the context of European Regional Agreements the Community block can amount to the preponderance of votes leading to resentment from minority states.[157]

The Vienna Convention on the Law of Treaties between States and International Organizations or between International Organizations has attempted to codify the position of international organizations in international treaty making and expressly recognizes the autonomy of the internal rules of the organization in question.[158] Essentially, according to Article 6 of this Convention, the capacity of an international organization is governed by the rules of that organization. In the case of the Community these are the rules examined earlier in this chapter.

Article 210 of the Treaty of Rome confers on the Community legal personality and it is this Article which is the starting point for the Community in respect of its treaty making capacity. Its competences are governed by the Treaty and the case-law of the Court on implied competences. These competences cover the whole range of the Community's objectives and insofar as the Community may take action to achieve any objective internally it appears it is able to conclude an agreement externally.

There is some debate over whether there exists an objective personality in an international organization which third states are obliged to recognize as the exercise of state sovereignty by members of the international community. If there is such a concept, the Community might claim participation in international conventions as of right. In practice, Community participation requires negotiation and agreement.[159]

Article 228 and the Community process for participation

The Community participates in binding international commitments through Article 228. Under this article the Commission is given the function of proposing Community participation in a particular agreement for which it requires a mandate from the Council.[160] The mandate authorizes the Commission to commence negotiation on behalf of the Community. The Commission has a general relations power which authorizes it to maintain relations short of those intended to lead to a binding commitment under Article 229. This power has not been the subject of much comment but appears to be the basis for participation in international relations which may arrive at documents which are not binding.[161]

The mandate to negotiate is adopted by the Council according to the voting system necessary for the adoption of internal rule on the same subject matter. Given what has already been said on the difficulties in ascertaining the appropriate legal basis for draft measures of the Community, deciding on the putative legal basis for a yet to be discussed international convention may present special difficulty. The mandate is not published but contains negotiating directives to the Commission which are binding on it.[162] There is a competence element to the mandate as it must contain the limits of the competence the Council is prepared to exercise in negotiations. The Council need not, but commonly does, limit the Commission to matters of exclusive competence, which in effect is the minimum legally possible. Within these limits the Council may even specify that nothing can be agreed outside the terms of existing legislation.

Obviously mandates may become outdated or irrelevant to the course of discussion and the procedures are so cumbersome for its review that mandates have delayed the negotiation of conventions while new versions are sought.[163] Article 228 provides for the possibility of general mandates in respect of particular conventions which will allow the Commission a certain freedom in pursuing negotiations which are more or less permanent.[164] A degree of flexibility is possible given the institution of committees to assist negotiations appointed by the Council to supervise the Commission in action.

As to actual representation, while the practice may appear confused, theoretically at least there are clear rules to determine who may speak on behalf of the Community.

First, where the Community participates as a full participant:

- if the agreement is an exclusive agreement the Community delegation will usually contain at least one Member State representative;[165]
- if the agreement is mixed there will be separate delegations who divide functions according to the division of powers;

and representation is divided accordingly.

By virtue of Article 228:

- the Commission represents the Community and must speak where competence is exclusive;
- the Commission may speak where the competence is purely potential, but usually does not do so given the ordinarily restricted terms of a mandate;[166]
- the Member States may speak in areas of potential competence but nonetheless states must attempt to adopt a common position upon which the delegation of the state holding the Presidency of the Council (which has no prescribed external relations functions) may speak;[167]
- Member States may speak in respect of retained competence.[168]

In practice, however, most areas of discussion cannot be categorized according to this simple formula and are partially Community and partially Member State questions (in effect mixed). The expedient solution is to allow the Presidency to speak in respect of an agreed common position though strictly this runs counter to Article 228 (at least insofar as the matter concerns exclusive competence). As a result the Presidency speaks according to the formula 'on behalf of the Community and its Member States'.

Where the Community has not attained full representation (as was the case in the Intergovernmental Negotiating Committee) the division of competences remains the same but of necessity the Presidency speaks in respect of exclusive and potential competences where the Commission might be expected to do so.[169] (See Figure 5.8.)

Decision making in the course of negotiation is often the subject of agreement in the Council between the Member States prior to the negotiation and should again follow the system which would be adopted for the adoption of internal rules. The Community and Member States attempt to resolve different issues each morning in coordination meetings prior to negotiations. These discussions cover both substantive and competence issues. Such informal arrangements often concern the Member States lest they concede a point which will prejudice the division of powers as decided upon in Brussels by the Council.[170]

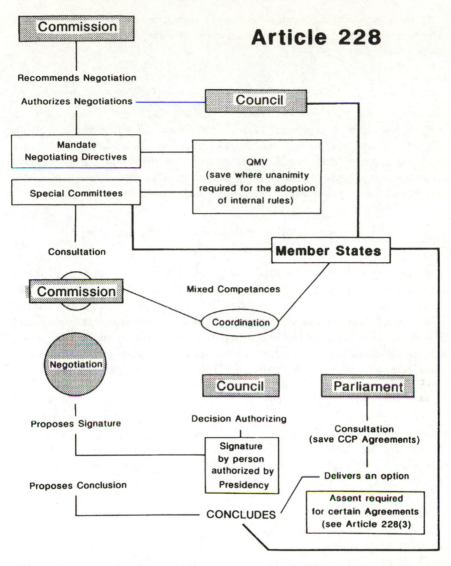

Figure 5.8 Article 228 and the exercise of external competence by the Community through a mixed agreement

In the absence of a common position it is not clear but in many cases it has been accepted that Member States must at least abstain from voting on these questions.

Community agreement and signature

Once the final text has been agreed, Article 228 has no provision on signature and the practice has been that the Council will agree a decision authorizing signature on behalf of the Community. The authorized signature is that of a person nominated by the President of the Council. Again delays in authorizing signature, which is the first stage in assent to be bound by an agreement, have caused some confusion in the Community ranks.[171]

Community conclusion and the effect of the agreement in Community law

Ratification of treaties is the final assent to be bound of states and is subject to national procedures. In international law ratification is formally done on delivery of instruments of ratification by a state to a named depository. The Vienna Convention on Treaties[172] distinguishes international organizations in that these parties to a convention do not ratify but deliver an act of formal confirmation to the depository. In practice the Community concludes an agreement by a decision and presents this to the depository with a declaration of competence where this is required.

Under Article 228 the conclusion of an international agreement is again by qualified majority vote unless unanimity is necessary for the conclusion of internal rules.[173] The instrument whereby agreements are concluded is ordinarily the decision. The Climate Change Convention was concluded by this procedure despite the objections of the UK which claimed unanimity was necessary. The UK formally abstained on the Council vote, as discussed further in the chapter that follows.

The statement that agreements are binding in Article 228 is strong support for the possible pre-emptive and even direct effect agreements adopted by the Community. The instrument of conclusion is the decision which has not yet been interpreted to give rise to direct effect. Nonetheless, the Court has confirmed that international agreements concluded by the Community form part of the catalogue of Community acts and may have direct effect. It has given direct effect to certain specialized agreements but has hesitated in respect of others. It has also stated that it is open to the Council to make provision as to the direct effect or otherwise of international agreements on conclusion.[174]

The possible direct effect of international agreements concluded by the Community suggests that they are generally binding on the Member States and therefore enforceable by the Commission. If this is the case it begs the question: to what extent are mixed agreements Community agreements within the meaning of Article 228?[175] Member States in adopting a decision on conclusion take some care to answer this question in a manner which avoids the possibility of direct enforcement.

First, in practice it is made clear that only exclusive competence is exercised thus preserving the division of competences in a mixed agreement and suggesting that only a portion of this agreement is in effect a Community agreement.[176]

Second, reflecting this, the Council ordinarily will state that the conclusion of a particular agreement is without prejudice to the existing division of competences.

It appears then that only the portion of a mixed agreement falling within the pre-existing exclusive competence of the Community is Community law.[177] Certainly it is not the practice of the Commission to enforce any element of international agreements concluded by the Community directly and instead it relies on enforcement through implementing measures.[178]

Nonetheless, a residual argument can be made that the Member States and the Community owe each other duties, each to see that their portion of the agreement is observed. Failure by the Member States to observe their portion of a mixed agreement may seriously prejudice the ability of the Community to perform what it has agreed. Particularly, international obligations, the fulfilment of which falls neither wholly within the Community's exclusive competence nor in the Member States' competence, at least require cooperation for compliance by both parties.[179]

The duty to facilitate the achievement of Community objectives and to refrain from actions which jeopardize their attainment in Article 5 may create obligations on the Member States which, though independent of divisions of competence or pre-emption rules, have a similar effect.[180] Competence may not therefore be the final word on Community enforcement. Distinguishing the legal effects of Article 5 which appear to encompass the law on competence and pre-emption and a more general duty of solidarity becomes difficult at this point. On the one hand the Rhine case demonstrates that the balance of competence by the Community in a particular agreement is ultimately a matter for the European Court and can be established, with reference to not only the pre-emptive effect of existing Community measures, but also the need to preserve the unity of the legal order. Similarly, the more recent ILO case confirms the duty of solidarity the sanctions for breach of which are as yet uncertain.

THE EFFECT OF COMMUNITY PARTICIPATION IN THE UN FCCC

As pointed out in Chapter 2, the commitment in the Framework Convention to adopt policies aimed at the reduction of anthropogenic greenhouse gas emissions is deliberately a weak one. The effect of the commitment in European law and, in particular, *vis-à-vis* the obligations owed to the Community by the Member States, is at best limited. In fact the Community commitment to stabilize emissions at 1990 levels by the year 2000 is a clearer and stronger statement than that contained in Article 4(2)(a) and (b).

Nonetheless ratification by the Member States and the Community has been the subject of close examination. States planning an increase in emissions could not sign or ratify the Convention. This is because, insofar as joint accession to the Convention can be interpreted to require an individual commitment to aim at stabilization by each of them, they are in breach of Convention. These states have been concerned therefore to establish that the commitment is joint and not severable and the Community will implement it through a system of burden sharing.

The Convention itself is ambiguous about joint implementation of the commitment and no special provision is made for the Community in this respect. From a Community point of view the declaration made at the time of ratification confirms that the commitment is intended to be joint. Greece, Ireland, Spain and Portugal have insisted on reference to burden sharing in all the Community instruments relevant to the commitment.[181] The monitoring mechanism,[182] which provides for Community reporting, itself incorporates the agreed aim of stabilization and provides an express mechanism for the balancing of burdens should this be required.

However, in the absence of a positive burden-sharing arrangement it is arguable that the Member States are as a matter of Community law obliged to take positive steps to ensure that the shared target is achieved insofar as this is one of the Community's international obligations. By reason of Article 5 joint ratification supports the existence of an individual legal obligation to take all steps necessary to achieve the Community obligation which is enforceable by the Commission. The putative obligation to take all steps to achieve what is a vague aim is not one which is easily enforceable but must be of some concern to states planning independent fulfilment of their obligations. Future protocols may not allow for such discretion in implementation nor obscurity as to the nature of the Member States' commitment.

CONCLUSION

The complexity of the Community system of law and the delicate institutional balance created by it lead to the danger that any possible positive action is lost in a cloud of procedural wrangling. In fact, the Community institutions and the Member States do manage to operate to good effect what for others may appear an incredibly cumbersome system.

The legal rules of the Community cannot supply results in the absence of political will. The judgments of the Court can, for the most part, only encourage Community participation in agreements for negative reasons. For example, the reason that competence has already been transferred to the Community by the Member States either through the Treaty itself or by the adoption of binding obligations in the internal sphere. Any expansion upon these spheres is dependent on the requisite agreement according to the procedures laid down in the Treaty. Even where a matter is one of exclusive competence new positions must be taken by agreement to see that it is actually exercised in international agreements.

The ability of the Community to act positively is enhanced by a gradual shift of power from the Council to the other institutions but perhaps more importantly by more extensive use of qualified majority voting. The ability of the Member States to impose a restrictive interpretation of the Community role through the Council is diminished.

The procedures in themselves, and the absence of a unified foreign policy, present greater difficulties. The sheer range of possible legal bases and voting procedures available for the adoption of measures and the apparent absence of simple objective criteria upon which the correct procedure may be chosen are inefficient. At least part of the problem is the absence of a clear identification in the Treaty itself of how the various legal bases relate to one another in the absence of which the Court must make do with manifest inconsistencies.

While many federal states operate subject to similar, but hardly more, complexity and uncertainty, none has attempted to do so externally. The abandonment of separate representation internationally by the Member States will certainly be the final step on the road to a federal Europe which the Intergovernmental Conference in 1996 will attempt to achieve.

Environmental treaty making is only part of such a common foreign policy and the inability of the Community to find common accord over other elements of foreign policy suggests that full integration will be very difficult to achieve. In the meantime, and in the absence of a unified foreign policy, third states are condemned to have to concern themselves with the workings of the Community's intricate internal machinery when dealing with the Member States or the Community, and the Community must itself contemplate the effects of the unique legal system it possesses on the agreements it adopts.

The example given by the Community in respect of enforcement and implementation of environmental law generally may be important but its response to the climate change issue leaves a lot to be desired. The failure of the Community to adopt a carbon tax is perhaps unsurprising but its failure to deal with its own internal version of shared but differentiated responsibility does not bode well for the adoption of a better definition internationally. The responsibility to show a lead in respect of reducing commitments is only one element of the Community's responsibility. If it is to show that it respects the difficulties experienced by the less developed nations in adopting their own commitments and has a serious intention of meeting its provisions it should be able to adopt a similar system in respect of its own members where the disparities in wealth are far less difficult to overcome.

To achieve such a commitment, the Community will have to adopt a strategy of coordinating its policies towards common goals centred on the sustainability transition. This suggests that policies being created for economic and social cohesion, which include trade, competition, transport and regional economic stimulation, will have to graft onto them policies that reduce greenhouse gas emissions. If the move towards economic integration does not carry explicitly a compatible move towards decarbonization and resource use efficiency, then the Community's external competence will have failed, and its political leadership will be undermined. The institutional test, therefore, lies in the strategic coordination of major economic and social initiatives aimed at greater quality of opportunity being meshed in the parameters of sustainable development, including the progressive reduction of greenhouse gases. So far there is no evidence that such thinking is being translated into inter-Directorate General cooperation. This remains the test of the Community's effective response to the UN Convention on Climate Change. Joint implementation is a radical and exciting notion. It should be applied in the convergence to sustainability, not to economic growth *per se* and social well-being in a disconnected manner.

NOTES

1 Professor Richard Macrory, Imperial College, London, and Martin Hession, Lecturer, Centre for Environment and Technology, Imperial College, London.
2 The European Community has legal personality and may participate in binding legal instruments. The European Union has no legal personality and may not. When considering international negotiations and participation in international agreements it is better to refer to the European Community.
3 'The right of entering into international engagements is an attribute of state sovereignty' Wimbledon Case PCIJ Reports 1923.
4 Perhaps usefully so. See Van Der Kerchove and Ost "Legal System Between Order and Disorder" Oxford University Press. The Preface to the French Edition quotes Le Moigne 1985: 168 and 'the capacity that legal systems should have, to tolerate – and even create – some ambiguities and equivocations, some redundancies and even some disorders within their internal as well as external articulations, so as to facilitate the conditions for some innovative moments of self organisation'.
5 The Single European Act of 1987 and the European Treaty of 1992 (Maastricht) are the major revisions. The Treaties are due for further revision in an Inter-governmental Conference in 1996.
6 26/62 *Van Gend en Loos* v. *Nederlandse Administratie der belastingen* 1963 ECR 1.
7 Per Lachmann 'International Legal Personality of the EC: Capacity and Competence' 1984 Legal Issues of European Integration 3, has argued that even where

competence has passed Member States retain capacity and distinguishes between the two (at p14 particularly). Probably the better view is that of John Temple Lang in 'The ERTA judgment and the Court's case-law on competence and conflict' 6 Yearbook of European Law 1987 183–220, who suggests that capacity and competence are inextricably linked p196 at n12.

8 Competence can be defined as the authority to undertake negotiations, conclude binding international agreements, and adopt implementation measures. Where competence is exclusive, it belongs solely to the Community to the exclusion of the Member States. Where it is concurrent either the Community or the Member States may act but not simultaneously.

9 'Practical men generally prefer to leave their major premises inarticulate' Oliver Wendell Holmes (1899) cited by Schwartzenberg Chapter 14.

10 Resulting implementation difficulties may be a concern, but third states will not concern themselves with matters of domestic controversy.

11 See Nollkaemper 'The European Community and International Environmental Cooperation: Legal Aspects of External Community Powers' 2 Legal Issues of European Integration pp55–91 (1987) at p71 et seq.

12 See accompanying Table of legal bases relevant to the environment (Table 5.1).

13 Hormones 68/86 Commission v. Council 1988 ECR 855, Veterinary Inspections 131/87 Commission v. Council 1989 ECR 3743, Generalised Tariff Preferences: Commission v. Council 1987 ECR 1493, and in respect of internal measures Case 300/89 Titanium Dioxide: Commission v. Council 1991 ECR 2857, Chernobyl – Greece v. Council 62/88 1989 ECR I 1526 and Case 155/91 Waste Framework: Commission v. Council 1991 ECR 2867 and Case 187/93 Waste Shipment – Parliament v. Council Judgment 28 June 1994 unreported. See also: The Emergence of Trade Related Environmental Measures (TREMS) in the External Relations of the European Community in Maresceau "The European Communities Commercial Policy after 1992: The Legal Dimension" pp305–306 1993 Kluwer. Demeret argues that measures restricting external trade for the purpose of protecting the environment ought properly to be adopted under Article 113; see p352, 353 et seq. and p385. Such measures have in fact been adopted on the basis of environmental policy.

14 That the Community has a broad potential competence to do so is confirmed in Article 130s of the Treaty, any exclusive competence based on the legal necessity of participation is much less clear.

15 The Luxembourg Conclusions 29 October 1990 defining a Community Commitment to stabilize as a whole became a matter of some controversy at INC 4 December 1991. In particular Canada criticized the EC approach which apparently dilutes the individual commitments of Member States. Greece, Ireland, Spain and Portugal are concerned to emphasize that their commitment is to a joint implementation of the stabilization target with burden sharing within the Community. Individually these states plan increased emissions bringing them into possible conflict with Article 4(2)a of the Convention (see below, the penultimate section).

16 As central to the definition of community exclusive competences is the possible impact of international measures on Community law.

17 See n23 below.

18 Preamble para 1.

19 For a discussion of the relationship between law and power in the international law context see Rosalyn Higgins "Problems and Process International Law and how we use it; The nature and function of International Law" Clarendon 1994.

20 See 294/83 *Parti écologiste 'les Verts'* v. *Parliament* 1986 ECR 1339 (para 23) and John Temple Lang "The Development of European Community Constitutional Law" 1991 Dublin University Law Journal Vol 13 pp36–54.

21 Articles 169 and 173 of the Treaty.

22 Higgins op. cit.

23 Fifteen judges appointed by common accord of the Member States.

24 Article 164.

25 For a description of this tendency Dominique Berlin Interactions between the Law Maker and the Judiciary within the EC, Koen Lanearts. Some thoughts about the interaction between Judges and Politicians in the EC 1992 Yearbook of European Law 17, Martin Shapiro 'The European Court of Justice' in "Europolitics", Alberto Sbragia (Ed), Brookings Institution Washington DC, and Hjalfe Rasmussen generally "On Law and Policy in the European Court of Justice" Martinus Nijhoff 1986.

26 In an EC context Dominique Berlin ibid at 19–21 and more generally Ely Democracy and Distrust.

27 Weiler Journey to an unknown destination: A retrospective and prospective of the European Court of Justice in the Arena of Political Integration 1993 31 Journal of Common Market Studies 417. Evidenced in the ILO and GATT cases recently, external relations is one area where community powers have been considerably revised or clarified in the face of Member State concern.

28 See for example Nicholas Emilou Subsidiarity – An effective barrier to the enterprise of ambition 1992 17 ELR 383.

29 Under Article 228 the Council, the Commission or Member State may apply for an opmion on the compatibility of an international agreement envisaged by the community with the Treaty.

30 For a theoretical approach to the nature and function of a legal system see Legal Systems between Order and Disorder op. cit.

31 Articles 145–154.

32 Qualified in this sense Germany, France, Italy and the UK – ten votes each, Spain eight votes, Belgium, Greece, Netherlands and Portugal five votes each, Austria and Sweden four votes each, Denmark, Ireland and Finland three votes each, and Luxembourg two votes.

33 A political agreement not to vote by qualified majority where a state claims a decision cannot be accepted for reasons of vital national interest. Arrived at after a French boycott of Community institutions and gradually abandoned. Note Community internal argument over the Blair House Agreement in the Uruguay Round of GATT. It must be noted that the agreement arrived at by the Commission was renegotiated following French pressure.

34 Shapiro op. cit., p148.

35 Maastricht Articles J–K.

36 Article 157: there are 20 commissioners; no more than two may come from any one state.

37 Unlike the Parliament the Commission need not establish any particular standing or interest to challenge measures through Article 173; see Generalised Tariff Preferences: *Commission* v. *Council* op. cit. and n13 supra.

38 Articles 137–144.

39 Article 173, Case 79/88 *European Parliament* v. *Council of the European Communities* 1990 ECR 2041, Case 187/93 *Parliament* v. *Council* (28 June 1994 Unreported), and see Bradley 'The European Court and the Legal Basis of European Community Legislation' 13 ELR 379 (1988)

40 See the section below on Community conclusion, p138.

41 Decision of the Council OJ L33 07/2/94 p11 Conclusion of the Framework Convention on Climate Change.

42 Text may be found attached to Decision on Conclusion.

43 Article 13 of the Vienna Convention. See John Temple Lang The Ozone Layer Convention a new solution to the question of Mixed Agreements 23 CMLR 157–156, at 160 and 167 and Simmonds The Community's Declaration upon signature of the UN Convention on the Law of the Sea 23 Common Market Law Review 521 544.

44 As of February 1996 Belgium has not done so.

45 Hession and Macrory 'The Legal Framework of European Community Participation in International Agreements' Vol 2 No 1 New Europe Law Review 1994 at p83 'Mixed agreements and Community Compromise: Four Modalities'.

46 Exclusivity of a particular competence can derive from the Treaty (see the section on exclusive policies below or in accordance with the third indent below) the pre-emptive effect of particular measures adopted within areas of potential competence (see the section on the shifting boundaries et seq., p. 123).

47 See the section on non-pre-emptive treaty provisions, p. 122.

48 See Duty of Solidarity below, if the competence is exercised internally legislative pre-emption gives rise to an area of exclusivity; see below, p. 129.

49 If the competence has not been exercised internally it may be exercised in the Agreement itself; see Rhine Case below.

50 Close 'Self Restraint by the EEC in the Exercise of its External Powers' 1 YEL 45 (1981).

51 See Neuwahl 'Joint Participation in International Treaties and the Exercise of Power by the EEC and its Member States: Mixed Agreements' 28 (1991) CMLR 717–740. The mixed agreement has been described both as a necessary evil Barav. 'The Division of External Relations Power between the European Economic Community and the Member States in the Case-law of the Court of Justice in Division of Powers between the European Communities and their Member States in the Field of External Relations', Timmermans and Völker (Eds 1981) at 29 and a faithful image of the federal international character of the Community by Joseph Weiler in 'The External Relations of non Unitary Actors: Mixity and the Federal Principle', in "Mixed Agreements" (O'keefe and Schermers Eds) 1983 Kluwer.

52 Through a specialized participation clause see Feenstra 'A survey of Mixed Agreements and their participation clauses' O'Keefe and Schermers "Mixed Agreements" 1983 at 207, John Temple Lang Ozone op. cit. n43.

53 For the development of this clause see Simmonds and John Temple Lang op. cit. and Rachel Frid 'The European Economic Community: A Member of a Specialised Agency of the United Nations' 4 European International Law Review 1993 pp239–255.

54 At INC 4 December 1991, India suggested an additional clause as a safeguard to the effect that the Member States should be present when the Community votes, perceived as unhelpful by the Commission.

55 Ruling 1/76 Nuclear Materials 1978 ELR 2158 at para 35.

56 Article 22(3) also requires that these organizations should inform the Parties through the depository of the Convention of 'any substantial modification in the extent of their competence'.

57 The Convention requires a declaration of competence. The Community's declaration refers to exclusive and concurrent competences and provides an annexe of adopted measures. As the Community's responsibility under the agreement is

limited to its exclusive competences the limits of exclusive competence are what is important.

58 See Rachel Frid op. cit. European International Law Review at p251 on the value of such declarations.

59 Or to adopt policies aimed at doing so, Article 4(2) of the FCCC.

60 These states which are planning an increase in emissions may well claim that they conform to Article 4.2(a) of the Convention as they have adopted policies aiming to stabilize jointly with the other Member States. How far they can be confident to do so in the absence of an agreement on burden sharing within the Community is a moot point.

61 See also Hession and Macrory 'The Legal Framework of European Community Participation in International Environmental Agreements' 1994 New Europe Law Review Vol 2 No 1 (Cardoza School of Law).

62 In the environmental field see particularly André Nollkaemper The European Community and International Environmental Cooperation: Legal Aspects of External Community Powers Legal Issues of European integration p55.

63 See Kutscher 'Methods of Interpretation as seen by a Judge at the Court of Justice', Judicial and Academic Conference 27–28 September 1976. Weiler 'The Transformation of Europe' 100 Yale LJ 2403, 2416 (1991).

64 Dominique Berlin op. cit. at p22.

65 See 22/70 *ERTA Commission* v. *Council* 1971 ECR 263 and Opinion 1/76 Re *Laying Up Fund for the Rhine Commission* v. *Council* 1977 ECR 742. In a sense the Community ends justify means, and this approach is open to criticism. Nonetheless it is implicit that the means adopted relate to the objectives in some sense. The test of proportionality requires that necessary and effective measures may be adopted to achieve Community objectives. Measures which clearly exceed the proportionality test are illegal – 15/83 *Denkavit* v. *Hoofdproduktschap* 1984 ECR 2171, 2175.

66 Article 190 requires that Community acts are reasoned and this must include a justification of the particular legal basis chosen. As a general rule all relevant bases must be cited save where the procedures required are inconsistent where a particular basis must be chosen. Generalised Tariff Preferences op. cit. see n12.

67 See in particular AG Lenz in Case 45/86 Generalised Tariff Preferences 1987 ECR 1493 for the Practice of the Commission and Council in ascertaining legal bases where two approaches can be demonstrated – instrumental and teleological – or as he puts it the objective and subjective methods para 62 p150 and case-law listed in n12 supra.

68 Nevertheless where one putative basis pertains to exclusive competence and another to concurrent competence it appears that those areas falling within the latter justify Member State participation. See Uruguay Round of CATT, OJ 1994 C 218/20.

69 This last objective was added at Maastricht and confirms the Community's external interest in the environmental field. While the Community has partaken in international agreements on the basis of the first three objectives the fourth objective tends to suggest that unilateral action in respect of global and regional problems is not part of the Community's valid objectives.

70 This provision was first introduced by the Single European Act 1987 as was the express legal basis for Community environmental measures. The Maastricht amendments have strengthened the requirements of environmental integration.

71 Many cases have concerned Article 100a and Article 130s, the former allowing for qualified majority voting at Council level, the latter unanimity; the disputes are likely to continue as the European Parliament has acquired the ability to veto measures adopted under 100a since Maastricht, while most measures adopted under Article 100 will follow the original Cooperation procedure with QMV in the Council.

72 Opinion 1/78 Natural Rubber Agreement 1979 ECR 2871 where Member State participation was authorized for limited purposes (para 60), but particularly the Opinion 1/76 Laying Up Fund for the Rhine 1977 ECR 741.

73 John Temple Lang 'Community Constitutional Law: Article 5 EEC Treaty' 1993 27 CMLR 645.

74 John Temple Lang 'The ERTA judgment and the Court's case-law on Competence and Conflict' 6 Yearbook of European Law 1987 183–220 characterizes the rule in Article 5 as one of cooperation rather than competence but that clearly the two are related (p196 para 8).

75 For positive duties see the section on scope and pre-emption below.

76 See generally Eugene Daniel Cross 'Preemption of Member State Law in the European Economic Community: A Framework for Analysis' 29 (1992) CMLR 447–472 for one synthesis of the rules on pre-emption, conflict and exhaustion, but also John Temple Lang 'The ERTA judgment and the Court's case-law on Competence and Conflict' 6 Yearbook of European Law 1987 183–220 at 197 for two interpretations of ERTA, and Neuwahl 'Joint Participation in International Treaties and the Exercise of Power by the EEC and its Member States: Mixed Agreements' 28 (1991) CMLR 717–740.

77 Bleckmann in Timmermans and Völker op. cit.

78 Whether conflict is necessary to give rise to pre-emptive effect is a matter of some discussion; see Eugene Daniel Cross 'Preemption of Member State Law: A Framework for Analysis' 29 CMLR 447 (1992) at 451 et seq. In addition what is meant by conflict is in itself difficult. Cross distinguishes between actual or potential conflict.

79 Such principles direct the exercise of a legislative power and are in essence procedural but may include objectives and principles with political rather than justiciable effects – in effect exhortatory statements rather than mandatory statements; the objectives and principles of environmental action such as polluter pays and the proximity principle in Article 130r(2). This latter principle has some legal effect as it has been used to justify (rather than overrule) national action to ban imports of waste into Wallonia (Re Imports of Waste: *Commission* v. *Belgium* 1993 CMLR 365). It therefore modifies the operation of Free Movement rules. Whether a full pre-emptive effect might be attributed to it is doubtful; it probably demonstrates an interpretative principle rather than a fully pre-emptive one.

80 See generally John Temple Lang 'The Court's case-law' op. cit. Case 22/70 ERTA (see below) speaks of measures which affect the operation or alter the scope of Community rules. But see Opinion 2/91 Re ILO Convention No 170 68 CMLR 800: a difficulty in determining whether a conflict exists is not an effect sufficient enough to create exclusive competence.

81 The interpretation of a measure to establish its scope and operation (effects) must be based on its objectives (the teleological approach).

82 Again there must be some interpretation of the objectives of the scheme; where a scheme is intended to be exhaustive any measure in some sense affects it. See John Temple Lang 'The ERTA judgment and the Court's case-law on Competence and Conflict' 6 Yearbook of European Law 1987 183–220 at 189 and cases

cited there in fn 8 and Re Imports of Waste: *Commission* v. *Belgium* 1993 CMLR 365 Re Directives 75/442 and 84/631 on hazardous and other waste respectively.

83 See Eugene Daniel Cross 'Preemption of Member State Law: A Framework for Analysis' 29 CMLR 447 (1992) Daniel Cross p453 on a proposed redefinition of Community pre-emption to include all instances of actual and potential conflict. We would suggest a working definition of pre-emptive effect to include where: 1. There is a direct conflict between texts – whether because the measure pre-empted is expressly forbidden, or the text of the measure directly contradicts a community provision. 2. There is an indirect conflict because the objects of the Community measure, the effects it is designed to achieve, are interfered with by the measure pre-empted.

84 Exhaustive provisions are pre-emptive because of implied conflict in the sense that any further legislation, even complementary legislation, would conflict with the measure's exhaustive character.

85 Scope and operation are probably only definable with reference to the objectives and the effects of a particular measure.

86 Re OECD Agreement on a Local Cost Standard 1975 ECR 1355, International Rubber, 1979 ECR 2871, 41/76 Donkerwolcke 1976 ECR 1921.

87 In effect the requirement of a unity of conditions is pre-emptive. However there are limits to the scope of the Commercial Policy International Rubber Agreement and Opinion 1/94 Re the General Agreement on Tariffs and Trade (Unreported 15 November 1994).

88 Nonetheless it appears clear that the Community can modify the effect of this ruling by expressly authorizing the conclusion of bilateral agreements by Member States. Further the strict limits of the Common Commercial Policy and its exhaustiveness remain closely argued issues in such areas as the GATT and international sanctions.

89 Kramer *et al.*, 1976 ECR 1279.

90 Fisheries: *Commission* v. *United Kingdom* 1981 ECR 1045.

91 Though the Community had here adopted framework legislation which gave flesh to the Community policy, Council Regulation 100/76 OJ L20 p1 and Council Regulation OJ L20 p19 thought the Court established effects on these rules were minimal [para 51].

92 Several fisheries conventions have been adopted by the Community as sole participant.

93 In addition the Community may delegate the exercise of competence to the Member States in certain cases as it has done in respect of certain bilateral trade agreements which receive express authorization. Alternatively where the Community has not achieved the right to participate the Member States act as trustees of the common interest.

94 The Ottawa Convention on Future Multilateral Cooperation in North West Atlantic Fisheries, The London Convention on Future Multilateral Cooperation in North East Atlantic Fisheries (Denmark for the Faroe Islands), The Reykjavik Convention for the Conservation of Salmon in the North Atlantic Ocean (Denmark for the Faroes).

95 But '*the exercise* of concurrent powers by the member states and the Community is impossible' Opinion 1/75 1975 ECR 1355 (our emphasis). See the section on the shifting boundaries below, p. 123.

96 See generally Kramer "Focus on Environmental Law": Chapter 3 'Objectives means and powers: Community Environmental Law' at page 67 *et seq*. 70–71.

97 See commentary in Nigel Haighs "Manual of Environmental Policy" Longman (Updated) on Community measures in respect of the following (9.3 Whales,

9.4 Seals, 9.8 Leghold Traps) and the following conventions: Convention on International Trade in Endangered Species, The International Tropical Timber Agreement and the Basle Convention on the Control of Transboundary Waste Movements, note in Demeret 'Environmental Policy and Commercial Policy: The Emergence of Trade Related Environmental Measures' (TREMS) in the External Relations of the European Community. In M. Maresceau (Ed) "The European Communities Commercial Policy after 1992: The Legal Dimension" 305–386 1993 Kluwer.

 98 The pre-emptive effect of several measures taken together may amount to such that no measure can be validly adopted in a particular area. The sphere of action therefore becomes exclusive by reason of occupation of the field.

 99 Dealt with at the section on Community conclusion below, p. 138.

100 Case 22/70 Re the European Road Transport Agreement: *EC Commission* v. *EC Council* 1971 ELR 60–79.

101 From the very restrictive to the very broad – see John Temple Lang The ERTA judgment and the Court's case-law on competence and conflict 6 Yearbook of European Law 1987 183–220.

102 This is confirmed by a close reading of ERTA itself.

103 In this context it is useful to note the Community's annexe attached to its declaration of competence to the Framework Convention. The list is dominated by non-binding acts which suggests the area of exclusive competence is limited.

104 The Court has always succeeded in finding a proper legal basis nonetheless and has never ruled there is no power to conclude the agreement (but see Opinion 1/91 Re Draft European Economic Area Agreement 1991 ECR 6079: 1992 1 CMLR 245).

105 Re OECD Agreement on a Local Cost Standard 1975 ECR 1355, Joined Cases 3, 4 and 6/76 Cornelis Kramer *et al.*, 1976 ECR 1279, Fisheries: Case 804/79 *Commission* v. *United Kingdom* 1981 ECR 1045–1080, ERTA Case 22/70 *Commission* v. *Council* 1971 ECR 263, Opinion 2/91 Re ILO Convention 170 on Chemicals at Work 68 1993 CMLR 800.

106 Community action which is contrary to human rights (though this is nowhere directly expressed in the Treaty) is illegal, Case 44/79 Hauer 1979 ECR 3727–3751. It appears that action which undermines the autonomy of the legal order is also disallowed (Opinion 1/91 Re Draft European Economic Area Agreement 1991 ECR 6079: 1992 1 CMLR 245 and Opinion 1/76 Laying Up Fund for the Rhine 1977 ECR 741).

107 In certain cases it may be invocable as between individuals, in other cases merely between individuals and emanations of the state. De Burca 'Giving Effect to European Community Directives' Modern Law Review 18992, 215.

108 Dominique Berlin op. cit. p24 *et seq.*

109 Hession 'The Role of the European Community in the Implementation of European Community Law' RECIEL Vol 2 Issue 4 1993.

110 It is not quite correct to say that a regulation has direct effect as the Treaty states that the regulation is directly applicable. Any terms of a directive which are insufficiently precise, unconditional or invocable by individuals are of course unlikely to have any real effect even if the Treaty guarantees their applicability. At the end of the day it is the substance of a measure which guarantees it having an effect on the results of a judicial decision.

111 See for example Timmermans 'The Division of Power between the Community and Member States in the Field of Harmonisation of Law – a case study' in Timmermans and Völker (Eds) "Division of Power between the European Communities and their Member States in the Field of External Relations"

(Kluwer 1981) p15 *et seq.*, Leenen 'Participation of the EEC in International Environmental Agreements', same volume.

112 The direct effect of a directive is limited in that provisions may be invoked by individuals to render national provisions inapplicable as against the Member State which has failed to adopt implementing measures but not as against other individuals.

113 Article 228 states, 181/73 Haegeman 1974 ECR 449 confirms that the ECJ has the jurisdiction to interpret these agreements.

114 See John Temple Lang 'The ERTA judgment and the Court's case-law on Competence and Conflict' 6 Yearbook of European Law 1987 183–220 at 190–193 and Re Imports of Waste into Wallonia: *Commission* v. *Belgium* 1993 CMLR 365, Directive 84/631 was found to have this effect, Reg 75/442 was not. It may be instructive to consider whether a rule of conflict or exhaustiveness was applied in this case.

115 The Wallonian Waste Case does demonstrate the difficulty of applying the framework of scope conflict and exhaustion to individual measures in that elements of each argument are demonstrated in the Judgment and Opinion of AG Jacobs. Whether conflict is necessary to establish pre-emption has been an issue. Eugene Daniel Cross 'Preemption of Member State Law' op. cit. The suggestion is that pre-emption exists even in the absence of conflict. The extent to which this is the case is a matter of some difficulty. John Temple Lang 'The ERTA Judgment and the Court's case-law on Competence and Conflict' 6 Yearbook of European Law 1987 183–220 (at 193 second paragraph), distinguishes exhaustion and exclusivity on the grounds that the former is preclusive because there is no room for further measures whether adopted by the Community or Member States; a true exclusive competence rule precludes measures by Member States regardless of whether the Community has acted. This raises the question whether ERTA is a rule of exclusive competence at all but merely one of exhaustion. Cross proposes a re-definition of pre-emption to include actual and potential conflict with Community law.

116 Paragraph 22.

117 Refer back to pre-emption difficulties above.

118 As has been suggested by many commentators Mastellone 'The External Relations of the EEC in the Field of Environmental Protection' Vol 30 ICLQ 1981 p104 at 113.

119 Opinion 2/91 Convention No 170 of the International Labour Organization concerning safety in the use of chemicals at work (1993); see Nicholas Emiliou Towards a clearer demarcation line: The division of external relations power between the Community and Member States.

120 Article 118a(3) but not it should be noted the reporting obligation introduced at Maastricht.

121 See paras 16–21.

122 Paragraph 22. It should be noted that one of these directives was adopted pursuant to Article 100a and laid down absolute standards. Article 100a allows more stringent standards to be applied in very limited circumstances. The Court did not consider the effect of this limited freedom.

123 Suggesting that potential conflict in this sense is not sufficient to trigger pre-emption. Had the ILO Convention established absolute standards this may not have been the case.

124 Opinion 1/76 Laying Up Fund for the Rhine 1977 ECR 741.

125 Paragraph 3 and para 4.

126 Unlike the CCP and Marine Fisheries Conservation Policies, the Transport Policy of which this Agreement forms a part is not exclusive *per se*. This underlies both the ERTA and Rhine judgments.

127 John Temple Lang 'The ERTA Judgment and the Court's case-law on Competence and Conflict' 6 Yearbook of European Law 1987 183–220 suggests a rule against interference with future Community decision making (p206) but the grounds upon which this might be said to arise are only vaguely defined. See also Hardy Opinion 1/76 of the Court of Justice: The Rhine Case and the Treaty-Making Powers of the Community 1977 14 CMLR 561.

128 Paragraph 11.

129 For this approach see Koen Lanaerts Regulating the Regulatory Process: 'delegation of powers' in the European Community ELR at p37 *et seq*.

130 John Temple Lang 'The ERTA Judgment and the Court's case-law on Competence and Conflict' 6 Yearbook of European Law 1987 183–220, at 206, suggests a putative rule against interference in future decision making. The existence of a Commission proposal on the matter is one element relevant to establishing the duty not to act. Whether this duty is one of competence is doubtful though perhaps this is less important than the fact that in the face of a proposal Member States are not free to take unilateral action to conclude an agreement without a formal rejection of the proposal at Council level. Otherwise the prerogatives of the institutions and the legislative scheme of the Community would truly be undermined. Whether this is expressed as a question of competence or as a duty of solidarity is unimportant.

131 Opinion 1/91 Re Draft European Economic Area Agreement 1991 ECR 6079: 1992 1 CMLR 245.

132 Paragraph 30 Opinion 1/91.

133 Pages 164 and 219.

134 Opinion 1/92 para 40.

135 The existence of a Commission proposal has been suggested as one factor. To give such a document pre-emptive effect at first appears excessive as it gives legal force to a measure which has not been finally adopted. Nonetheless to preserve the Community's legislative procedures and the Commission's prerogatives such a rule is supportable. The Court of Justice may then be called upon to rule as to the necessity of Community participation, see Hardy 597 *et seq*., or the Council may decide to reject the proposal. Whether the Council may do so is a question of the justiciability of the necessity concept.

136 Paragraph 12 of Rhine Case.

137 Joined Cases 3, 4 and 6/76 Kramer op. cit. at paras 30–33* para 45 – 'A duty to use all political and legal means at their disposal to ensure participation by the Community'.

138 Ibid para 45 and Fisheries Case 804/79 *Commission* v. *United Kingdom* op. cit.

139 ILO Opinion 2/91 op. cit. [paras 12 and 36–38] and Kramer [para 45] and Nuclear Materials (Euratom) Ruling 1/76 [paras 34 and 36].

140 This has been suggested in the context of particular votes at international institutions. The recognition of a reporting obligation in Article 130t may not have the effect of reinforcing this duty in respect of more stringent measures but reinforces the duty of consultation.

141 Close Self Restraint by the EEC in the Exercise of its External Powers 1 YEL 45 (1981).

142 Though the competence of the Community in respect of extraterritorial fisheries has been a matter of some comment Lhoest 'The European Community and Antarctica: Competence of an Extraterritorial Nature in The Antarctic

Environment and International Law' (Verhoven, Sans and Bruce, Eds). See Fisheries Case 309 79 op. cit. at para 33.

143 See Article 102 of the Act of Accession of the UK Ireland and Denmark, Joined Cases 3, 4 and 6/76 Cornelis Kramer *et al.*, 1976 ECR 1279 and Fisheries: *Commission* v. *United Kingdom* 1981 ECR 1045.

144 Birds Directive 709/409 OJ L103 25.4.79.

145 3626/82 OJ L384 31.12.78 as amended.

146 Article 15 (but note there is a reporting requirement in respect of additional species).

147 Adopted using Article 235 (130s precursor); recent amendments have been under 130s.

148 Similar arguments were raised in respect of Water Quality Standards and The Paris and Barcelona Conventions, causing difficulties in respect of the 7th Meeting of the Paris Commission June 1985.

149 5th Conference of the Parties 1985 com 85 729 see André Nollkaemper 'The European Community and International Environmental Cooperation: Legal Aspects of External Community Powers Legal Issues of European Integration' p55 at p71 (Buenos Aires Meeting).

150 The situation created by the Court in Fisheries whereby unilateral action became impossible and Member States could only act in consultation and ultimately with the Commission's agreement may form a framework for legal consequences of a duty of unity in external relations. Though this concerned an exclusive competence and rules for the exercise of Community competence by the Member States, the exercise of Member State competences by the Member States in mixed agreements might be subject to similar strictures.

151 Charter of the United Nations Article 2.

152 Paris Convention on the Prevention of Marine Pollution from Land-based Sources (Decision 75/437 1194 25.7.75) and Barcelona Convention on the Protection of the Mediterranean Sea against Pollution (Decision 585/77 1240 19/9/777 p3), for example.

153 Rachel Frid op. cit.

154 The block vote is an issue for third parties who may insist that the EC should be treated no differently as any other party where it acts in its own name.

155 Note 86 supra see "Haigh Manual of European Environmental Policy" Longman Chapter 13.

156 See n139 and n140 supra.

157 Swedish/Norwegian Statement op. cit. 7th Meeting of Paris Commission 1985 PARCOM 7/4/1 expressing concern.

158 Manin 'The European Communities and the Vienna Convention on the Law of Treaties between States and International Organizations or between International Organizations' 1987 CMLR 457–481.

159 Seyrsted.

160 This was not expressly the case prior to the Maastricht amendments but it occurred in practice.

161 Such as an agreement on a non-binding energy charter, or perhaps Agenda 21.

162 For obvious reasons the mandate is confidential. The European Parliament in particular is concerned about the secrecy involved see for instance Resolution of E OJ C66 15/3/82 p68.

163 Jachtenfuchs 'The European Community and the Protection of the Ozone Layer' J. of Common Market Studies 28 261 at 265.

164 Maastricht introduced this innovation but the Commission had proposed General Mandate in respect of three Conventions COM (86) 5673 prior to this.

165 The so-called bicephalous delegation, see Groux and Manin 'The European Communities in the International Legal Order and Groux Mixed Negotiations' in O'Keefe and Schermers Mixed Agreements op. cit.

166 Member States leave the exercise of such competence to implementation measures internally.

167 By virtue of a duty of unity in respect of mixed agreements. Member States may agree that the Presidency should speak outside the Community framework under the Common Foreign Policy of the European Union.

168 In this way many Member States may choose to add to an agreed Community position.

169 Where the competences are exclusive it does so as trustee of the common interest; see the section on scope and pre-emption above.

170 Reported to be a problem at INC 4 of FCCC.

171 Anecdotal evidence regarding the signing of the Montreal Protocol.

172 Between States and International Organizations and between International Organizations.

173 The European Parliament now has the right to be consulted in respect of all agreements save those adopted under the Common Commercial Policy. There may be a time limit for opinions set by the Council according to the urgency of the matter. For Association Agreements (Article 238) agreements establishing institutional frameworks by organizing cooperation procedures, agreements with important budgetary implications and those amending acts where the Parliament has a final veto, Parliament must assent before they may be adopted.

174 Bresciani 87/75 1976, Pabst 17/81 (Association Agreements), Kupferberg 1982 (Free Trade Agreement) and *Polydor v. Harlequin* 270/80, see Van Themaat The Impact of the Court of Justice of the EC on the Economic World Order 82 Michigan Law Review p1422 at 1428–9 and Bourgeois Effects of International Agreements in European Community Law: Are the Dice Cast? Vol 82 Michigan Law Review 1250, Hession 'The Role of the EC in the Implementation of International Environmental Law' Vol 2 Issue 4 Review of European Community and International Environmental Law p341.

175 This at least is the case where the Council does not expressly exclude the exercise of potential competences, but if the rule in the Rhine Case has been broken or even where statements preserving competence are continued only in the minutes of the Council the Court is bound to interpret the division according to the principles described.

176 See Close 'Self Restraint by the EEC in the Exercise of its External Powers' 1 YEL 45 (1981) and Neuwahl 'Joint Participation in International Treaties and the Exercise of Power by the EEC and its Member States: Mixed Agreements' 28 (1991) CMLR 717–740 at 725, 726, 729 et seq. and, particularly, p731, to the effect that the mixed agreement is designed to obviate the necessity of the Community exercising its potential competence in an international agreement and to prevent any pre-emptive effect arising.

177 See Bourgeois 'Effects of International Agreements in European Community Law: Are the Dice Cast?' Vol 82 Michigan Law Review 1250. The article does not deal with added difficulties determining the extent to which a mixed agreement is a Community agreement, but see Van Thematt op. cit.

178 Kramer in 'The Implementation of Community Environmental Law' 1991 German Yearbook of International Law 431.

179 As Rachel Frid points out in respect of the FAO 'activities within the FAO are not designed to accord with the distribution of community competence'.

180 See the section on scope and pre-emption above and in this context the conclusions of Phillip Allot in 'Adherence to and Withdrawal from Mixed Agreements' in Mixed Agreements op. cit. at p120 (Conclusion 5).

181 Conclusions of the Joint Environment Energy Councils of 29 October 1990
and 13 December 1991 include the wording: 'the council notes that some
member countries according to their programmes are not in a position to commit
themselves to this objective [stabilization of CO_2 emissions at 1990 levels by the
year 2000] . . . [and] may need targets and strategies which can accommodate that
development, while improving the energy efficiency'.

182 Decision 93/389 of June 1993 OJ L 167 09/07/93 p31 see Article 2.1.

6

CLIMATE CHANGE POLICIES AND POLITICS IN THE EUROPEAN COMMUNITY

Nigel Haigh

The legal arrangements that have evolved to allow the European Community (EC) to participate in international conventions in addition to its Member States have been described in the preceding chapter. There is no concealing the fact that these are frequently obscure and ambiguous. At a simple level the arrangements can be seen as an aspect of the continuing power struggle between the EC institutions and the Member States. At a deeper level, they are an extraordinary attempt to share a commodity – sovereignty – which in principle is indivisible. The Member States remain sovereign states with the power and the desire to continue pursuing their own international relations. Simultaneously, as the policies of the EC develop, the Member States are in the process of diminishing their own powers by ceding more sovereignty to the EC in order to achieve objectives that are beyond the reach of any one state on its own. In this respect the softening of the concept of sovereignty that is effectively taking place within the EC may well have lessons for all nations dealing with global environmental problems.

This chapter traces the development of EC policy for climate change including its involvement with the Framework Convention on Climate Change. It starts with a discussion of the different institutions of the EC, and of the making of EC policy, and places EC climate policy in the context of pre-existing EC policy for the environment and for energy. It concludes with observations on the importance for the success of the UN FCCC of a Community that actually meets its collective target, and provides a sound political and economic basis for further reductions of greenhouse gases beyond 2000.

THE INSTITUTIONS OF THE EU AND EC

As its name implies, the European Union, of which the European Community is the central pillar, is very much more than an international organization created among national states for the purposes of pursuing an

155

agreed objective. The Treaty of Maastricht, which came into effect in November 1993, is the most recent of a number of treaties entered into by the Member States who have created an organization which possesses many of the attributes of sovereignty while enabling the Member States to retain essential aspects of their own sovereignty.

Particularly when dealing with international conventions, there is a possibility of incorrectly using the term 'European Union' when the term 'European Community' is appropriate, so it is as well to be clear about the difference. The 'European Economic Community' was created by the Treaty of Rome in 1957. The Treaty of Maastricht created the 'European Union' supported on three 'pillars' and simultaneously amended the most important and pre-existing pillar – the Treaty of Rome – in a number of respects including changing the name 'European Economic Community' (EEC) to 'European Community' (EC). In addition to the EC, the other two pillars supporting the EU are: common foreign and security policy; and home affairs and justice policy. All policies previously carried out by the EC such as agricultural policy, transport policy and environmental policy continue to be pursued by legislation adopted under the Treaty of Rome so it is perfectly correct to call them 'EC policies', but as the EC is part of the EU it is also correct to call them 'EU policies'. But since the EU, unlike the EC, does not have legal personality it is the EC and not the EU that signs and ratifies international conventions. Confusing? That is the price for accommodating national sensibilities.

Accordingly it is the EC that has signed and ratified the Climate Change Convention and although EC policies relating to climate change are also EU policies, for consistency this chapter will refer to EC policies for climate change and not to EU policies.

The single characteristic of the EC, which sets it apart from other international organizations, is its possession of institutions able to propose and adopt legislation which can bind the Member States without further review or ratification by national governments or parliaments or by the head of state.

These institutions (before the changes resulting from the accession of Austria, Finland and Sweden) are as follows:

- A Commission of 17 members which has the sole right to propose EC legislation. The Commission is supported by the Commission services divided into a number of Directorates-General.
- A Council composed of a minister from each Member State which has the power to adopt legislation proposed by the Commission. Legislation must sometimes be adopted unanimously but sometimes it may be adopted by qualified majority voting (QMV).
- A Parliament with 567 directly elected members which participates in the legislative process in accordance with various provisions of the Treaty.

These provisions differ according to the subject under consideration (consultation procedure, cooperation procedure, co-decision procedure).

- A Court of Justice composed of 13 judges with jurisdiction over a range of matters relating to the functioning of the EC including the implementation by Member States of the obligations placed upon them by EC legislation.
- A Court of Auditors composed of 12 members which examines the accounts of all revenue and expenditure of the EC.

The EC 'legislature' is therefore composed of the Commission, the Council and the Parliament acting together in accordance with the Treaty. Only the Commission may propose legislation and only the Council can finally adopt it. The Parliament, depending on the subject matter under discussion, has a greater or lesser influence, including in some circumstances the power of veto. There are three types of binding legislation that the Commission may propose (Regulations, Directives and Decisions) in addition to non-binding Recommendations and Opinions. Regulations are directly applicable in the Member States exactly like national legislation. Directives are binding as to the results to be achieved but leave the Member States the choice of form and method for implementing them. Directives in general therefore require national legislation for them to become effective although the Court has developed the doctrine of 'direct effect' which can apply in certain circumstances. Decisions are also binding in their entirety on those to whom they are addressed and the Decision is the chosen instrument for the EC to conclude international conventions.

Where EC legislation can be adopted by qualified majority voting in the Council, a Member State may have to accept obligations to which its government is opposed. It is in these circumstances that the EC most clearly assumes a supranational character. Under such conditions the Parliament is also given a greater role than merely to give an opinion, thus providing some greater legitimacy to the ultimate decision to overrule a particular Member State. The Parliament's greater role is manifested in the 'cooperation procedure' or the 'co-decision procedure' which gives the Parliament certain extra powers to modify proposals and in the extreme to veto them (Wilkinson, 1992). The Treaty of Maastricht extended the number of occasions when qualified majority voting applies, so that it has become the standard procedure for environmental legislation. Nevertheless, the Treaty specifically applies exceptions for which unanimity is required. These are particularly relevant in the field of climate change since they include 'provisions primarily of a fiscal nature' and 'measures significantly affecting a Member State's choice between different energy sources and the general structure of its energy supply'. Despite this, as we will see below, the EC adopted a decision to ratify the Climate Change Convention by qualified majority.

THE MAKING OF EC POLICY

A major difficulty in discussing EC policy for climate change, or indeed for any other subject, is the ambiguous way in which the term EC policy is often used. As we have seen, the EC is a 'Community' composed of Member States whose tasks are carried out by a number of institutions. The institutions of the EC are not themselves the EC, and a proposal made by the Commission does not become EC policy until it has been adopted in a formal text by the Council following the proper procedures laid down in the Treaty. Similarly, a mere opinion of the Parliament is not EC policy, nor yet is a political declaration by the Council until it is embodied in a legally binding text. Such opinions or declarations may be precursors to EC policy and give an indication of possibilities. Since the Commission is the initiator of EC legislation in which policy is embodied, it is easy for the policies which it is advancing to be confused with EC policy. But strictly this is incorrect. Although the Commission put forward a proposal for a carbon and energy tax, such an idea would have only become EC policy if the Council had adopted a formal text on the subject. Thus the position of EC policy is rather different from national policy. In most nation states, a minister, having secured agreement in cabinet, can state something as government policy and expect that it will be treated as such even if it has not been approved by the national parliament. This is because of a presumption that any government in power is likely to be able to have any necessary legislation passed in parliament.

By contrast the EC has no person or body with authority comparable with a prime minister or national cabinet. The Commission has much less power over the Council than most national governments do over their own legislature, as is shown by the fact that many proposals from the Commission are never adopted. It is perfectly proper to talk of the 'policy of the Commission' once it has made a formal proposal but that does not mean it is or will become EC policy.

Commission proposals are adopted for transmission to the Council by all the Commissioners meeting together as a college. They are developed by the relevant Directorate-General following consultations with any other Directorate-General likely to be affected. In this respect the process of EC policy making is comparable with that within national administrations, where, with regard to environmental matters, environmental ministries are often weak in comparison with longer established ministries that are usually supported by powerful economic interests.

The provision in the Treaty that environmental protection requirements must be integrated into the development and implementation of other EC policies is therefore very important, for in principle it strengthens the hand of the Directorate-General for the Environment when dealing with other Directorates-General. So long as DG XI (Environment) was confining itself, as it did in its early years, to proposing legislation, e.g. for water pollution or nature protection, that might cause pain in the Member States but did not affect the priorities of the other Directorates-General, the latter were not particularly concerned. But the moment DG XI begins to become involved with, say, transport on the grounds that EC policies for road building will increase emissions of greenhouse gases the likelihood of conflicts arises. In the resolution of such conflicts the 'integration' provision of the Treaty comes into play.

EC energy policy

The EC's involvement with energy is still not formally provided for in the core Treaty although the associated Treaties that created the European Coal and Steel Community and the European Atomic Energy Community have influenced the coal industry and many aspects of the nuclear industry since the origins of the EC. Attempts to form an EC energy policy have had to confront the very different pattern of energy supply in Member States, some of which are richly endowed with energy resources while some have none – apart from the sun. Some Member States depend heavily on nuclear power (France) while some are totally opposed to it (Denmark). Following the oil price rises of the 1970s, the EC began developing policies for energy efficiency in the interests of reducing energy imports, which is a common concern of all Member States. In 1986, for example, the EC adopted a target of improving the efficiency of final energy use by at least 20 per cent by 1995 for reasons unrelated to climate change. When in 1988 it became clear that this target would not be met without more vigorous action being undertaken a new range of measures was initiated which came to be associated with climate change policies. These include the SAVE and ALTENER programmes on energy efficiency and renewable energy respectively (see below) and a range of research and development programmes (notably JOULE and THERMIE).

EC environmental policy

EC environmental policy formally began in 1973 with the adoption by the Council of the First Action Programme on the Environment. This was despite the absence of any express authority in the Treaty for such a policy. It was not until the Single European Act amended the Treaty of Rome in 1987 that express authority for an environmental policy was provided, thus effectively legitimizing the extensive body of environmental legislation that had by then been adopted under a rather elastic interpretation of the original Treaty. The title on the environment introduced by the Single European Act established certain principles of environmental policy, including the principle that preventive action should be taken; that environmental damage should as a priority be rectified at source; that the polluter should pay; and that environmental protection requirements should be a component of the EC's other policies. This last principle is of very considerable importance for climate policy, and was strengthened by an amendment made by the Maastricht Treaty to become 'environmental protection requirements must be integrated into the definition and implementation of other Community policies'. The Maastricht Treaty added the precautionary principle which is also of considerable importance for EC climate policy. (For a discussion of the wider significance of the principle, see O'Riordan and Cameron (1994).)

The Single European Act also introduced the 'principle of subsidiarity' expressly for environmental policy and the Treaty of Maastricht then extended the principle to all EC policies. It confines action by the EC to those matters whose objectives cannot be sufficiently achieved by the Member States and that can be better achieved by the EC. This has had implications for EC climate policy as we shall see below.

The Treaties and the principles provide the basis for policy. But to be effective EC policy must be embodied in items of legislation placing obligations on the Member States: indeed, an extensive body of EC environmental legislation exists to promote this (Haigh, 1992a). To give an indication of the legislation it plans to propose in the years ahead the Commission has drafted action programmes on the environment. Five such programmes have been approved by the Council in the last 20 years. The resulting EC legislation has significantly altered the way all Member States have to behave, though not all like to admit this. EC environmental policy has thus developed from being a fringe political activity, for which there was no formal Treaty provisions, to being one of the policy arenas of the EC best known and accepted by the general public (Worcester, 1994).

A topic that was successfully handled by the EC, and that has provided some lessons for the climate change debates, is acid rain. In the mid-1980s when the issue was finally forced onto the agenda of reluctant countries, the EC agreed on legislation which should result in an EC-wide reduction in sulphur dioxide emissions from large combustion plants of 58 per cent by the year 2003 compared with a 1980 baseline. This experience gave the Commission the confidence that it could tackle a major environmental issue transcending national frontiers with considerable implications for the cost of energy. A useful precursor to the subject of climate change was thus provided. Simultaneously in the mid-1980s the EC was involved in the negotiation of the Vienna Convention on the Protection of the Ozone Layer, and its associated Montreal Protocol, an involvement which not only influenced the final form of the Montreal Protocol but which also delivered intact a bloc of 12 countries as signatories to the Convention. This outcome would by no means have been a foregone conclusion without the EC (Haigh, 1992b).

By the end of the 1980s, therefore, when the issue of climate change reached the political agenda, the EC was in a position where it had acquired sufficient experience of environmental policy making to feel able to take on a leading role. Had the negotiations in the 1980s over acid rain resulted in failure, the way in which the EC approached the climate change issue would have been very different.

THE DEVELOPMENT OF EC CLIMATE POLICY

EC policy for climate change can conveniently be regarded as having evolved in three phases which ended with ratification of the Convention by the EC in December 1993. We are now effectively in a fourth phase.

First phase – 1988–1990

It was the European Parliament, in a Resolution in 1986, that was the first of the EC institutions to recognize the need for EC policy on climate change (Official Journal of the EC 1986). This followed the Villach conference, as introduced in Chapter 1, which concluded that the time was ripe for policy initiatives on climate change.

It was not until November 1988, however, that the Commission issued the first of its communications on the subject (European Commission, 1988). Its Fourth Action Programme on the Environment, adopted in late 1987 and covering the years 1987 to 1992, had made no mention of climate change except as a subject for further research (Official Journal of the EC 1987). The Commission's communication of November 1988 reviewed the scientific findings and possible actions to be taken without making any precise recommendations beyond proposing a work programme. In the same month the IPCC was established. The earliest EC developments can therefore be seen to have largely coincided with developments elsewhere.

The seriousness of the subject was confirmed at the highest EC level in June 1990 when the European Council (composed of heads of state and government) meeting in Dublin called for early adoption of targets and strategies for limiting emissions of greenhouse gases (*Bulletin of the EC*, 1990, 19). This paved the way for political agreement at a joint Council of Energy and Environment Ministers in October 1990 that CO_2 emissions should be stabilized in the EC as a whole by the year 2000 at 1990 levels. The wording of the Council's conclusions included the following:

EC Member States and other industrialized countries should take urgent action to stabilize or reduce their CO_2 and other GHG emissions. Stabilization of CO_2 emissions should be in general achieved by the year 2000 at 1990 levels, although the Council notes that some Member countries according to their programmes are not in a position to commit themselves to this objective. In this context countries with, as yet, relatively low energy requirements, which can be expected to grow in step with their development, may need targets and strategies which can accommodate that development, while improving the energy efficiency of their economic activities. The European Community and Member States assume that other leading countries undertake commitments along the lines mentioned above and, acknowledging the targets identified by a number of Member States for stabilizing or reducing emissions by different dates, are willing to take actions aiming at reaching stabilization of the total CO_2 emissions by 2000 at 1990 levels in the Community as a whole.

This declaration was precipitated by the Second World Climate Conference in November 1990 and enabled the EC to take a strong and leading role, particularly in relation to the United States. This political agreement was not at that time embodied in any legally binding EC text and was in any event qualified by the assumption that other leading countries undertook commitments along similar lines. The qualification was not spelled out in any detail but has always been understood to include at least the United States, the world's largest emitter of CO_2.

The political commitment to stabilize was taken in a conscious attempt by the EC to show itself as a leader on the subject. But it was also based on a certain pragmatism. Several Member States had by then adopted national policies to curb, stabilize or reduce greenhouse gas emissions. By making assumptions on what might happen in Member States which had adopted no national targets it appeared that stabilization was an attainable goal for the EC within ten years.

This first phase of EC policy making on climate change, which can be said to extend from the recognition of the problem by the Commission to the adoption of a political target, i.e. the period from 1988 to 1990, developed remarkably rapidly despite inevitable resistances from some quarters. The political decision by the EC influenced the course of negotiations on the UN FCCC. Despite its non-binding character, Article 4(2) of the Convention would certainly have been much weaker without the EC's prior position. The Article was a compromise negotiated between the United States and EC Member States. The political weight of a combined EC put a reluctant United States under pressure.

Second phase – 1990 to Rio

The second phase of EC policy making can be said to span the period between the political agreement in 1990 and the signing of the Convention at Rio in June 1992.

The Commission, doubtless encouraged by the success of the first phase, made no secret of its desire to see the EC maintain its leadership role. It had been encouraged to do this by the declaration on the 'Environmental Imperative' issued by the European Council at Dublin in June 1990 which stated:

> The Community and its Member States have a special responsibility to encourage and participate in international action to combat global environmental problems. Their capacity to provide leadership in this field is enormous.
>
> (*Bulletin of the EC*, 1990, 19)

Given this encouragement, it is no surprise that the Commission started by adopting an approach in which the major elements were to be decided at EC level. With hindsight it can be seen that this approach was over-ambitious and thereby delayed by perhaps a year the turning of the political commitment into a legally binding text.

Immediately after the joint Energy and Environment Council of October 1990 an internal draft of a Communication was prepared by the Directorates-General for Environment and for Energy (DG XI and XVII). It stated that stabilizing CO_2 emissions actually meant reducing emissions growth by some 10 to 20 per cent. Energy efficiency actions already under-way, like the SAVE programme, seemed unlikely to secure such results. Accordingly, an adequate strategy was to consist of several measures complemented by economic and fiscal incentives. Action in the transport sector was also recognized as being necessary, including speed limits for cars and improved traffic and transport management. By May 1991 a further draft identifed four major elements of climate policy, namely (a) a regulatory approach, (b) fiscal measures, (c) burden sharing among Member States and (d) the scope for complementary action at the national level.

The concept of burden sharing, which the Netherlands in particular pressed for during its Presidency in the second half of 1991, was derived in part from the example of the Community's policy response to the issue of acid rain which it developed in the mid-1980s. Under a directive adopted in 1988 Member States were allocated different targets for reducing emissions of sulphur dioxide from their existing power plants. The experience of the long drawn-out negotiations on this directive should have given some warning of the difficulties inherent in this approach which could better be described as horsetrading rather than the sharing out of a burden based on objective criteria. The idea of a directive allocating different targets for CO_2 emissions was eventually abandoned and fiscal measures, i.e. a carbon energy tax, then became the cornerstone of the Commission's proposals. One objection, not often publicly mentioned, to the idea of national targets for CO_2 emissions being set at EC level was the transfer of powers that this would have implied over a matter which touched so many aspects of national life.

Some Member States were reluctant to grant the EC so much competence, an issue discussed more fully below.

The deadline that came to concentrate the minds of the Commission and the Member States was the United Nations' Conference on Environment and Development (UNCED) to be held in Rio in June 1992. If the EC was to maintain its leadership role which it had forged for itself at the Second World Climate Conference in November 1990, it was necessary to have some policy flesh to put on the bones of the political decision to stabilize CO_2 emissions. It would be necessary to demonstrate how the EC was to achieve this. One course would have been to translate the decision on stabilization into a legally binding instrument. This could also have required Member States to develop their own programmes with their own targets. Such an instrument could have been agreed fairly quickly after October 1990. Not surprisingly, given the Commission's traditional desire to expand EC competence, it chose the path of trying to develop a complete and ambitious package of measures to be agreed together. The most ambitious of these, namely the carbon energy tax, proved to be the most contentious, and delayed all the others. As finally announced in June 1992 these measures consisted of four elements:

- a framework directive on energy efficiency within the SAVE programme;
- a decision on renewable energies – ALTENER programme;
- a directive on a combined carbon and energy tax; and
- a decision concerning a monitoring mechanism for CO_2 emissions.

These four elements are described more fully below.

Third phase – Rio to ratification

The third phase spans the period from the signing of the Convention at Rio in June 1992 to its ratification by the EC in December 1993. During this period the Council had to decide whether to accept the four detailed proposals that the Commission had put forward just before Rio, the most important of which in the eyes of the Commission was the proposed carbon energy tax.

The carbon energy tax

Liberatore (1995, 61–69) explains how the tax idea was conceived in the Commission and ended in the Council of Environment Ministers. The story began with a Commission-initiated research programme from 1979 to 1986, during which time the evidence on global warming was collected. However, the Fourth Environmental Action Programme of 1987 contained no specific measures. These were developed through an interservice group of officials representing the Directorates-General on tax, competition, environment, energy, regional policy and research. The individual responsible for promoting the international leadership role for the Commission was its committed but flamboyant Commissioner for the Environment, Carlo Ripa di Meana. He promoted the cause of the tax, initially simply as a penalty, but subsequently as an innovation in fiscal policy, in meetings in Washington in 1990, and in Bergen (the pre-Rio conference on sustainable development for the northern rich countries) in the same year.

Three elements combined to advance discussions, namely the interservice group initiative, a belief in the need to show leadership (urged on the Commission by the Germans and the Dutch in particular), and a joint meeting of energy and environment ministers prior to the Dublin Summit. The key to innovative reform lay in the connection between energy futures and climate change on the one hand, and economic development and energy efficiency measures on the other. This coupling of 'no regret', namely interlinking the economic gains of energy conservation, stimulated the belief that the tax could be a valuable trigger in economic renaissance. In this endeavour, the European Parliament also became involved, holding hearings and initiating debate. It is of interest to note that the interservice group largely left out transport and agriculture in its deliberations. It focused on energy too early on, then tax reform, a highly contentious issue, and failed to make an effective case for even informal integration. The client groupings were simply not ready for realignment.

The proposed tax was to be shared equally between carbon content and energy content on the grounds that a pure carbon tax would have favoured nuclear energy. For this reason alone the joint carbon and energy tax was always opposed by France which favoured a straight carbon tax. The tax was to be progressive starting at the equivalent of 3 US dollars per barrel of oil and rising by 1 dollar per barrel per year until 10 dollars had been reached. The tax had opponents within the Commission, within industry, and among the Member States. Agreement within the Commission to the proposal was only secured by exempting energy-intensive industries and making the proposal conditional on comparable action by other OECD countries. For a time there were some hopes that the United States might introduce an energy tax at a level that could be said to be comparable. When that hope was not realized it was clear that either the conditionality element would have to be dropped or the tax would have to be abandoned. In the Council the tax was opposed by France, by the less developed Member States who argued that their development would be impeded, and by the UK. The UK's objection was not to the idea of fiscal measures but to the principle that taxation is a matter that should be the responsibility of the Member States. The UK also argued that it did not need this tax to meet its own target. Since, even under the revised rules for decision making in the Council, fiscal matters can only be decided by unanimity, it was improbable that an EC carbon energy tax would be agreed. In December 1994 the

Council eventually determined that there would be no tax set at EC level but Member States were encouraged to develop their own taxes. This was regarded as a severe setback by the 'northern' Member States (Germany, the Netherlands, Denmark, Sweden, Finland), and the idea has certainly not gone away.

Renewable energy

The decision on renewable energy (ALTENER) is a programme of EC funding amounting to about 35 million ECU over five years. This is estimated to achieve a reduction of CO_2 emissions of 180 million tonnes by 2005. The agreed funding was smaller than originally envisaged by the Commission but otherwise proved uncontentious and has been adopted. Whether it will meet the target set for it is a moot point.

Energy conservation

The directive on energy saving measures (SAVE), as adopted, places an obligation on Member States to introduce national programmes relating to a number of matters. On the grounds of subsidiarity it was much weakened from the original proposal made by the Commission as described below. This is in addition to the laying down of EC standards for traded products such as central heating boilers and the requirement for energy labelling of products.

The monitoring mechanism

The last of the four elements is much more important than its title of 'monitoring mechanism' might imply. Not only is the Commission to evaluate data on greenhouse gases reported by the Member States but, as described more fully below, the Member States are to 'devise, publish and implement national programmes for limiting anthropogenic emissions of CO_2' in order to contribute to commitments in the UN FCCC and the EC's own stabilization target. Because the decision is more precise about the nature of the national programmes than is the Convention, and since the EC target is also sharper than in the Convention, the decision must now be regarded as the cornerstone of EC climate policy.

If the EC carbon energy tax, which was seen by many in the Commission as the cornerstone, has been abandoned at least for the time being, then the monitoring mechanism forces responsibility back on the Member States, not only to set their own targets but to devise national measures to ensure that national targets are achieved. An important role for the Commission is to ensure that the national programmes add up to the EC stabilization target, and to sound the alarm if they do not. One of the regrettable features of the

166

saga of the proposed carbon energy tax is that it delayed the monitoring mechanism by about a year. As it is, the national programmes so far sub-mitted to the Commission are inadequate and do not yet enable firm con-clusions to be drawn as to whether the stabilization target will be met.

Ratification

A curious side-show in the process leading to ratification of the Convention was an attempt by six countries, led by the Netherlands and Germany, to link ratification to adoption of the EC carbon energy tax. The UK called this bluff by refusing to change its view on an EC tax, and by implying that it would not be particularly concerned if the EC did not ratify the Convention and that it would ratify on its own. This it did, quickly followed by Germany and then the others. The idea of the EC not ratifying caused concern to those countries like Spain that were intending to offset their planned increase in CO_2 emissions against reductions by other countries (e.g. Germany, the Netherlands). In the event a face saving formula was found which enabled the EC to decide to ratify. The formula left the decision on the tax to the Council of Finance Ministers (with whom the decision rested anyway) and emphasized the possibility of national taxes which might amount to something equivalent to the proposed EC tax. Had the Netherlands not withdrawn its opposition to EC ratification, then it is possible that some Member States, such as Portugal and Spain, would not have ratified on the grounds that on their own their national plans could not be said to live up to the requirements of the Convention. In the event all Member States, as well as the EC, ratified the Convention (though, tech-nically, Belgium has still to do so). The EC will accordingly have to submit a report to a future Conference of the Parties on how it will fulfil the obligations in the Convention.

Fourth phase: follow-through

We are now in a fourth phase where the attention is on how far the national programmes will achieve the stabilization target. The following sections review the various mechanisms by which national actions are coordinated at EC level.

The 'monitoring mechanism'

The monitoring mechanism was adopted by the Council as Decision 93/389 in June 1993 (Official Journal of the EC 1993b). Although the Decision does not expressly say so, the monitoring mechanism consists of the Commission consolidating and evaluating certain information submitted by the Member States. The Decision also requires Member States to 'devise, publish and implement national programmes for limiting their anthropogenic emissions of CO_2 ' in order to contribute to (a) the qualified target for stabilization of CO_2 emissions by 2000 at 1990 levels in the EC as a whole (the qualification being the assumption that other leading countries undertake commitments along similar lines) and (b) the fulfilment of the commitment relating to CO_2 in the Convention by the EC as a whole. The programmes are to be updated periodically with a frequency to be decided by the Commission. The contents of the programme are specified more precisely than the equivalent requirements in the Convention and from the first update each national programme is to include:

- the 1990 base year anthropogenic emissions of CO_2;
- inventories of national anthropogenic CO_2 emissions by sources and removal by sinks;
- details of national policies and measures limiting CO_2 emissions;
- trajectories for national CO_2 emissions between 1994 and 2000;
- measures being taken or envisaged for the implementation of relevant EC legislation and policies;
- a description of policies and measures in order to increase sequestration of CO_2 emissions;
- an assessment of the economic impact of the above measures.

One important difference from the Convention is the additional requirement to provide 'Trajectories of national CO_2 emissions between 1994 and 2000'.

The Decision requires Member States annually to report to the Commission on their CO_2 emissions and removal by sinks for the previous calendar year, which again is a more precise obligation than in the Convention. Within three months of receiving the information, the Commission is to produce inventories for the whole Community and is to circulate them to all Member States.

The Decision also requires the Commission to evaluate the national programmes in order to assess whether progress in the Community is sufficient to ensure fulfilment of the commitments. Annually after the first evaluation, the Commission, in consultation with the Member States, is to assess progress, on the basis of the national programmes and inventories and is to report to the Council and the Parliament. With the help of a committee, the Commission is to establish procedures for the evaluation of the national programmes and the frequency with which they are to be updated.

National programmes

The first evaluation of the existing national programmes under the 'monitoring mechanism' was issued by the Commission in March 1994 (European Commission, 1994). This evaluation was based on the national programmes available to the Commission in early August 1993 not all of which were then complete. The Commission lists the following goals set out in the programmes:

Belgium	5%	Reduction from 1990 by 2000
Denmark	5%	Reduction from 1990 by 2000
Germany	25–30%	Reduction from 1987 by 2005
Greece	25%	Increase from 1990 by 2000
France	13%	Increase from 1990 by 2000
Ireland	20%	Increase from 1990 by 2000
Italy		Stabilization on 1990 level by 2000
Luxembourg		Stabilization on 1990 level by 2000
The Netherlands	3–5%	Reduction from 1989/1990 by 2000
Portugal	30–40%	Increase from 1990 by 2000
Spain	25%	Increase from 1990 by 2000
UK		Return to 1990 level by 2000

From this list it can be seen that all countries except Germany have given CO_2 emission figures for the year 2000.

A comment is necessary about the difference between the word 'stabilization' and the words 'return to'. 'Stabilization' implies that emissions will not rise above the given figure after the year 2000 whereas the words 'return to' carry no such implication.

This list can be rearranged in Table 6.1 under three headings relating to the target for 2000.

The assumption is made in the table that Germany will reduce its emissions by 2000 from 1990 but the exact amount has not been projected by the German government, which, however, remains committed to its national target of reducing by 25 per cent by 2005 from a baseline of 1987. Germany's CO_2 emissions decreased by 14 per cent between 1987 and 1993 but this reduction was due to the economic restructuring of the former East

Table 6.1 The CO_2 reduction programmes of EC Member States, 1990–2000

Reduction		Stabilization or 'return to'	Increase	
Belgium	−5%	Italy	Greece	+25%
Denmark	−5%	Luxembourg	France	+13%
The Netherlands	−3% or −5%	UK	Ireland	+20%
Germany	−?		Portugal	+30% or 40%
			Spain	+25%

Germany as will be noted in the following chapter. Some of this reduction was already achieved before 1990 so that Germany's possible total reductions between 1990 and 2000 remain problematic.

From a knowledge of national emissions in 1990 it is possible to calculate what the emissions might be in the year 2000 (Table 6.2) assuming that national goals are fulfilled.

Since Germany is by far the largest emitter of CO_2, responsible for nearly a third of all EC emissions in 1990, the assumption made about German emissions in 2000 will dramatically affect the total EC result. On the conservative assumption made in the table that Germany stabilizes its emissions by 2000 at 1990 levels it can be calculated that the EC as a whole will show a 4 per cent increase in 2000 compared with 1990. If the assumption is made that Germany reduces its emissions in 2000 by 12.5 per cent compared with 1990 (i.e. half the 25 per cent reduction proposed by 2005 compared with 1987) its CO_2 emissions in 2000 become 879.4 metric tonnes, with the result that total EC emissions in 2000 are approximately equal to the 1990 total. The EC stabilization goal would then be achieved, assuming that the other national goals do not change.

The accession to the EC of Austria, Finland and Sweden does not significantly alter this conclusion since these countries add only 5 per cent to the total EC emissions of CO_2 in 1990. Finland and Sweden have a target of stabilization by the year 2000 and Austria has a target of a 20 per cent reduction by 2005 from the 1988 level. All three countries will have difficulties in achieving their targets because they have low per capita emissions (in the case of Sweden because hydro- and nuclear power provide almost all electricity). If anything they will add to the EC's difficulties rather than ease them.

Table 6.2 National CO_2 emissions in million metric tonnes

	1990	2000
Belgium	112.0	106.4
Denmark	53.01	50.36
Germany	1005.0	1005.0 (assuming no change)
Greece	73.7	92.13
France	365.7	413.24
Ireland	30.8	36.96
Italy	402.4	402.4
Luxembourg	12.5	12.5
The Netherlands	157.3	152.58 (assuming −3%)
Portugal	39.9	55.86 (assuming +40%)
Spain	210.7	263.38
UK	579.2	579.2
Total EC (12 Member States)	**3042.21**	**3170.01 = 4% increase on 1990**

The Commission will come forward with further evaluations based on revised national programmes and inventories of emissions and as time goes by a more accurate assessment of what might happen in 2000 will emerge. But from the evidence so far, several countries, including the four so-called 'cohesion countries', are assuming that they will increase their emissions considerably. France is projecting an increase of 13 per cent because it is assuming increased consumption of fuel in the transport sector, with little scope for compensating reductions in emissions from electricity production since it is largely derived from nuclear power stations. The cohesion countries all project increases of 20 per cent or more. This explains why they, and particularly the largest of them, Spain, were insistent that the EC should ratify the Convention since if their increases are subsumed in the EC total, and the EC as a whole achieves stabilization, then their own increases can be presented as part of 'joint implementation' by the EC as a whole. On their own they felt that they would not be able to fulfil the requirements in the Convention to endeavour to return emissions by 2000 to 1990 levels. The position of these countries is further discussed below.

The cohesion countries

Spain, Portugal, Greece and Ireland all have a national GDP which is below 90 per cent of the EC average and are therefore eligible for payments from the cohesion fund which was agreed in the negotiations leading to the Maastricht Treaty.

The cohesion fund, set out in Regulation 1164/94 (Official Journal of the EC 1994b), is intended to strengthen economic and social cohesion within the EC through the provision of finance for projects in the poorer Member States. A total of over 15 000 million ECU is provided until the end of 1999 for projects relating to environmental protection and transport infrastructure, each of which must cost at least 10 million ECU.

The group of four countries can be regarded as a distinctive category for climate change purposes for the following reasons:

- their per capita CO_2 emissions are relatively low;
- their combined contribution to the EC total of CO_2 emissions is fairly small (about 12 per cent);
- there is a presumption that their relatively low GDP will increase relatively rapidly to achieve the goal of economic convergence;
- as a result of the above, and based on the principle of burden sharing, there is an assumption that they will not be constrained to CO_2 targets as demanding as those of other Member States.

The programmes so far prepared conform to the last expectation in that they all forecast substantial increases in emissions by the year 2000. Indeed, these forecasts in several cases exceed the trajectory of emissions in recent years, as Figure 6.1 illustrates.

Many of the data on which the four programmes are based arise from national energy plans, whose main purpose is clearly the development of energy resources to meet economic growth. Both the Spanish and Portuguese reports emphasize that environmental protection was an important consideration in the development of the plans, but energy emissions seem in most cases to have been calculated from existing energy plans, rather than as a factor in developing energy plans.

The four cohesion countries present in microcosm the dilemma of selecting appropriate targets for the developing countries under the Convention. Relative to the richer EC Member States the cohesion countries have low per capita CO_2 emissions, but in a world context the cohesion countries are developed countries with relatively high per capita CO_2 emissions. If the EC as a whole accepts that joint implementation can entail the cohesion countries having unrestrained or only moderately restrained increases in CO_2 emissions, it is difficult not to follow the same argument for developing

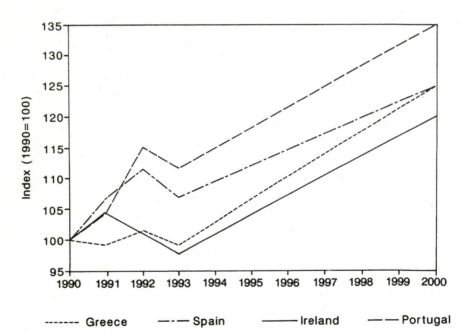

Figure 6.1 Actual and planned CO_2 emissions from the cohesion states 1990–2000

Note: Actual emissions are derived from fossil fuel use for years 1990–1993 (Eurostat). Planned emissions for years 1994–2000 are interpolated from each state's own year 2000 estimates. No account is taken of non-fuel emissions or energy fixation

countries. A difference of course is that the cohesion countries represent a small proportion of the total EC population whereas the developing countries are a large proportion of the world population.

Burden sharing

Although the phrase 'burden sharing' was not used in the conclusions of the Council in October 1990 which first set out the EC's stabilization target, the concept was there, and the phrase is part of the language in which EC climate policy is collectively discussed. The national targets for curbing CO_2 emissions constitute a form of burden sharing that has been called the 'bottom-up' approach since the choice of national target has been made by the country itself. If the national plans of the Member States happen to result in the EC's stabilization target being achieved then the 'bottom-up' approach creates no problems. But this will not be so if it becomes clear that EC stabilization will not be achieved. In 1991 the Commission had indeed begun work on a draft directive that would have attempted to allocate CO_2 emission reductions among the Member States on the model of Directive 99/609 concerned with emissions of sulphur dioxide from large combustion plants. This approach met opposition and was abandoned. The experience of the Large Combustion Plants Directive suggests the difficulties entailed in trying to set legally binding targets for a number of countries each having quite different circumstances. Even if the 'top-down' approach has been abandoned for the time being it remains implicit in the setting of an EC target (i.e. stabilization) since, if it looks as though the target will not be achieved by the 'bottom-up' approach, some process of negotiation between the Member States and the Commission will have to take place in which pressure will be applied on those Member States that are perceived by the others as not having adequately shouldered their share of the EC burden. A version of the 'top-down' approach may well therefore re-emerge but possibly in an informal way. The negotiations may need to be based on some attempt at an objective evaluation of what fair burden sharing should entail.

The ALTENER programme

The ALTENER programme (Official Journal of the EC 1993c) of EC funding for renewable energy sets three objectives which Member States are to take into account in their own programmes:

- increasing the renewable share of energy supply from nearly 4 per cent to 8 per cent between 1991 and 2005;
- trebling the output from renewable sources other than large scale hydroelectric plant;

- securing for biofuels a 5 per cent share of the fuel consumption of motor vehicles.

An indicative budget of 40 million ECU is suggested for the first five-year programme, subject to budgetary constraints. Funding is to be mainly in the form of joint funding with private or public sector sources. The main forms of activity to be supported are:

- pilot actions on solar collectors and water heaters;
- pilot schemes on biofuels for vehicle fleets;
- pilot studies on least-cost integrated planning and demand management;
- improving financial arrangements for renewables projects;
- local plans and pre-feasibility studies;
- pilot projects for photovoltaics on buildings;
- windfarm planning;
- bioclimatic systems in building design.

The SAVE programme

Even before the EC's commitment on CO_2 emissions, the Council had encouraged the Commission to pursue energy efficiency programmes in the interests of reducing energy imports. In a Resolution of 1986 the Council adopted for the Community a target of improving the efficiency of final energy demand by at least 20 per cent by 1995. By 1988 it was clear that this target would not be met without more vigorous action being undertaken, as the improvement in energy intensity of final demand was clearly weakening. A specific programme (SAVE – Specific Actions for Vigorous Energy Efficiency) was therefore developed as a key part of the EC's climate change policy. Various elements of EC policy already in place or under development were subsumed within the SAVE programme, and it did not therefore develop in a particularly logical or orderly fashion.

The SAVE programme is now set out in Council Decision 91/565 (Official Journal of the EC 1991) and Council Directive 93/76 (Official Journal of the EC 1993d). The Decision provides an indicative budget of 35 million ECU for the five-year period of 1991 to 1995. Funding is mainly in the form of joint funding with private or public sector sources. The main forms of activity to be supported are:

- technical assessments of data needs for setting technical standards and specifications;
- support for Member States' initiatives on energy efficiency in respect of training and information schemes and pilot schemes in all sectors including transport;
- measures to help foster better coordination of national and supranational initiatives on energy;

- measures to implement the programme for improving efficiency of electricity use.

Each year the Member States submit to the Commission a list of the proposed measures under the programme, and the bodies which are to undertake them. The Commission then submits a draft list which takes account of the views of its advisory committee to the Council for approval.

The Directive requires the Member States to establish national implementation programmes for the following six fields through the use of specific laws and regulations, or economic and administrative instruments:

- energy certification of buildings;
- the billing of heating, air conditioning and hot water costs on the basis of actual consumption;
- third party financing for energy efficiency investments in the public sector;
- thermal insulation of new buildings;
- regular inspection of boilers;
- energy audits for undertakings with high energy consumption.

The scope of the programmes is to be determined by the Member States themselves on the basis of potential energy efficiency improvements, cost-effectiveness, technical feasibility and environmental impact. No quantified targets are set.

The final form of both the decision and the directive was weaker than originally proposed. For example, the original Commission proposal for the directive included a measure to ensure regular inspection of cars, but this was dropped from the final version. Energy audit requirements were also limited to energy-intensive enterprises only. Furthermore, all the detailed requirements of the original proposal were removed, leaving responsibility for the content of programmes entirely in the hands of the Member States, to be determined by the criteria noted above. In essence, these changes were determined largely by the operation of the subsidiarity principle which is further discussed below.

The European Parliament and others were highly critical of this weakening of the proposals, having sought a more detailed and binding measure which would act as an effective framework for the SAVE programme. Overall, then, both the scope and content of SAVE were pared down very significantly in the course of the legislative process. The Commission initially estimated that the programme would contribute 25 per cent of the CO_2 savings required of the EC but the effect of the programme as it now stands is likely to be far less than this. Critics question whether this now constitutes the 'vigorous' programme which its name implied.

One area which has remained for decision at EC level is the setting of energy efficiency standards for traded products and the labelling of products with information on energy consumption. The EC clearly has a role here since differing national standards would create barriers to trade. Standards have been set in directives for hot water boilers, and labelling requirements have been agreed for ovens and refrigerators. The same does not apply to standards for the insulation of buildings, which of course are not traded.

Michael Grubb (1995, 172–173) has looked at all the EU measures to improve energy efficiency and reduce CO_2 emissions, and has found them sadly wanting (see Table 6.3). Originally designed to cut CO_2 by 11 per cent by 2000, the best, he claims, that can be hoped for is 3–4 per cent.

Table 6.3 Evaluation of EC CO_2 saving measures

Measure	% CO_2 in year 2000	
	Initially projected	*Likely*

SAVE: series of binding directives on energy efficiency
    Outcome: directives delayed and weakened;
    many initiatives left to Member States
	3%	1%

CARBON ENERGY TAX: $3/bbl starting 1993
    to $10/bbl
    Outcome: original concept abandoned; new
    proposal for selective broadening of excise taxes
	3–5.5%	<1%?

ALTENER: integrated programme for promoting
    renewables
    Outcome: targets agreed; minimal funding
	1%	<1%?

THERMIE: technology demonstration and diffusion
    programme
    Outcome: incorporated in 4th Framework
    Programme
	1.5%	1.5%

NATIONAL MEASURES: Outcome: monitoring decision	[National programmes]	

Source: Grubb (1995, 173)

Competence

A particular reason why some Member States can be expected to resist a 'top-down' approach to burden sharing, although this may not be stated publicly, is the loss of sovereignty entailed. Even if it is possible to agree a fair allocation of reductions among the Member States, the setting of these percentages figures in a directive has considerable implications for the future, because competence is transferred to the EC from the Member States for the setting of further targets. Such a transfer of competence has been accepted for chlorofluorocarbons (CFCs) in the context of protecting the ozone layer and for sulphur dioxide in the context of combating acid rain. But both those subjects have significantly different implications compared with curbs on CO_2. CFCs are traded products and for that reason alone the EC is bound to be involved in any controls over their use. Any loss of sovereignty is therefore confined to a limited range of substances, but because production of many of these substances is being phased out, any loss of sovereignty is now only theoretical. Sulphur dioxide is not a product in trade but is the unwanted by-product of burning fossil fuels containing sulphur. Its reduction can be achieved by several means which entail costs but do not deny the possibility of continuing to use certain fuels.

Carbon dioxide, by contrast, is the product of innumerable activities and curbs on its emission affect great areas of national life including industrial production, electricity consumption, domestic heating and public and personal transport. A ceiling set on national emissions of CO_2 therefore has far wider repercussions. Even if a Member State is comfortable with a particular target which it may have chosen itself, or which may have been agreed in a process of negotiation with other Member States, it may well be cautious about relinquishing the power to set subsequent, and more stringent, targets. And as we have seen from the previous chapter, once the EC acquires competence for making rules internally, it simultaneously acquires the competence to agree rules on that subject in any international conventions.

Qualified majority voting (QMV)

The question of loss of national sovereignty is a concern of all Member States though some express the concern more openly than others. It is a concern even when decisions are made unanimously in the Council. But the concern becomes much more acute once there is the possibility of a Member State being outvoted in the Council and having to accept some obligation, such as an emission target, against its wishes. Before the Single European Act all EC environmental legislation had to be adopted unanimously. The Single European Act introduced qualified majority voting (QMV) for legislation affecting the internal market, and the Treaty of Maastricht, by amending Article 130s of the Treaty of Rome, made QMV the normal method of adopting all environmental legislation albeit with some exceptions. These exceptions are fundamental to climate policy because they include 'provisions primarily of a fiscal nature', which therefore covers the proposed carbon energy tax, and also 'measures significantly affecting a Member State's choice between different energy sources and the general structure of its energy supply', which must include many measures to curb CO_2 emissions. However, measures dealing with the energy efficiency of traded products such as gas boilers can certainly be adopted by QMV under the Article in the Treaty concerned with the single market (Article 100a) and therefore already some climate-change-related measures are subject to QMV. The introduction into the Maastricht Treaty of the exceptions relating to 'measures significantly affecting a Member State's choice between different energy sources' shows how sensitive at least some Member States are to the prospect of being outvoted on such issues. Nevertheless, there is the possibility of the Treaty being amended yet again following the intergovernmental conference in 1996 which could result in all environmental measures being adopted by QMV although fiscal matters could well still require unanimity.

An indication of possible conflicts ahead is provided by the adoption of Council Decision 94/69 (Official Journal of the EC 1994a) concerning the conclusion of the Convention during which the UK abstained. The Council legal services had advised that the Decision should be adopted by QMV under Article 130s(i) while the UK argued that the Convention significantly affects a Member State's choice between different energy sources and that a Decision to ratify the Convention must therefore do the same. The UK's abstention appears to be in the nature of a marker for the future and the UK is unlikely to seek annulment of the decision by the Court.

Subsidiarity

Under the principle of subsidiarity as now formulated in the Treaty, the EC:

> shall take action only if and in so far as the objectives of the proposed action cannot be sufficiently achieved by the Member States and can therefore, by reason of the scale or effects of the proposed action, be better achieved by the Community.

This formulation of the principle of subsidiarity does not deal with whether action should be taken within the Member State by the central government or by local or regional governments since that is a matter for the Member States themselves to decide. Nor does it deal with the question of whether action is better taken by international machinery, such as that created under the Climate Convention, rather than by the EC. Any broad view of subsidiarity must deal with these questions. Such a view was contained in one of the principles set out in the First Action Programme on the Environment of 1973 (Official Journal of the EC 1973) which states that 'in each category of pollution, it is necessary to establish the level of action (local, regional, national, Community, international) best suited to the type of pollution and to the geographical zone to be protected'.

The following two types of question therefore arise in the context of climate change. If the machinery of the Convention, to which all Member States are Parties, is adequate to deal with the climate issue, what is the justification for the EC duplicating that machinery? Similarly, if national (or local) action on some matters such as the insulation of housing adequately deals with that matter, what is the need for the EC to set insulation standards? The issue of subsidiarity is therefore very relevant for climate change policy and has indeed influenced decisions. Opposition to the carbon energy tax by the UK is not so much an objection to the idea of such taxes but to them being set at EC level rather than at national level. Equally, the SAVE Directive was amended in the Council, as we have seen, by the removal of several detailed requirements such as EC-wide insulation standards for each climatic zone. Instead the SAVE Directive merely requires Member States to establish national programmes for a number of fields including the thermal insulation of buildings. The discussion about subsidiarity will therefore certainly continue in the field of climate change as it does in other areas, and the resolution of any particular conflict is likely to be pragmatic.

POLICY INTEGRATION

None of the four measures proposed by the Commission just before the Rio conference in 1992 is specifically directed at particular policy areas such as transport, industry or agriculture, although all are affected. The Treaty now requires that environmental protection requirements must be integrated into the definition and implementation of other Community policies. This means that all EC policies shall now contribute to climate change policies. The principal EC document that sets out the framework for doing this is the Fifth Action Programme on the Environment that was drafted by the Commission and whose general approach was approved by the Council in February 1993 (Official Journal of the EC 1993a).

The Fifth Programme is very different from the earlier programmes which largely consisted of a list of proposals for environmental legislation that the Commission would bring forward. The Fifth Programme took its inspiration from the Dutch National Environmental Policy Plans whose themes have been the integration of the environmental dimension into all sections of public policy. The choice of the title of *Towards Sustainability* underlines that the document is of a new kind:

- it sets out to provide a longer term, and more comprehensive, strategy for Community action involving most of the major actors;
- in addressing the overarching theme of sustainability it proposes some fundamental adjustments to current economic and social trends and greater integration of environmental objectives into other policy areas;

179

- it selects five sectors of the European economy for special attention;
- it is explicit in identifying different actors as having responsibilities for the execution of the Programme;
- it considers ways in which responsibilities can be divided amongst the different actors and offers some interpretation of the concept of subsidiarity;
- its proposals are presented in a more structured and coordinated fashion, and related to specific time frames and targets up to the year 2000;
- it is accompanied by the *State of the Environment* report which provides some data to underpin or at least illustrate the arguments advanced;
- it proposes some modest institutional changes, including a process whereby the implementation of the Programme is to be reviewed in 1995;
- it includes a brief chapter on costs, although this avoids any analysis of the costs or benefits of implementing the Programme.

The Programme identifies actors, target sectors and key environmental issues. The first of the environmental issues listed is climate change. The five target sectors chosen are manufacturing industry, energy, transport, agriculture and tourism.

The Programme suggests a broader range of policy instruments than reliance on legislation, such as economic and fiscal instruments and voluntary agreements, and also sets out two guiding principles, namely the precautionary approach (although it is not defined), and the concept of shared responsibility. Shared responsibility is interpreted as meaning that all the major economic and political players in society have a role in implementing the Programme including the general public, both as citizens and consumers.

The Programme is a wide ranging document which sets ambitious targets intended to steer the EC on the path towards sustainable development. One result is that the Commission has adopted internal procedures which include the appointment of senior officials with special environmental responsibilities within each Directorate-General.

The way the mechanisms will work in practice remains unclear, but the political conflicts involved are considerable, particularly in the transport sector. Here the EC is promoting a trans-European road network which is bound to result in increasing road traffic and hence to increasing CO_2 emissions. This conflict mirrors comparable conflicts within each Member State and its resolution will be long drawn out. What can be said at present is that climate change arguments reinforce at a strategic level other environmental arguments about road traffic which may have greater force at local level, such as air pollution or destruction of habitats. For that reason, they can be applied in strategic discussions such as those concerned with trans-European road networks. In the absence of an EC climate policy it would be much more difficult to bring environmental arguments to bear on EC level discussions about road building. The same will hold for other policy areas.

EC INVOLVEMENT IN THE CLIMATE CHANGE CONVENTION

The previous chapter discusses the legal aspects of EC involvement with the Convention and shows that its decision to become a party was a political choice and not the result of a legal obligation. We have also seen how the Council's political decision to stabilize CO_2 emissions in the EC as a whole by the year 2000 at 1990 levels gave an impulse that led two years later to the signing of the Convention with stronger provisions than might otherwise have been expected. Had no Convention been agreed the EC would have then had to consider whether on its own it should have pursued a policy for climate change. A powerful argument would have been advanced that it should do so as an encouragement for the rest of the world. Once the Convention was signed the argument was different. One view is that action by the EC was no longer necessary, and that there was no real need for the EC to be a party to that Convention. Clearly the cohesion states did not share this view. They have had a very good reason for wanting the EC to be a party since otherwise it is doubtful that they fulfil the requirements of the Convention given that they are projecting significant increases in CO_2 emissions by the year 2000. The 'cohesion countries' are relying on the EC as a whole to stabilize its emissions so that they do not have to. The uncertainties concerning this form of 'joint implementation' are discussed in the previous chapter.

The EC (as opposed to its Member States) in fact played only a limited role in the negotiations leading to the Climate Change Convention. One key issue was whether the EC should insist on the Convention including a firm commitment of 'stabilizing' CO_2 emissions by the year 2000 at 1990 levels. The United States was firmly opposed to such a commitment and the EC was thus faced with two possibilities: to press for a strong Convention in the knowledge that the United States would probably not then sign, or to agree to a form of words which would enable the United States to sign while still taking account of the EC's own political commitment. There were some EC Member States which would have accepted a stronger Convention without US participation, while others argued that without US participation the Convention would hardly be worth while, and other countries would be given an excuse not to sign.

During these discussions the UK Secretary of State for the Environment, Michael Howard, allegedly with the encouragement of some other Environment Ministers from EC Member States, travelled to the United States and agreed a form of words with US officials which forms the basis of Article 4(2) of the Convention. Whether this can be regarded as an EC contribution to the framing of the Convention is a matter of opinion. Formally it was not since no formal Council decisions were taken on the subject, but without the machinery provided by the EC for discussion between ministers it may not have happened.

Before the first Conference of Parties in March 1995, the Council of Environmental Ministers urged the CoP to establish a protocol for setting targets and timetables to reduce greenhouse gases beyond 2000. The proposed dates were 2005 and 2010, though no figures were provided. The EC may therefore claim some influence on the modest outcome of the CoP.

The ability of the EC to contribute to the evolution of the Convention at future Conferences of the Parties will depend on its ability to meet its own target. If it cannot do that, the confidence with which it can take a leading role will be much diminished. Given Germany's leading role both in the Secretariat and in sharing the burdens of the EC's total CO_2 contribution, much will depend on that country's response. We examine that in the chapter that follows.

CONCLUSIONS

Given that climate change is a global issue and that global machinery to handle it has been created under the Convention to which all Member States are party, two questions have to be asked: what contribution can the EC make to combating climate change in addition to that made by Member States, and what extra value is added by the EC being a Party to the Convention?

A complete answer cannot yet be given to either question despite the fact that the EC provided an impetus in 1990 that led to the Climate Convention being stronger than it would otherwise have been. Indeed without the EC's initiative it may never have been agreed. The political decision – albeit qualified – taken by the joint Council of Environment and Energy Ministers in October 1990 to set the target of stabilizing CO_2 emissions in the EC as a whole by 2000 at 1990 levels provided the basis for Article 4(2) of the Convention, which would otherwise have been weaker. The EC has also ensured that all Member States have become Parties to the Convention, which could not have been a foregone conclusion otherwise.

The EC was not able to sustain its early leading role in setting the stabilization target and experienced delays in coming forward with measures at EC level to fulfil the obligations it set itself in 1990 and those of the Convention. In the absence of a carbon energy tax that the Commission initially saw as the cornerstone of EC climate change policy the so-called 'monitoring mechanism' has now assumed that role. It not only requires the Commission to evaluate data on greenhouse gases reported by the Member States, but requires that Member States 'devise, publish and implement national programmes for limiting anthropogenic emissions of CO_2' in order to contribute to commitments in the Convention and to the EC's own target.

The 'monitoring mechanism' thus clearly places responsibility back on the Member States not only to set their own targets but to devise national measures to ensure that they are achieved. This merely repeats the obligation in the Convention but additionally the Commission has the role, though by no means its only one, of ensuring that the national programmes add up to stabilization in the EC as a whole and of sounding the alarm if they do not.

Many of the existing national programmes so far are inadequate and do not enable firm conclusions to be drawn as to whether the stabilization target will be met, but there is a real possibility that it will not be. The 'monitoring mechanism' further provides the basis for a continuing dialogue between the Member States and the Commission in which they can apply pressure on one another to improve their performance. The EC machinery ensures that this dialogue can be conducted much more frequently than under the Convention. Until the EC is confident that it is comfortably meeting its own self-imposed target of stabilization it will be difficult for it to play the leading role in the evolution of the Convention that it played when the Convention was being drafted.

The EC does, however, have these further roles:

- It can set standards for traded products and relating to labelling requirements. These standards need to be set for the single market to operate and the EC's climate policy provides extra pressure for them to be set at a high level.
- It can supplement national efforts on energy saving and relating to renewable energy particularly by providing finance. This role is so far rather limited.
- Most importantly it can ensure the integration of environmental protection requirements into the definition and implementation of other EC policies and indeed it has a duty to do so.

The very existence of an EC policy for climate change should in the long run influence other EC policies, even if the results are not yet very visible. The EC, for example, disburses large sums of money under the 'cohesion fund' to the four poorest Member States for transport infrastructure projects. EC climate policy provides a strategic argument that should influence the spending of this money on forms of transport that emit less rather than more CO_2 . The same pressure should also apply to other policies if the Fifth Action Programme on the Environment is to fulfil its promise.

In institutional terms, the EC experience reinforces the view that policy integration across Directorates-General that heretofore tended to promote their core mandates will be difficult to achieve, despite the duty in the Maastricht Treaty. The presence of 'environmental officials' in each of the relevant Directorates-General has yet to show much influence. Policy coordinating machinery remains rudimentary. The specific energy programmes of

SAVE and ALTENER are modestly funded, with limited effect on national energy policies. Ecolabelling and appliance standards are still at an early stage. The idea of a carbon energy tax has collided with the desire of Member States to retain control of fiscal policy. The climate warming debate is still very young in the Community. The elements of reform are there – in tax policy, policy integration, burden sharing and the purposeful application of the precautionary principle. Until Member States take global warming seriously and substantively seek to mend their own institutional ways, the Commission can promise and prepare, but no effective institutional transformation can be expected.

ACKNOWLEDGEMENTS

I am grateful to David Baldock, Malcolm Fergusson, Konrad von Moltke and Jon Birger Skjaerseth for contributions to this chapter.

REFERENCES

Bulletin of the EC (1990) Declaration by the European Council on the environmental imperative, Vol. 23, No. 6, 16–18.

European Commission (1988) *Communication from the Commission to the Council: The Greenhouse Effect and the Community*, COM (88) 656, Brussels.

European Commission (1994) *First evaluation of existing national programmes under the monitoring mechanism of Community CO_2 and other greenhouse gas emissions*, COM (94) 67, Brussels.

Grubb, M. (1995) Climate change policies in Europe: national plans, EU policies and the international context, *International Journal of Environment and Pollution* 5(2/3), 164–179.

Haigh, N. (1992a) *Manual of Environmental Policy: the EC and Britain*, Harlow: Longman.

Haigh, N. (1992b) The European Community and international environmental policy, pp. 228–249 in A. Hurrell and B. Kingsbury (eds) *The International Politics of the Environment*, Oxford: Oxford University Press.

Liberatore, A. (1995) Arguments, assumptions and the choice of policy instruments: the case of the debate on the CO_2/energy tax in the European Community, pp. 55–72 in B. Dente (ed.) *Environmental Policy in Search of New Instruments*, Dordrecht: Kluwer.

Official Journal of the EC 1973 C112 20.12.73.
Official Journal of the EC 1986 C255 13.10.86.
Official Journal of the EC 1987 C328 7.12.87.
Official Journal of the EC 1988 L336 7.12.88.
Official Journal of the EC 1991 L307 8.11.91.
Official Journal of the EC 1993a C138 17.5.93.
Official Journal of the EC 1993b L167 9.7.93.
Official Journal of the EC 1993c L235 18.9.93.
Official Journal of the EC 1993d L237 22.9.93.

Official Journal of the EC 1994a L33 7.2.94.

Official Journal of the EC 1994b L139 25.5.94.

O'Riordan, T. and Cameron, J. (eds) (1994) *Interpreting the Precautionary Principle*, London: Earthscan Publications.

Wilkinson, D. (1992) Maastricht and the environment: the implications for the EC's environmental policy of the Treaty on European Union, *Journal of Environmental Law* 4(2), 221–239.

Worcester, R. (1994) European attitudes to the environment, *European Environment* 4 (6), 3–8.

7

CLIMATE CHANGE POLITICS IN GERMANY

How long will any double dividend last?

Christiane Beuermann and Jill Jäger

INTRODUCTION

The issue of climate change is characterized by three main dimensions which must be borne in mind when analysing the development of climate change politics.

First, the global dimension. From the point of view of atmospheric concentrations, it is not important where measures to reduce greenhouse gas emissions are implemented. Every nation contributes to the problem and because of the generally small shares of responsibility (even in most industrialized countries), every country's action is needed to solve it. For the same reason, the action of one country is not restricted to its territory. But, on the other hand, the future impact of a specific measure cannot be related to this measure. For example, it is not possible to predict or quantify the mitigated increase in temperature or sea-level rise as a result of the efforts of single countries.

Second, the time dimension. The impact of today's action lies in the future. The long term cause and effect relations of both natural processes and the political process compared with the short term dependence or orientation of political decision making processes are an obstacle to effective policy making. Efforts undertaken in the present seem to have no results within the time period upon which decision makers base their calculations on how to keep or gain power and influence. For example, the legislative cycle (which in Germany is four years) forces governments to present positive results of their policy at the end of the legislative period to gain an element of political credibility.

Third, the uncertainty dimension. In addition to the uncertainty aspects of the first and second argument, natural scientists are not able accurately to predict regional climatic changes or say exactly when the signal of global warming will be detectable. The 'extreme climate events' and slight average global temperature increase observed in the last hundred years are still within the range of natural variability.

The analysis that follows introduces the political style and the development of institutions in Germany, along with a short overview of the German greenhouse gas emission inventory. The main part of the analysis focuses on how climate change diffused from the scientific onto the political agenda. As soon as it became a high priority political issue, a reduction target for CO_2 emissions was adopted in a short period of time. Given the institutionalized structures of problem solving and decision making processes at the federal level, a closer look is taken of interministerial responses and on the relationship between the international climate policy of Germany and its national policy. Subsequently, the adequacy of the measures in the CO_2 emission reduction programme is evaluated in terms of its effects on other policy sectors taking the German transport policy as an example. The dependence of climate change politics on economic circumstances and on different interest groups is given attention. The localization of implementation is reviewed in terms of municipal level responses.

NATIONAL POLITICAL STYLE AND THE DYNAMICS OF INSTITUTIONAL CHANGE IN GERMANY

Reunification of Germany in October 1990 led to changes in policy style in Germany. These changes resulted internally from the attempts to reintegrate the *Länder* of the former German Democratic Republic (GDR) and externally from the need to adjust the role of a larger Germany within the European context.

Important aspects of German policy style remained unchanged. As Dyson (1982) pointed out, German policy style reflects historical experience and the cultural attitudes that have emerged from that experience. The collapse of the Weimar Republic, the experience of the Third Reich and of the devastation of war led West German political leaders to spell out the normative framework of policy to a greater degree than in other West European societies. Dyson discusses two important aspects of the West German policy style: the *intellectual style*, which is very objective, and the *negotiation style*, which emphasizes the importance of consensus in policy. Dyson detects a shift of policy style during the 1960s from regulation and status preservation to concertation and status preservation.

Katzenstein (1987) argues that the political culture of West Germany gives rise to a consensual policy making style which results in incremental change. According to Katzenstein certain aspects of Germany's political culture (i.e. coalition governments, cooperative federalism, and parapublic institutions) 'create a dense network of multiple dependencies which inhibits all actors, including the federal government, from taking bold steps in new directions'. Cavender Bares (1993) concludes, however, on the basis of an analysis of policy making in response to acid deposition, stratospheric ozone depletion and the increased greenhouse effect, that policy development in response to environmental risks has experienced discrete jumps in policy. Cavender Bares suggests that the nature of policy making in Germany in response to global environmental risks differs from other policy realms through its strong reliance on scientific information, international influences and the presence of an institutionalized green movement in Germany.

In their comparison of acid rain politics in Germany and the UK, Boehmer-Christiansen and Skea (1991) conclude that the political and administrative structures in Germany served to encourage the early emergence of the environment as a major political issue. They point out the important role of the party political system in Germany, where the political condition of coalition and federal politics make the German system more sensitive to changing public moods. Furthermore, they suggest that the degree of centralization of political power significantly affects the nature and quality of decision making. The less centralized federal structure in Germany tends to lead to complex decision making processes involving a great deal of bargaining in the search for a wider consensus. Boehmer-Christiansen and Skea point out that most of the implementation of policy in Germany is done at the *Länder* level, which means that the federal level can focus on strategic issues. This is partly a function of the German constitution which affords *Länder* representation in the upper house, or Bundesrat. If a majority of the *Länder* are governed by opposition parties, then the Bundesrat may take their side. This can result in fairly complex bargaining.

Policy linkages and the views of the main political parties

In principle, all German federal ministries and the Office of the Chancellor have had some interest in environmental policy since the development of the first environmental policy programme at the beginning of the 1970s. Initially the main responsibility for the development of environmental policy was with the Federal Ministry of the Interior (BMI). In 1986 there was a major change in the assignment of responsibilities for environmental issues and the Federal Ministry for the Environment, Nature Conservation and Nuclear Safety (BMU) was created in the aftermath of the Chernobyl nuclear accident (see Weale *et al.*, 1992).

The BMU is still one of the small ministries that will remain in Bonn after the majority of the administration and the government has moved to Berlin. Other ministries have well-defined interests in environmental issues, e.g. the Federal Ministry for Agriculture, Nutrition and Forestry (BML) and the Federal Ministry for Research and Technology (BMFT), which supports the development of environmental technology. In 1994, BMFT absorbed the Federal Ministry for Education and Science.

Of major importance in the response to the increasing greenhouse effect is the Federal Ministry of Economic Affairs (BMWi), whose positions reflect the interests of industry, including the coal and steel industries and the utilities. This ministry is responsible for the development of energy policy. The Ministry of Finance (BMF) also has an interest in the economic impacts of environmental policies. It is the only ministry to exercise the power of veto in decisions of other ministries. The Foreign Ministry (AA) is involved in all environmental issues that have an international and transboundary nature. Most of the other ministries also play important roles (e.g. Ministry of Transport – BMV), or have at least some responsibilities in climate-change-related areas.

Environmental policy development in Germany is also influenced by other groups. The environment ministries of the *Länder* play an important role and there are regular meetings of environment ministers. In addition, the Federal Environment Agency (UBA) drafts legislation, observes and follows up scientific, technological and political developments, and designs public and other information campaigns. The Federal Environment Ministry also receives advice from an Advisory Council on Environmental Questions. Enquete Commissions (discussed in more detail later) make recommendations to the government, and a number of permanent Bundestag committees also participate in the discussion of environmental policy. Further, a Scientific Council on Global Change was established in 1992. This council reports directly to the Federal Chancellor and the Cabinet.

There are five main political parties represented at federal level in the German Parliament. The Christian Democratic Union (CDU) and its sister party in Bavaria, the Christian Social Union (CSU), are 'conservative' and have formed a coalition government with the smaller, liberal Free Democratic Party (FDP) since 1982. The Social Democratic Party (SPD) has been in opposition since 1982. The Green Party has now joined with the reformers' party of former East Germany (Bündnis 90). In 1994, this coalition gained 7 per cent of the vote to guarantee representation in the Bundestag. In a number of *Länder* governments there have been SPD/Green coalitions in recent years.

All of the traditional parties began to embrace environmental policies after the political upheavals at the end of the 1970s, which led to the establishment of the Green Party in West Germany and its election to the Federal Parliament in 1982. Both the reasoning behind the support for environmental policy and the intensity with which the party programmes reflect environmental issues differ, of course, from one party to another. However, given the structure of coalition government and a consensus seeking policy style, environmental 'thinking' has pervaded many areas of policy making.

Greenhouse gas emission inventory

According to the report of the Federal Environment Ministry (BMU, 1993) the 1990 anthropogenic emissions of the most important gases involved in the increase of the greenhouse effect were as shown in Table 7.1.

Table 7.1 Anthropogenic emissions of greenhouse gases in Germany

Gas	Emissions
CO_2	1031 Mt/a
CH_4	6000 kt/a
N_2O	220 kt/a
CFC	37 kt/a

Source: BMU (1993, 10)

Following the Federal Environment Ministry (ibid), of the total emissions of CO_2 in 1990, the producers were:

Traffic and transport (18.8 per cent)
Commercial, institutional and private combustion (19.9 per cent)
Industrial combustion installations (23.5 per cent)
Public power plants and district heating installations (35.6 per cent)
Non-energy activities (2.2 per cent)

Of the total emissions of CH_4 in 1990, the producers were:

Waste management (36.7 per cent)
Agriculture (34.2 per cent)
Extraction and distribution of fossil fuels (25.0 per cent)
Others (4.2 per cent)

Of the total emissions of N_2O in 1990, the producers were:

Production processes (45.5 per cent)
Agriculture (34.1 per cent)
Public power plants and district heating installations (5.9 per cent)
Traffic and transport on roads (4.1 per cent)
Others (10.5 per cent)

Given the target of reducing CO_2 emissions by 25 per cent of their 1987 level by the year 2005 (the development of this goal is discussed in detail later in this chapter), it is of interest to look at the developments in CO_2 emissions since 1987. The development of emissions between 1987 and 1993 is shown in Table 7.2.

Table 7.2 Development of CO_2 emissions in Germany, 1987–1993

CO_2 emissions	Year			
	1987	1990	1991	1993[1]
in Mt				
Former West Germany	723	709	745	723
Area of former GDR	345	303	220	179
Germany	1068	1012	965	911
per capita (t CO_2)				
Former West Germany	11.4	10.8	11.2	10.9
Area of former GDR	20.5	18.5	13.7	11.2
Germany	13.4	12.3	11.7	10.9
per unit GNP (g CO_2/DM)				
Former West Germany	322	288	280	275
Area of former GDR	1026	1205	808	
Germany		376	339	315

[1] The figures for 1993 are provisional.
Source: BMU (1994)

The CO_2 emissions in East Germany were halved after 1987 as a result of the drastic economic changes after 1989 and the related strong reduction of primary energy consumption. In addition, the rapid change of the fuel structure in East Germany reduced CO_2 emissions (in 1989 300 Mt of lignite were mined in East Germany, in 1993 only 116 Mt). The share of lignite in primary energy consumption sank from about 70 per cent in 1989 to less than 50 per cent in 1993. In West Germany the emissions in 1992 were 15 Mt (2 per cent) higher than in 1987. A slight decrease of about 1 per cent between 1992 and 1993 is believed to be due to the economic recession.

The per capita CO_2 emissions in Germany are about three times the global average and about twice the average for the EU countries as a whole.

The Federal Environment Ministry estimates that about 1.5–2 billion tonnes of carbon are stored in forests in Germany. Further, they conclude that in the existing 10.8 billion hectares of forest carbon levels are growing by about 5.4 million tonnes of carbon per year. However, the capacity of forests to act as a reservoir is limited and it is estimated that forest management can reduce national CO_2 emissions by between 5 and 10 million tonnes of carbon (2–4 per cent of annual national emissions of CO_2).

DEFINITION OF POLITICAL PRIORITIES

Information dissemination

As indicated in the first chapter, it is important that scientific knowledge awakens public interest and becomes involved in the political decision making process. Therefore the media play an important transmitter role.

Before the time period we characterize as the first phase of German climate change politics (1987–1990), there was of course some research on the issue in Germany. Single entrepreneurial scientists, such as Hermann Flohn, focused attention very early on the problem of climate change. Nevertheless, there was not enough interest in the issue to shift it onto the political agenda until the mid-1980s.

Reviewing articles in the two German journals *Bild der Wissenschaft* and *Der Spiegel* shows that articles about global environmental issues have a long tradition in these journals. The count of articles about the issues 'ozone depletion', 'acid rain (*Waldsterben*)' and 'climate change' shows (Figure 7.1) that between 1970 and 1978 all three issues were covered on a low level.

The number of articles on environmental issues generally increased from a low level at the beginning of the 1970s to a high level in the late 1980s and beginning of the 1990s. At the same time, the structure of these reports changed from mediating the public opinion to policy to event-related reporting (Voss, 1990, 190) and to the concentration on single issues.

Comparing the quantities of reports with rankings of environmental problems by people (Commission of the European Community, 1992) shows differences: while in 1991 media attention focuses on climate change, the most important issue by far to the public is ozone depletion/the ozone hole. The problems of climate change and forest die-back were felt to have almost the same priority. In contrast, the differences between these two issues in levels of media reporting are enormous.

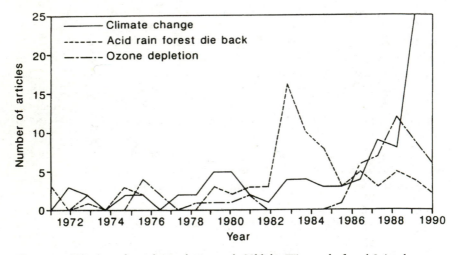

Figure 7.1 Number of articles in the journals *Bild der Wissenschaft* and *Spiegel* concerning the issues of acid rain, ozone depletion and climate change between 1971 and 1990

Source: Jill Jäger, unpublished

Achieving a reduction target

In the past, emissions of some greenhouse gases were influenced indirectly and unconsciously by policies focusing on other issues, especially on traditional air pollution. The underlying principles of the air protection law enforced in 1974 (Bundesimmissionsschutzgesetz, BImSchG) are both the 'polluter pays' and the 'precautionary' principles. In §3 BImSchG, the law refers to *Immissionen*. This is a specific German term meaning all emissions of substances which adversely affect the environment of a defined area independent of where the sources of the emissions are located. The term 'environment' includes human beings, animals, plants and inanimate things (Hansmann, 1992). Regulation under the BImSchG starts with the emissions of pollutants and emission-related procedures in both the industrial and individual sectors (including the transportation sector). In fact, the law and its executive ordinances resulted in approved end-of-pipe technology by setting limits for substances recognized as pollutants. Therefore, the main greenhouse gases were not an issue of regulation though all greenhouse gases are theoretically covered by the law: §3 BImSchG defines emissions regulated by the law as emissions from all kinds of facilities which cause changes in the air's natural composition.

Because the energy-related CO_2 emissions cannot be mitigated by end-of-pipe technologies economically and because the BImSchG does not turn processes towards low greenhouse gas emissions, this is generally not a promising approach for climate policy. In the political arena, this is reflected by the fact that a tightening up of measures and laws was only demanded where that seemed to be useful (e.g. the BImSchG has been complemented by a decree to limit emissions of volatile halogenated hydrocarbons). Amendments and further developments of the BImSchG did not become the core of climate change politics.

The political response to the problem of the increased greenhouse effect started with a comprehensive analysis of the issue. The existing forms of discourse of the Federal Cabinet with the German Bundestag and the German Bundesrat led to an extensive information policy concerning both the destruction of the ozone layer and the increased anthropogenic greenhouse effect. This is shown by more general discussions, questions and applications raised by opposition parties and the corresponding answers of the government.

This can be a protracted process, partly because ruling parties need time to respond to complex cross-policy arenas such as climate change.

As a result of the increasing discussion, the Enquete Commission *Vorsorge zum Schutz der Erdatmosphäre* (Preventive Measures to Protect the Earth's Atmosphere) was constituted in October 1987 (BT.-Drs. 11/533, 11/678, 11/787). An Enquete Commission is composed of politicians and scientific experts. Knowledge is transferred between them in the sense of interpreting the science and formulating political needs. Therefore, the Commissions have the potential to be the most important platform for discussions about the transformation of science into political action. The work of the Enquete Commission of the 11th German Bundestag is judged to be extremely successful and to have shortened the phase of issue framing considerably.

Given the background information of the Enquete Commission's work, at the ministerial level in early 1990 the political activities gained a new quality. BMU was asked by the Chancellor to prepare a Cabinet Decision for a CO_2 reduction target after informal communications by BMU officials with colleagues at the Chancellor's office (Müller, 1993a). Backed by the Chancellor's assignment, in four weeks BMU worked on a feasibility study using the knowledge from the Enquete Commission and the Federal Environment Agency (Umweltbundesamt – UBA) and concluded that a reduction of CO_2 emissions by 30.5 per cent would be possible. Both the Enquete Commission and the UBA commissioned detailed studies to support this position (Schaefer *et al.*, 1990). Therefore, a 25 per cent reduction goal seemed realistically achievable.

On 13 June 1990 the government agreed to a goal of 25 per cent reduction of the CO_2 emissions in the former West Germany based on 1987 levels. On 7 November 1990 the Federal Cabinet reaffirmed this target for the former West Germany and stated that larger reductions were expected in

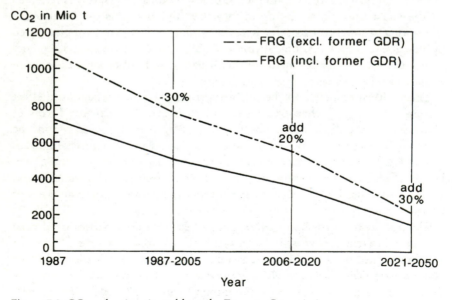

Figure 7.2 CO_2 reduction timetable at the Enquete Commission

Source: German Bundestag, 1991, Vol. II, 88

the area of the former GDR. On 11 December 1991 the Cabinet reaffirmed previous decisions and stated that Germany would aim to reduce CO_2 emissions by 25–30 per cent by the year 2005 based on 1987 values. The Enquete Commission in its 1990 report had a goal of 30 per cent CO_2 emissions reduction. The reduction path recommended by the Commission is shown in Figure 7.2.

As a further reaction to the final report of the first Enquete Commission, in September 1991, the German Bundestag agreed strictly to apply the precautionary and polluter pays principles and to integrate environmental protection in all political areas (BT.-Drs. 12/1136). This application is stressed in every environmental resolution because these principles are the basis of German environment policy since 1971. They are used to explain why Germany puts emphasis on the limitation of greenhouse gases and not on adaptation research and measures.

DEVELOPMENT OF STRATEGIES

Given the ambitious German target, discussion focused on how to fulfil it. To understand the resulting measures, it is necessary to have a closer look at the actors involved, their relations and negotiation strategies.

Ministerial responsibilities

Though competences for environmental protection had been centralized in the BMU after the Chernobyl nuclear accident, many important policy sectors concerning the environment still remained in other ministries, such as energy policy, which falls under the power of the BMWi. Other ministries such as the BMV, the Federal Ministry for Building and Town Planning (BMBau) and the BML also play important roles in decision making on issues that affect greenhouse gas emissions (Müller, 1990, 167).

The influence of a ministry depends on available resources (Weidner, 1989, 18). Just comparing the staff shows that BMU is one of the smallest and thus in administrative terms less powerful ministries (Table 7.3). Nevertheless, compared with the 1993 budget, in the 1994 budget, BMU's expenditure increased by 5 per cent while other ministries had to cope with cut-backs (e.g. BMWi −11 per cent, BML −4 per cent, BMFT −1 per cent). BMV (+23 per cent), whose policy does not reflect the problem of climate change, as described later in this chapter, and BMBau (+32 per cent) gain large increases. Besides that, the main difference between BMU and BMWi, BML and BMV is that the latter three, while representing the German government, also take into account the interests of industry and others. BMU does not have such connections but stands between all lines: other ministries, industry and environmental groups.

Table 7.3 Comparison of federal ministries

Ministry	Total personnel costs (in 1000 DM)	Other administration expenditure (in 1000 DM)	Total expenditure (in 1000 DM)			Civil servants
			1993	1994	Difference	
BMF	3,281.967	1,196.936	5,814.374	5,899.911	+85.537	1594
BMWi	581.609	283.722	5,436.837	14,145.230	−1,817.407	1141
BML	438.303	152.811	14,271.075	13,326.419	609.376	734
BMV	2,097.970	2,434.256	43,871.946	53,808.262	+9,936.745	775
BMU	02.338	285.395	1,262.953	1,331.375	68.979	567
BMBau	115.284	136.560	7,988.932	10,537.608	+2,548.676	344
BMFT	90.796	36.125	9,610.982	9,468.132	−142.850	489
BMI	3,247.371	1,020.694	8,789.388	8,527.167	−262.221	1079
BMZ	58.122	21.676	8,423.881	8,365.214	−58.667	354
AA	1,134.801	261.304	3,632.539	3,803.824	+171.285	1248

Source: Bundeshaushaltsplan für das Haushaltsjahr (1994)

This constellation considerably weakens the effectiveness of political programmes of the BMU because of interministerial conflicts. While BMWi initiatives are supported by intense lobbying of industry, BMU has to compensate its lack of clout by intense public relations (Müller, 1990, 168). This is reflected by the statements of Minister Töpfer in the press reports (BMU-Pressemitteilungen), which use every possible occasion to stress the importance of Germany as a leader in international climate policy and therefore the necessity to act nationally.

This process is similar to that experienced in the UK, as noted in the chapter that follows. The climate change debate forced open some form of interministerial dialogue, stimulated in part by more imaginative and comprehensive participatory processes. Norway, and to some extent Italy, followed a broadly similar pattern. Policy coalitions trigger wide ranging consultative arrangements.

However, there was always the large risk that the results of environmental policy would be disappointing compared with the original demands. In fact, the reputation of Minister Töpfer suffered because of this (IPOS, 1992, 25). Figure 7.3 shows the results of an opinion poll, where the work of BMU was on average valued at −0.3 points on a range from −5 to +5. It should be noted, however, that besides other influencing factors, the valuation was highly connected with the party allegiance preferences of the respondents, so these conclusions may not be fully representative.

The interministerial working group

For implementing the CO_2 reduction target and analysing the related topics, an interministerial working group (Interministerielle Arbeitsgruppe – IMA)

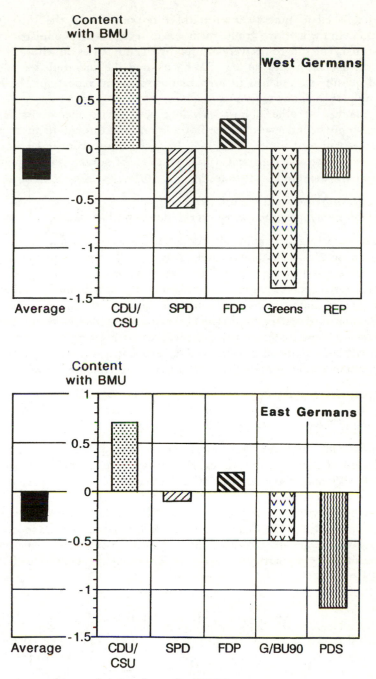

Figure 7.3 Confidence placed in the work of BMU

Source: IPOS (1992)

was established in June 1990 with BMU responsible for the IMA 'CO_2 reduction'. In the working group, members of several federal ministries were represented. The subgroups were chaired by the ministries, in whose competence the specific issues lie (e.g. BMWi chaired the subgroup for energy-related questions), and had to prepare the subgroup reports of the IMA. BMU did not chair any subgroup.

Programmes of other ministries concerning environmental issues can generally only be influenced by the BMU in an IMA through its institutionalized mechanisms (Müller, 1990, 169). In such an interministerial working group, the responsible department has two strategic advantages independent of who chairs the several subgroups:

- the *Initiativrecht* (right of initiatives) and
- the *Problemdefinitionsmacht* (power to define problems).

In the preparation of political programmes, the responsible department alone can determine contents. Afterwards the other participating departments can suggest changes to the submitted programme. This process of interministerial coordination seldom leads to programmes completely different from those made by the initiating department. The mechanism of coordination is the so-called *Negativkoordination* (negative coordination): the different departments restrict themselves to averting negative consequences of a programme for their 'own' department.

The potential areas for conflict in the IMA 'CO_2 reduction' were:

- the political struggle for power between BMWi and BMU;
- the distribution of administrative competence between BMU and BMWi;
- economic versus environmental interests.

The difficulties between BMU and BMWi concerned the choice of words: BMWi had enormous problems calling the CO_2 reduction programme a 'goal'. BMWi preferred the term 'orientation', which would be much less committing. Nevertheless, in need of a fixed target BMU 'played the numbers game with the other ministries' (Müller, 1993a). This is reflected in official statements by the Federal Cabinet declaring that it consciously had not passed a binding quota because first of all the preconditions and consequences of policy, in particular energy policy, should be determined (BMU, 1990, 24).

The second Enquete Commission

Because of the successful work of the first Enquete Commission, there was an overwhelming consensus to constitute a new one in the 12th legislative period. In contrast to the first Commission, however, the working structures of the second were very controversial and mistrust and frustration dominated the process (Ganseforth, 1993). How is this development from the first to the second Enquete Commission to be explained? There are several kinds of difference.

The Chairperson The first Commission was chaired by CDU member Bernd Schmidbauer. He was able to free himself from party and interest ties. Early on the opposition suspected he would support industry's interests, but subsequent negotiations showed that this was not the case. He played a most important integrating and confidence building role in the Commission. Additionally, the opposition felt that in the beginning of the Commission's work the other CDU members were dominated by Schmidbauer; hence the unanimous decisions. Later, the other CDU members' influence grew, and more disagreement arose.

With the new chairman of the second Enquete Commission Klaus Lippold (also CDU Member of Parliament) and a new deputy chairperson from the SPD the style of leadership changed. The new chairman was seen to be very close to party lines and as a lobbyist. As a consequence it was more difficult to reach agreement, as subsequent interviews revealed.

The issues The first Enquete Commission dealt mostly with the natural science background to the problem of the increased greenhouse effect. The scientists were able and willing to present difficult scientific knowledge in a way that non-experts could understand. This dialogue was partially restricted in the second Enquete Commission, especially for the transport sector. Another point is that politicians suggested that the scientific experts are less interested in questions of implementation (second Enquete) than in scientific basics (first Enquete).

Working process How the Enquete Commission's members worked was strongly dependent on the above-mentioned preconditions. Because of the confidence placed in the first Enquete Commission, different opinions were not obstacles for cooperation, as consensus building mechanisms were used. Another reason for the differing success of the work of the two Enquete Commissions is that new political issues of high domestic priority emerged, such as the recession, unification and increasing unemployment. These distracted public attention from the Commission's work, possibly another factor for its success.

While the Commission appeared initially to be praised, the cosiness of agreement was criticized by some of the opposition, particularly at grassroots level. Oppositions like to maintain some questioning positions (Ganseforth, 1993).

Negotiation structures

The task of the IMA 'CO$_2$ reduction' was to analyse which measures should be implemented. It had to consider the overall economic and socio-political consequences, the financial effects to be expected, and the priorities to be set. In a second step, the IMA was supposed to estimate how long it would take to implement measures, which interdependences and which potentials for conflicts with other political sectors existed and which measures would have to be coordinated internationally.

The first report of the IMA, presented on 7 November 1990, noted the changed conditions following German unification and recommended the reduction of emissions in the Neue Bundesländer (former GDR) by a larger amount because the development of an equal standard of living in the Neue Bundesländer compared with the Alte Bundesländer (former FRG) would provide opportunities to reduce CO_2 emissions drastically over the next 15 years. The CO_2 reduction programme (CO_2-Minderungsprogramm) which resulted from the IMA's work stated the necessity of international cooperation and precautionary measures. It also formulated demands for a tightening up and further development of existing arrangements (e.g. the Montreal Protocol). First priority for measures focusing on the greenhouse effect should be given to economic instruments. The most important part of the programme was the aim to reduce the energy-related CO_2 emissions.

The IMA was instructed by the German government to extend the programme. Subsequently, in September 1994 in the third report 109 measures to reduce CO_2 emissions were listed (Beschluß der Bundesregierung, 1994, 86–136). Moreover, a first evaluation of reduction potentials of other greenhouse gases than CO_2 was provided (Table 7.4) more as an orientation on what seems to be possible than a recommendation to adopt reduction goals as in the case of CO_2.

Phasing out of CFCs

The regulation of CFCs was primarily motivated by the enforcement and implementation of the Montreal Protocol. Therefore, additional action beyond the ozone-related measures did not seem to be necessary in the context of climate change.

Table 7.4 Reduction potentials for other greenhouse gases than CO_2

Greenhouse gas	Scenario 1990–2005
Methane (CH_4)	−48%
Nitrous oxide (N_2O)	−25%
Nitrogen oxides (NO_x)	−25%
Carbon monoxide (CO)	−51%
Non-methane volatile organic compounds (NMVOC)	−43%

Source: Beschluß der Bundesregierung (1994, 45)

In 1987, the CFC using industries committed themselves to reduce CFCs in spray cans by 75 per cent by the end of 1988 and 90 per cent by the end of 1989 and to report on CFC consumption. The attempt to use the controversial instrument 'voluntary restrictions of industry' was part of the consensus-oriented strategy of the first Climate Enquete Commission. Compliance with the agreements was to be monitored and non-compliance answered with sanctions. The reporting commitments were not fulfilled resulting in increased demands for a ban based on the Chemical Law in 1989 in the German Bundesrat (BT.-Drs. 11/8166). Referring to the precautionary principle, the obligations of the draft law were stronger than those in the Montreal Protocol and its London Amendment in 1990. Both the production and the consumption of the substances covered by the law were regulated. On the same day, when, after modifications, the ban was passed by the Federal Cabinet, a modified voluntary agreement was announced. On 1 August 1991 the ban came into force (BR.-Drs. 18/91).

The discovery of the dramatic increase of chlorine in the stratosphere above the northern hemisphere by NASA in February 1992 stimulated a new policy push: the Bundesrat (BR.-Drs. 103/92) demanded an amendment of the CFC–Halon ban for earlier phase-out of production and consumption. At the international level government was asked to push at the Fourth Conference of Parties to the Montreal Protocol for shorter EU deadlines and a worldwide ban on ozone depleting substances in 1995. Exceptions for developing countries should be reversed and financial aid be extended. Initially, this engagement was rejected by the government (BR.-Drs. 485/92). As another consequence, Germany's three CFC producers, Hoechst AG, Solvay (formerly Kali-Chemie AG) and Chemiewerk Nünchritz (East Germany), committed themselves to an earlier phase-out of production (i.e. in 1993) than agreed, if the CFC consuming industries would reduce usage of CFCs (BMU Pressemitteilung 75/92). This agreement had widely been accepted by affected CFC consuming sectors. In December 1992 at the EU Environment Council, negotiations resulted in shorter EU deadlines, in fact by 1 January 1995. Afterwards, the federal government, in particular the Federal Environment Minister, presented the German phase-out of CFCs as a 'success story' and as a consequence of Germany's role as a leader in CFC reduction politics. In his opinion, this led to a considerable acceleration of the phase-out of the production of CFCs regulated under the Montreal Protocol and its amendments by the Parties of the Protocol (BMU-Pressemitteilung 44/93).

The phase-out deadlines in Germany are shown in Table 7.5.

German climate policy in the international context

The responsibility for national and international action lies in two different departments of the BMU. National policies are the responsibility of the

Table 7.5 German phase-out deadlines for ozone depleting compounds

Function	Substance CFCs	R 22 HCFC	Methyl chloroform	Carbon tetra-chloride	Halons
Aerosols	August 1991	August 1991	August 1991	Not used	Not used
Refrigeration			Not used	Not used	Not used
Equipment					
Large scale	January 1992	January 2000			
Large scale					
mobile	January 1994	January 2000			
Small	January 1995	January 2000			
Foams			Not used	Not used	
Packaging					
material	August 1991	August 1991	August 1991		
Dishes	August 1991	August 1991	August 1991		
Construction	August 1991	January 1993	August 1991		
Insulation	January 1995	January 2000	January 1995		
Others	January 1992	January 2000	January 1992		
Cleaning agents and					
solvents	January 1992	Not used	January 1992	January 1992	Not used
Extinguishers	Not used	Not used	Not used	Not used	January 1992

Source: Cutter Information Corporation (1993, 3)

Department for 'Nature Conservation and Ecology' and international climate policy of the Department for 'Basic and Economic Issues of Environment Policy, International Cooperation'. International efforts are sometimes used to move national policy forward. For example, as the Rio conference was approaching international negotiations were seen as an opportunity for progress in the negotiations at the national level to maintain credibility (Müller, 1993a).

During the negotiations for a Framework Convention on Climate Change (FCCC), Germany showed support for a Climate Convention and first Protocols on greenhouse gases, especially on CO_2 emissions, to be signed during UNCED in 1992. In its report on the results of the UNCED, BMU placed particular emphasis on joint implementation of commitments in Article 4(2)a, saying that this was a German proposal. However, as a whole, the FCCC and its commitments in Articles 4(2)a and b did not meet German expectations. Showing Germany's strong interest in the follow-up process, Chancellor Kohl at UNCED offered to host the first Conference of Parties of the FCCC (CoP) in Berlin. This interest was also underlined by Germany's commitment to the 'prompt start' in anticipation of Article 12

FCCC. As one of these 'prompt measures', the preliminary National Climate Report was published in August 1993 (BMU, 1993). The CO_2 reduction programme is a key element. Another element is the financial support by the Federal Ministry for Economic Cooperation (Bundesministerium für wirtschaftliche Zusammenarbeit, BMZ) for the preparation of country studies in developing countries. The final National Climate Report was published in September 1994 (BMU, 1994).

The national German reduction goal conflicts with international commitments, as pointed out in the previous chapter. This is because the year 2005, not 2000, is chosen as the national target deadline. Together with the uncertain course of the German reduction timetable, the contribution of the German CO_2 emissions to the total CO_2 emissions of the EU – and for that reason the ability to achieve the EU stabilization target given other European emission reduction timetables – remains unclear. To illustrate this, several theoretically possible courses of CO_2 reduction in Germany are presented in Figure 7.4.

Because of the connections between climate change politics and other domestic 'problems' (recession, structural change, unemployment), the Environment Ministry stated that there would be no change of the target year because that might lead to a national renegotiation of the target (Müller, 1993b). Under the German EU Presidency in the second half of 1994 a new initiative was announced by the Federal Environment Minister

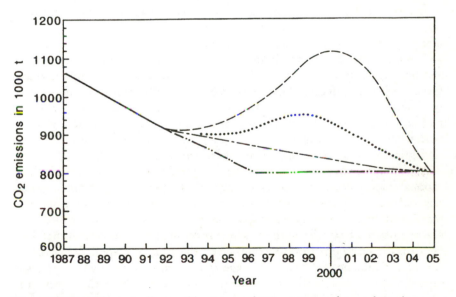

Figure 7.4 Some *theoretically* possible courses of CO_2 emissions for reaching the German CO_2 reduction goal

to implement the CO_2/energy tax (BMU-Pressemitteilung 44/94). Despite support from a group of like-minded countries – Denmark, the Netherlands and the new Scandinavian Member States – this initiative failed. The politics of national sovereignty and fixed subsidiarity have won for the time being.

EVALUATION OF THE RESPONSE STRATEGY

The formulation of the measures of the CO_2 reduction programme is subject to substantial criticism from almost all non-governmental groups and the opposition parties. Both the implemented measures and also the delay in implementing crucial ones, e.g. the amendment of the law on the energy industry or a tax on energy, are criticized. Moreover, the effectiveness of single measures for reducing greenhouse gas emissions cannot be verified because projections of the reductions expected from each measure were officially never made. The positive points of an effective climate policy were watered down during the negotiation processes of each measure as in the case of the Thermal Insulation Ordinance or the Ordinance on Heat Use. The setting of legislative frames at the federal level is oriented to the needs of the local authorities than to other actors at the federal level. Local authorities interested in climate-change-related issues consider that to be one of the main obstacles for an effective climate policy. Other criticism results from not having broader possibilities to promote least emission solutions. For example, the 1991 Act on the National Grid (*Stromeinspeisungsgesetz*) improved the profitability of renewable energies but the chance to promote progressive, efficient, electricity generating technologies at the same time was not taken. Table 7.6 provides an overview of the structure of the 109 measures of the CO_2 reduction programme.

The majority of measures implemented to date can be characterized as 'subsidies', promoting useful activities or investment by federal financial aid. Other measures affected only small groups (utilities, CFC producers) or were incidental to any CO_2 reduction. At the end of the 12th legislative period, none of the unpopular measures which would have increased the federal revenue had been implemented, although several federal resolutions gave priority to the use of economic instruments.

A possible explanation is that the behaviour of market actors is less rational than in economic theory (Müller, 1994). The macroeconomic advantages of economic instruments, such as more flexibility, more efficiency, more innovation, may be feared by microeconomic actors because they create more uncertainties in the strategic planning processes and in the behaviour of direct competitors. Second, given the general responsibility of the Minister of Finance for taxes, the expected steering effect of an eco-tax was regarded as being an unreliable generator of revenue. Since there is a clear differentiation in Germany between sources of revenue in general taxes and earmarked taxes (*Sonderabgaben*), there was no chance that BMU could

Table 7.6 Structure of the CO_2 reduction programme

Of the 88 measures enforced by September 1994 were	% of enforced measures
R&D	22
Information	11
Other incentive measures	34
Regulation	17
Fiscal measures	7
Other	9

Of the 21 measures not enforced by September 1994 were	% of not enforced measures
R&D	–
Information	–
Other incentive measures	–
Regulation	29
Fiscal measures	24
Other	14
Not further specified	33

Source: Own compilation. Based on: Beschluß der Bundesregierung (1994, 86–136)

draft a new general energy tax. BMWi, being responsible for energy policy, would have rejected any such proposals. This reveals both the role of law and administrative politics in the handling of a cross-departmental policy measure in the name of reducing climate change impacts. Unless there is a clear macroeconomic objective, or unless interest group coalitions are properly coordinated around innovative fiscal measures, such institutional departures are unlikely. In the UK, Norway and Italy, the conditions were suitable, but not, apparently, in Germany.

Projections of total CO_2 reduction by 2005

In July 1992, Minister Töpfer announced the results of a study by the UBA, which estimates potentials for CO_2 reductions (BMU-Pressemitteilung 71/92 and UBA, 1992, 40, 132, 137). UBA concluded that the protection of the environment supports economic growth in Germany and that this would be the main growth sector in the future, especially with respect to the EU internal market. Töpfer also pointed out that the national implementation of the reduction target would be an important contribution to international action. The estimates were based on the announced (not the implemented) measures of the CO_2 reduction programme and concluded that a reduction of German CO_2 emissions by approximately 31 per cent could be achieved by the year 2005. This fits exactly with the CO_2 reduction goal. A division in

Neue and Alte Bundesländer shows the differences: in the Alte Bundesländer the reduction would be around 28 per cent; in the Neue Bundesländer it would be around 36 per cent due to larger potentials of reduction.

The 1987 to 1992 figures show no significant change for the Alte Bundesländer (Figure 7.5). In particular, the CO_2 emissions of the traffic sector were increasing (UBA, 1993a, 156). For the Neue Bundesländer, there is a strong decrease, primarily due to the economic collapse following unification. All together, there is a decrease in total CO_2 emissions between 1987 and 1992 of 14 per cent from 1086 Mt/a to 933 Mt/a (BMU, 1993, 84). Other less optimistic estimates from Prognos AG (1992, 435) conclude that CO_2 emissions will only be reduced by 10 per cent by the year 2005 compared with 1987 levels.

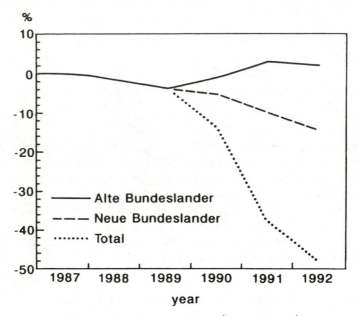

Figure 7.5 Energy-related CO_2 emissions (base year 1987)

Source: UBS (1992, 154)

The transport policy

In Germany, in the early 1990s about 19 per cent of the total CO_2 emissions were from the transport sector. Since greenhouse gas emissions from this sector have continuously increased in the past (see Figure 7.6), it is of interest to see whether the issue of climate change is reflected in the federal transport policy and whether this policy is adequate to achieve the reduction goal for the year 2005.

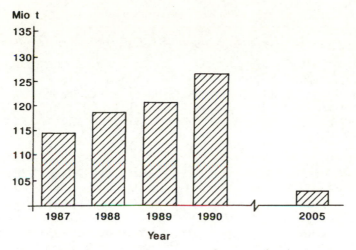

Figure 7.6 Development and projection of energy-related CO_2 emissions in the transport sector

Source: UBA (1992, 156)

Table 7.7 shows the development in transport figures from 1960 to 1990 (only Alte Bundesländer). The emission-intensive motorized individual transport as well as road haulage and air transport increased steadily while at the same time other transport modes decreased continuously, especially public transport and railways.

Approximately 45 per cent of total motorized individual transport is due to leisure activities. This share remained almost constant for 30 years (see Figure 7.7), while the total amount of person kilometres increased by almost three and a half times.

Figure 7.8 illustrates that the estimated trend figures for the transport sector are not compatible with the recommendations of the Enquete Commission to reduce transportation-related consumption of end energy and related CO_2 emissions by 30 per cent by the year 2005. On the contrary, end energy consumption is estimated to increase by about 60 per cent in this period.

In the past transportation policies concentrated on both adapting the existing infrastructure to increasing transport figures and limiting system-related negative effects of the transport sector itself. There is no evidence of changes in orientation of transport policy towards mitigation policies. Despite this not very encouraging situation, many experts conclude that a reduction of transport-related CO_2 emissions by the recommended amount is possible. But demands for more or less ambitious restructuring measures in the transport sector have generally not been picked up by the Federal Minister for Transport. It has not even been possible to enforce a speed limit for highways.

Table 7.7 Transport figures

Transport capacity (motorized transport) in billion person kilometres – Pkm (%)							
Year	1960		1970		1980		1990	
Motorized individual transport	161.7	(63.8)	350.6	(76.8)	470.3	(78.6)	593.8	(82.1)
Taxi/hired cars	0.8	(0.3)	1.7	(0.4)	2.2	(0.4)	2.5	(0.3)
Public transport	48.5	(19.1)	58.4	(12.8)	74.1	(12.4)	65.2	(9.0)
Rail traffic (short distance)	19.3	(7.6)	15.6	(3.4)	14.7	(2.5)	17.2	(2.4)
Rail traffic (long distance)	21.7	(8.6)	23.6	(5.2)	26.4	(4.4)	27.4	(3.8)
Air transport[1]	1.6	(0.6)	6.6	(1.4)	11.0	(1.8)	18.4	(2.5)
Total	253.5	(100.0)	456.5	(100.0)	598.6	(100.0)	724.5	(100.0)

Transport capacity (goods traffic) in billion tonne kilometres (%)							
Railways	53.1	(34.4)	71.5	(33.2)	64.9	(25.4)	61.8	(20.6)
Inland navigation	40.4	(28.5)	48.8	(22.7)	51.4	(20.1)	54.8	(18.3)
Pipelines	3.0	(2.1)	16.9	(7.8)	14.3	(5.6)	13.3	(4.4)
Air transport[1]	0.0	(0.0)	0.1	(0.05)	0.3	(0.1)	0.4	(0.1)
Long distance road haulage	23.7	(16.7)	41.9	(19.5)	80.0	(31.3)	120.4	(40.1)
Short distance road haulage	21.8	(15.4)	36.1	(16.8)	44.4	(17.4)	49.4	(16.5)
Total	142.0	(100.0)	215.3	(100.0)	255.3	(100.0)	300.1	(100.0)

[1] Only CO_2 emissions emitted by air transport above German territory. Otherwise the figures would be approximately ten times higher

Sources: DIW (1992); Schallaböck (1993b).

This is an example of policy distintegration. Just as is argued in Chapter 12 that climate change policies rely on piggy-back linkages to other policy initiatives, so the reverse is also possible. If a stable policy core is resistant, the piggy-backing is not possible, and climate change futures suffer as a consequence. Lip service and denial, as suggested in Chapter 4, can block institutional transformation.

The BMV's transport policy was not really affected by the recommendations of the first Enquete Commission, who recommended a ranking of transport-related measures giving first priority to the avoidance of transport. They suggested further measures to change the modal split, technical optimization of transport modes and flows, and changes in transport behaviour in general. This list of priorities was confirmed by one of the regular conferences of the Federal Environment Minister and his colleagues of the Bundesländer.

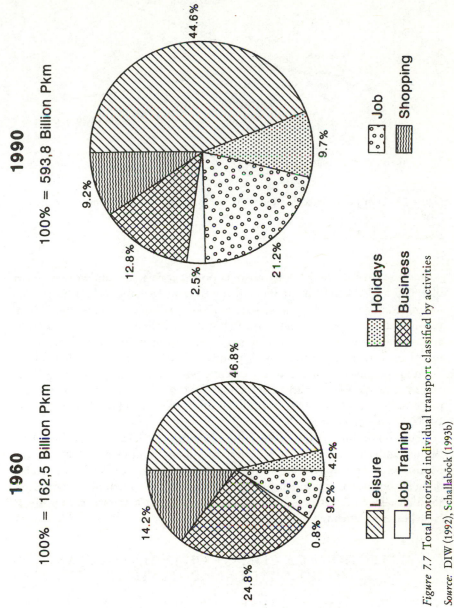

1960

100% = 162,5 Billion Pkm

1990

100% = 593,8 Billion Pkm

Figure 7.7 Total motorized individual transport classified by activities

Source: DIW (1992), Schallaböck (1993b)

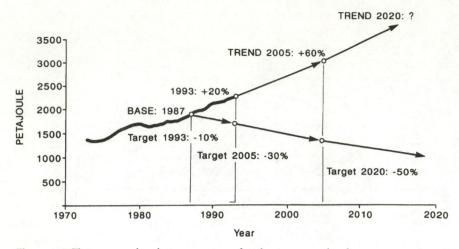

Figure 7.8 Transport-related consumption of end use energy: development, trends and targets

Source: Schallaböck (1993a)

In the mid-1990s the BMV's transport policy consisted mainly of transport planning and the reform of the railway system (Schallaböck, 1993b). The federal transport planning was characterized by three elements: the federal planning of transport infrastructure (Bundesverkehrswegeplanung, BVWP) and, following unification, measures to speed up planning procedures and a list of infrastructure projects with priority. The aim of these policies is to build up a transport infrastructure to cope with more traffic over longer distances. The BVWP earmarked a total of DM 200 billion for infrastructure projects by the year 2010, to provide an additional 5000 km of railways and 10 000 km of roads.

The overall integration, especially economic integration, of the Neue Bundesländer is of high priority. Therefore, many politicians – not only the Ministry for Transport – are convinced that other factors, such as environmental considerations which by law are part of the planning procedures, should be given lower priority until the most urgent infrastructure projects have been implemented.

REFLECTIONS ON CLIMATE CHANGE POLICY AND SOCIO-POLITICAL ADAPTATION

Industry: policy of voluntary restrictions

Following the same lines of argument as in the case of CFCs, voluntary restriction was presented by industry associations as the most effective and market-sensitive strategy for reducing CO_2. Fiscal solutions were rejected because they were regarded as public income services and could take funds away from businesses, funds which otherwise would realize climate protection measures. On the other hand, regulation alone is seen to be unsuited for handling climate change because it relies on vague terms like 'standard of technology' which do not cover entrepreneurial, product and production-related characteristics (VIK, 1993, 15; BDI, 1992, 187). Therefore, in November 1991 industry offered voluntary measures for climate protection connected with opportunities for international joint implementation (e.g. VDEW, 1993). This means that in cooperation with industry, government would formulate attainable reduction targets and industry independently would decide where and how these targets could be achieved. During the negotiations, however, the enthusiasm of industry for action was dampened as recession mounted. At the beginning, industry did not feel able to make binding agreements to reduce CO_2 or to increase energy efficiency, as long as there was no binding agreement that other burdens would not be implemented at the national or international (EU) level. For that reason, the initially favourable discussions between industry and BMU, BMWi and BMF lost their momentum (VIK, 1993, 18). By October 1993, the consultations were hijacked by discussions on the attractiveness of Germany for both domestic and foreign investment (site discussion) and the recession in general.

Once again we see a combination of reverse piggy-backing, consolidation of established interests around core positions, the absence of precaution, and only cosmetic compliance. These are essentially first-order, almost non-shifts in, institutional responses as laid out in Chapter 4.

Environmental organizations emphasized from the beginning the negative aspects of voluntary restrictions, as shown by several examples, e.g. the announcement of the beverage industry to use returnable bottles in 1977, which was not met in the past (Sauter, 1993, 2). Only when faced with economic losses because of increased environmental consciousness of the consumers or with regulation (e.g. the restriction on the use of asbestos) do voluntary agreements work (Wicke, 1993, 272, 275). In the view of one of the major German environmental non-governmental organizations (Bund für Umwelt und Naturschutz Deutschland – BUND), the urgent need for solutions to the climate change problem requires the application of regulation to forbid single substances (e.g. CFCs) where necessary, but mainly the implementation of economic instruments. Concerning the costs, BUND argued that voluntary measures would also cause financial burdens if the measures were to be effective. The use of economic arguments by an environmental NGO is quite interesting and shows the changes these groups have made in responding to industry's statements. We will note how this aggregation of policy interests is also found in the UK and in Norway.

As mentioned above, the discussion in Germany is linked with the questions of joint implementation. Protection of the climate indicates the need to increase development aid even though Germany fails by a long way to meet the 0.7 per cent of GDP goal for aid spending. For that reason, joint implementation was judged to be necessary, shifting the tasks from state level to those businesses which are willing to practise joint implementation under the precondition that there are no obligations for new taxes.

Non-governmental organizations

The role of German non-governmental organizations is not given much attention here because, for a long period, climate change was not an issue for them. This was due to fears that nuclear energy as a CO_2-free source of energy would be promoted. In addition the German NGOs mainly focused on local or national problems. This changed once it was recognized that energy conservation combined with the phase-out of nuclear energy could be feasible and that climate change issues were being negotiated internationally. During and after the preparation for the UNCED, a multitude of fora and working groups developed partly because of financial aid by the Federal Environment Ministry.

Two initiatives indicate a new quality of NGO activities leading to shifting coalitions with fundamentally different interest groups which have few areas of shared interest. First, the BUND has formed a coalition with an association of young entrepreneurs (Bundesverband Junger Unternehmer – BJU) to support the implementation of an ecological tax reform. This initiative is described in more detail below. Second, the DIW has carried out a study for Greenpeace Germany analysing the macroeconomic effects of an ecological tax reform especially by quantifying the impacts on the unemployment level. DIW concludes that such a reform would create 300 000 to 800 000 additional 'green' jobs (Greenpeace, 1994). Other studies support this positive net effect (Jarras and Obermair, 1994).

The positions of the large associations of industry, such as the Confederation of German Industry (Bundesverband der Deutschen Industrie – BDI) and the German Chamber of Commerce (Deutscher Industrie und Handelstag – DIHT), give the impression that there is one cohesive interest group. For the coal and steel sector this is not surprising, because they would be among the main 'victims' or losers of an economic restructuring. There are smaller associations which support the implementation of ecological thinking in entrepreneurial decisions, such as the Federal German Working Group for Environment-Conscious Management (Bundesdeutscher Arbeitskreis für Umweltbewußtes Management – BAUM).

A newer initiative, which met with much public response, was organized by BJU and BUND. This alliance was built to demonstrate that there is not necessarily a contradiction between economic and ecological interests, and that the differences between the new (young) generation of entrepreneurs and conservation groups existed only in the minds of politicians and established interest groups. The latter tend to be dominated by older decision makers associated with medium size enterprises, who still give less priority to environmental issues (BJU, 1993). The initiative points out that environmental protection is not an issue in itself but is a precondition for future economic development. Action must not be delayed for simple financial reasons, for the difficulties and potential follow-up costs will increase. Given the German debate on whether the country will continue to be a preferred site for industry, disadvantages as a result of the restructuring process were also discussed. In contrast to the traditional associations, they were given less weight than the future beneficial effects. It was stressed that an ecological structural change from pollutant- and resource-intensive to labour-intensive sectors would be induced by measures such as an ecological tax reform. Therefore, or so it is believed, foreign ecologically oriented businesses would be attracted to Germany. Innovation would subsequently be steered to future competitive advantage. The response to this initiative was enormous, probably due to the aggressive marketing strategy used (Schön, 1993). Within four months after the BJU presented its statement, a report about it and the environment commission's chairman was published in the weekly newspaper *Die Zeit* (Vorholz, 1993), focusing public attention on the whole initiative and causing a large number of requests for interviews in other newspapers and on television. The chairman of the environment commission felt that BJU's statement caused surprise and learning. This surprise was an intended part of their strategy. Clearly there are limits to the alliance, since the BJU is convinced that a ministry for the environment is not necessary and the industry sector itself can solve the problems in a more efficient way – if there is an appropriate framework. Environment problems are considered to result from misguided economic policy. For that reason, lobby activities concentrated on the Federal Ministry for Economic Affairs. The alliance with an NGO underlines that the business sector is not one united group, but it is an open question how influential the initiative will become.

Economic factors

At the beginning of the German unification process, environmental protection seemed to be one of the main points for future politics. At that time most of the politicians thought that the economic rebuilding in East Germany would not be very expensive (IÖW Informationsdienst, 1992b, 1). Since Germany was facing a recession, the assumed contradiction between the environment and the economy was stressed more often. Expensive environmental strategies were acceptable because environmental protection

213

creates disproportionate burdens for the poor. Environmental protection competes with other concerns of society for financial transfers so that the scope for redeployment of labour and capital becomes narrowed (IÖW Informationsdienst, 1992a, 1). In the short run, a delaying strategy might protect endangered industries from additional costs and ease their struggles for improved international competitiveness. In the long run, only strong leadership secures advantages in international competition and induces the necessary structural change. Industries adversely affected by climate change policies and identified as influencing political decision making processes reacted to environmental protection discussions (not measures) with reference to 'endangered site Germany' (DIHT, 1993, 23). Reviewing this discussion, we conclude that in difficult economic situations environmental protection becomes less accepted the higher the protection level is, because of its cost character and the distributional questions. Then, demands for a break in established environmental policy arise, and new decision sequences may be created.

In the opinion of the German Environment Minister in 1993, environmental protection will strengthen Germany's economy and must be seen as an investment in future development (BMU-Pressemitteilung 23/93, 5). There was no necessity to lower environmental demands because of economic difficulties. A study carried out by two of the main German economic research institutes (DIW, Rheinisch Westfälisches Institut für Wirtschaftsforschung – RWI) on behalf of the UBA concluded that high German environmental protection levels are one of the less important factors for determining the site of new investment (UBA, 1993b). The burden of environment protection expenditure in different economic sectors suggests that these costs are almost negligible, especially compared to labour costs (DIW 1993, 201). Moreover, it has been shown that businesses increasingly receive environment related subsidies (UBA, 1993b, 60). The difficulties in quantifying them result from the large number of institutional levels from which subsidies can be obtained (the EU, the federal government, the Bundesländer). Compared with other countries the expenditure for environmental protection in Germany is by far the highest in absolute figures ($16.37 billion in 1991). But calculated as a percentage of GDP, the range does not differ very much among the main trading partners of Germany. Figure 7.9 shows two contrary trends: in group 1, expenditure as a percentage of GDP increased from a medium level in 1980 to the top level in international comparison in 1991 (except for France). In the bigger group 2, expenditure of the countries leading in 1980 steadily decreased.

Assuming that there is a relation between high environmental protection costs in a country and a weak position in international competition, products which would cause high environmental pollution without environmental protection measures cannot be exported by these countries because of the higher prices due to environmental charges. However, in 1988,

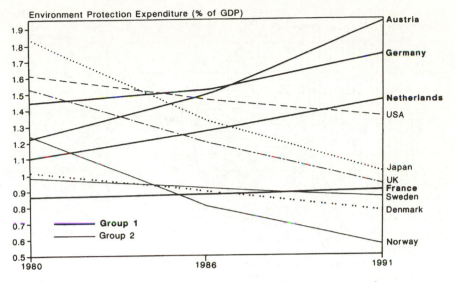

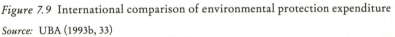

Figure 7.9 International comparison of environmental protection expenditure

Source: UBA (1993b, 33)

Germany was the leading nation in exporting those products, followed by the United States (Figure 7.10).

The high level of environmental protection expenditure of private businesses in the United States and Germany correlates with the number of patent applications of these countries in Europe (Figure 7.11). This may be an indicator for a high level of private R&D activities concerning environmental issues. But while in 1986 the proportions of the total applications did not differ very much, in 1991 the United States dominated while, except for the UK, all other proportions decreased.

LOCAL CLIMATE INITIATIVES

To judge the opportunities and the possible developments of climate change policies at the level of the municipality, we select energy policy as an example. The legal situation is controversial since there is doubt over whether energy provision is a public sector responsibility such as waste disposal and the provision of public transport (e.g. several articles in Ipsen and Tettinger, 1992). If the answer is yes, following Article 28.2 of the German constitution (GG), the municipalities can decide if they themselves want to start providing energy, continue existing programmes or instruct utilities.

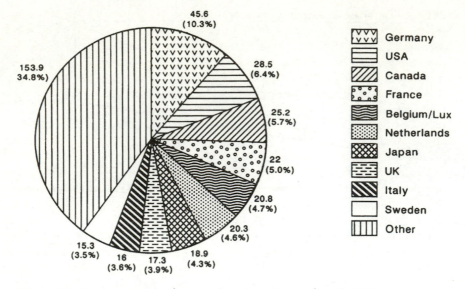

Figure 7.10 German exports of environment-intensive products in 1988
Source: DIW (1993)

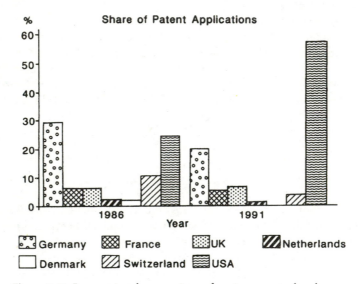

Figure 7.11 International comparison of environment-related
patent applications

Source: UBA (1993b, 108)

Another question is whether it can be deduced from Article 28.2 GG that municipalities own the local energy production or the distribution facilities (Schmidt-Jortzig, 1992, 91). Is there a constitutional right for municipalities to install combined heat and power? As yet this is still legally vague, though it is becoming politically more relevant.

Apart from the constitutional aspects, legislation and the structure of the German energy industry are important. The existing law on the energy industry (*Energiewirtschaftsgesetz*) has controlled the structure of electricity and gas provision since 1935. Then, a centralization process in the energy utility sector legalized monopolistic structures (Kohler, 1984, 17). As a result, the utilities (*Elektrizitätsversorgungsunternehmen* – EVU) were able to control electricity surpluses by deterrent pricing and exclusive control of the supply network. Until the 1980s, this was the main obstacle to institutional innovation in the energy sector.

Given that these practices are not compatible with the regulations limiting restriction of competition (*Gesetz gegen Wettbewerbsbeschränkungen* – GWB), efforts to demonopolize the EVU sector were introduced by the 4th (1980) and 5th (1989) reforms of the GWB. While the treaties between the EVU and municipalities (*Konzessionsverträge*) were signed for long periods (30 to 50 years), the modern treaties are generally limited to 20 years. Treaties signed before the amendment of 1980 are included in special regulations, so that 82 per cent of all treaties will expire by 1999 (Köpke, 1992, 150). In addition, the demarcation treaties between the EVU, which define the regional monopolies, are now limited to 20 years and after that have to be negotiated. These alterations mark a decisive turning point for municipalities to influence energy provision, and decentralize energy provision. In the opinion of the Federal Cartel Office (Bundeskartellamt), the aim was to strengthen competition between the EVU for service areas and not to improve the municipalities' positions in negotiations (BT.-Drs. 10/243, 85).

It is obviously not the case that all local governments realize the chances for and the potential benefits of restructuring. For example, the municipalities got used to the secure source of revenue *Konzessionsabgabe*, a special tax paid by the EVU to the municipalities for letting them exclusively provide energy. In addition, the EVU and decision makers are personally and institutionally intertwined (Köpke, 1992, 153). Other general obstacles are a lack of know-how in energy infrastructure provision, and restrictions on municipal finances. For that reason, there is a strong correlation between progressive concepts at the local level and environmental consciousness. What we see here is the emergence of institutional reform, mostly of a legal kind, to permit local collectives to run 'whole energy management' systems. The changes in the law were not designed with this in mind; hence the piggyback opportunity. So the key is leadership, interest group alliances and a supportive industrial policy that could transform the quality of energy provision at the local level in Germany.

The climate action plan of Heidelberg

After a meeting of the first Enquete Commission where the necessity for new municipal energy and transportation/traffic concepts to meet the CO_2 reduction goal were considered, the local authorities in Heidelberg decided to analyse the opportunities for action. This was led by the city's Mayor, Frau Beate Weber, a former Chair of the Environment and Consumer Affairs Committee of the European Parliament. In June 1991, the town asked the Institute for Energy and Environment Research (Institut für Energie- und Umweltforschung – ifeu) to analyse municipal opportunities to reduce CO_2 emissions. The town was only able to finance the study because of funding by the EU (DG XVII). This proved to sceptics that climate change was an issue that the town had to address. Another purpose, and something essentially new, was public participation in the planning processes from the very beginning. A forum with members from different groups of society was established (Würzner, 1993). The resulting study (Schmidt et al., 1992) concentrated on CO_2 emissions caused by transportation/traffic and the energy sector (in particular electricity and heating), though the whole industrial sector was left out. This was justified because of the special economic characteristics of the town with its major service sector and very small industrial base. Rumour has it that the local business community was not cooperative. Table 7.8 shows a selection of measures in the transportation/traffic sector recommended by the climate protection study of ifeu.

The study concluded that about 90 single or combined programmes should be initiated. These would lead to a reduction of CO_2 emissions by 28 to 33 per cent by the year 2010 compared with 1987–1988 levels (Schmidt et al., 1992, 1005–1062). It was stressed that most of the measures would have positive effects beyond CO_2 reduction. This payoff was necessary

Table 7.8 Selected opportunities for CO_2 reduction in the transportation sector of Heidelberg

Description	Example
Economic or fiscal measures	Rationing of parking spaces, pricing of public transport
Regulation	Traffic abatement, reduction of car parks
Technical measures	Speed limit, P + R system
Infrastructure	Extension of public transport and cycleways
Organization	Car sharing, shortening of fixed times in public transport
Information and PR	PR concerning bicycling or public transport

Source: Schmidt et al. (1992, 1152, 1153)

because the public was not generally aware of the climate change issue and of the relevance of, for example, the traffic sector. A major educational effort linking climate change reduction benefits to other city activities was set in place. One year after the study was finished, several recommendations concerning the energy aspects were acted upon. The implementation of traffic measures is complicated, because traffic is seen as a more political issue. Opportunities for fresh approaches were seized upon. The local *Stadtwerke* began to think about concepts such as least-cost planning which had not been considered before. However, by 1994 implementation was endangered by the town's deteriorating finances. The goal of reducing CO_2 emissions by 30 per cent by the year 2005 will probably not be achieved.

Generally, it can be concluded that some local authorities in Germany are in a complex process of learning and understanding that they can play key roles in the response to the issue of climate change. However, as municipal finances get worse through the amendment of federal legislation this will probably become the main obstacle for the implementation of climate protection measures in the near future. Several municipalities are thinking of selling parts of their local utilities to regional utilities for financial reasons (e.g. Bremen, Hannover, Düsseldorf, Frankfurt, München, Berlin). However, 'finance' is a double-edged argument: there are many cost-effective options which save money and reduce greenhouse gas emissions.

Climate Alliance of the Cities

In addition to individual municipal action, several European local governments and the indigenous people of the rain forests formed an alliance in 1989 called *Klima-Bündnis der Städte* (Climate Alliance of the Cities). From 180 member cities in August 1992 the number grew to 371 members in July 1993 including 124 German, 40 Austrian, 117 Dutch, 37 Italian, 6 Swiss and 2 French municipalities. The indigenous people are represented by COICA (Coordinatores des las Organisaziones Indigenes de la Cuenza Amazonica) and other similar organizations. The special aim of this alliance is mutual support to protect the earth's atmosphere. For that reason, in 1990 the European cities committed themselves to reduce their CO_2 emissions by 50 per cent by the year 2010, to stop using CFCs immediately and not to use tropical woods any longer (Klima-Bündnis der Städte, 1990). Another goal is to make the exchange of information and experience easier and to strengthen it. For that reason, the climate alliance is building up a computer-aided data base called *Klima-Bündnis-Infothek*, which concentrates on individual, different municipal concepts, new planning methods, and offers consultation and development programmes (Alber, 1993, 14).

Working in a similar direction but based on its own research programmes, the International Council of Local Environmental Initiatives (ICLEI), which was founded in 1990, is another, initially Canadian-based, international union of local governments. It is supporting municipal administrations to improve their ability to mitigate environmental problems, to take measures against existing environmental problems and to improve natural or built environmental quality on the local level (ICLEI, 1993a, 1993b). Internationally, it organizes the exchange of information and experiences of other municipal initiatives and mediates joint or exemplary projects. One main element of its international research programme was the 'Urban CO_2 Reduction Project' (Torrie, 1993), in which 14 municipalities of the United States, Canada, Europe and Turkey, including the German cities of Saarbrücken and Hannover, participated. The aim of the project was to quantify and compare urban energy use patterns and to analyse the specific local opportunities for reducing energy intensity and CO_2 emissions. In 1993, these activities resulted in the 'Convention of European Municipal Leaders on Climate Change' including the campaign 'European Cities for Climate Protection'.

The German National Climate Report of August 1993 and the Report in 1994 explicitly state that the reduction of CO_2 emissions by 25–30 per cent can only be achieved if all federal levels are involved and actively contribute. The Bundesländer and the municipalities are required to complement and support the federal government's approach. In 1992 the *Land* of Northrhine–Westphalia produced its own climate report, following similar procedures adopted by the federal government (Landesregierung Nordrhein–Westfalen, 1992). The report also stresses the larger potentials for reduction in the Neue Bundesländer. But in contrast to other statements, for the first time it is recognized that these potentials might not be easily tapped because of the financial difficulties in these Länder. Concerning local and regional energy supply concepts, attention is drawn especially to the engagement of municipalities in international projects (BMU, 1993, 135). The conjunction with or participation in similar OECD project groups is emphasized, showing that these concepts are not national single-handed efforts but internationally (OECD-wide) agreed future developments. The importance of these initiatives is underlined by reference to municipal workshops and handbooks. Several remarks in the German National Climate Report attempt to prove that the Federal Republic of Germany follows a 'comprehensive' approach: for example, the federal government shows its interest and support to connect the international and municipal level by referring to the fact that the results of an international conference on the promotion of municipal environmental protection in Berlin have been incorporated in Agenda 21. Nevertheless, there is no official estimate of the extent to which municipal initiatives contribute to the reduction of CO_2 emissions or emissions of greenhouse gases in general. However, those municipalities which are developing climate protection concepts are reviewing very critically the general setting by the federal level.

CONCLUSIONS

The German climate policy can be divided into two phases: the discussions and negotiations which led to the adoption of the ambitious German CO_2 reduction target in 1990, and the implementation measures to achieve this target.

1987–1990: the boom in climate change politics

Given the longer scientific interest in climate change, the question is: what were the driving forces and circumstances which pushed the issue onto the political agenda? Why was it possible to adopt the ambitious reduction target for CO_2 emissions of minus 25 per cent by the year 2005 based on 1987 levels in such a short time period only in Germany and not in many other countries? Reviewing the German experience, we find that the delay between scientific statements and political action was shortened by at least two factors: information dissemination by the media, and the work of the Enquete Commission.

In the starting phase the obstacles to political agenda setting were overcome by focusing and raising public attention by repeated media coverage. Since the early 1980s the steady, but low level coverage of environmental problems by the media changed into a cyclical coverage starting with acid rain and turning to climate change in the late 1980s. Conspicuously, the issues of ozone depletion and climate change were often mixed up. This is reflected by a simultaneous development in the number of press reports from 1986 on. Additionally at that time, public interest was raised by the presentation of scientific results in international (e.g. IPCC) and national (Enquete Commission) fora. This very important phase can be interpreted as an interactive self-reinforcing group process in which the activity of one element causes more activity in the others.

In the German case alarmed public opinion stimulated by press reports and inquiries by the parliamentary opposition probably played the most important role in shifting the climate change issue onto the political agenda. The establishment of the Enquete Commission in 1987 indicated the overall importance of the issue and kept it continuously in the minds of government and other decision makers in the following years. Given the Enquete Commission's role as a transmitter of scientific knowledge between scientific researchers and politicians, the success of the 'Climate Enquete' was a result of several factors that were neither inevitable nor predictable.

1 The commitment and impartiality of the people involved – especially the Commission's chairperson.
2 The consciousness of the Commission's members that priority should be given to analysing the issue and to announcing the most efficient and effective solutions overriding party political interests.

3 The relative ease of reaching a scientific consensus because there were no major scientific critics in Germany. Questions of implementing measures (political issues) were much more contentious.

4 The German economy was booming and German unification had not yet taken place so financial questions were less important.

5 Special national and international attention to and respect for the Commission's work and interest in the issue in general. International fora such as the IPCC strengthened the position and reputation of the Commission. That was one of the reasons why politicians trusted in the results and recommendations of the scientists. At the same time, the ambitious recommendations of the Commission as a parliamentary institution were used to increase the reputation of German environmental policy and to underline Germany's claim to be a leader in international climate policy. Critics of 'traditional' government politics who agreed with the Commission could no longer be ignored.

The special difficulty facing this analysis is that individual influencing factors such as willingness for consensus or for commitment to dialogue cannot easily be institutionalized. These individual factors require great care over the selection of both political and expert members. Nevertheless, analysing the work of the Enquete Commissions and their special circumstances helps to show which factors are necessary for effective policy making. It shows how important it is to intensify communication between the different actors.

During this generally fruitful phase of raising consciousness and discussion, a number of new initiatives spread out at the local level forming focal points for new municipal development. The recommendations and influence of the Climate Enquete Commission were very crucial in this instance because they motivated local actors and induced political response. At the local level the climate change issue was used by key actors to shift local environmental or other problems related to climate change in one way or another onto the agenda. It is clearly understood by some of them that the implementation of a sustainable development path is necessary but that it is unimportant precisely how to sell it – as climate change politics or employment politics, etc. – so long as the specific environmental issue remains near the core of political decision making. Despite this multitude of encouraging new concepts, this was still an initial phase: the number of these initiatives was small and highly dependent on actors in key positions. Remember too that there were actors who delayed or impeded action.

1990–1994: disillusionment or realism in climate change politics?

The subsequent institutional process of setting up an interministerial working group (IMA) was characterized by the problems of the environment ministry concerning responsibility and power to get programmes accepted by other ministries (BMWi, BMV, BML) and the difficulty in finding appropriate responses to the target. This process contained no new and innovative elements which focused on the characteristics of the climate change problem or which were designed to reduce the 'institutionalized' problems. The measures in the CO_2 reduction programme are designed in a very general way. There are no projections of how much any measure will contribute to the reduction of CO_2 emissions or even which sector will contribute to which share of the total reduction. There was no change of the programme between the first version and the version which is integrated in the National Climate Report. While this Report is the response of the German government to its obligations under the EU monitoring directive, the government is not fulfilling all of these obligations. There appears to be a contradiction between Germany's international claim to be a leader in climate change politics at the EU level and in the CoP process and the development of national climate policy. Between August 1993 and September 1994, the CO_2 reduction programme was inflated from 30 to 109 measures counting every government activity which could be somehow interpreted as a climate protection measure. At the same time, the focus of the work of the IMA shifted from CO_2 to other greenhouse gases. Consequently, statements are more vague and less ambitious.

Reviewing the policies of other ministries that concern climate-change-related issues such as transport policy or energy policy, we find that climate change has not become one of the bases on which policy makers and administrators found their decisions. Traditional thinking about the policy needs of the economy has not been overcome. This non-integration at present becomes all the more obvious by analysing the dependence of political activity in the climate change field on the overall economic situation. Under changed economic frames (recession, German unification) the costs of environmental protection in general but including climate-change-related measures are stressed by some industries and the ministries with traditionally good connections to them (BMV – automobile industry; BMWi – large industry associations, especially the coal and steel industries). It is unlikely that any new policy arena will be accepted by these ministries, if their clients do not accept it. Nevertheless, we observe some movement: younger entrepreneurs have formed progressive new alliances with environment NGOs. Though they do not seem to have much influence, it is a first step in a promising direction. There is also some discussion of the implementation of voluntary restrictions on CO_2 emissions by the government and the large German industry associations. At present these appear to be aimed more at hindering or delaying decision making.

The influence of financial and legislative shortcomings on institutional innovation is also reflected by the difficulties faced by progressive municipalities when seeking to implement their local climate initiatives. This points to the necessity for combined action at government level.

We conclude that from an institutional perspective it was possible in Germany to settle the need for climate change politics within other policy areas. But, it has not become a priority when compared with other 'traditional' policy areas such as transport or energy. There are no new institutional innovative moves or elements at the government level: the government's response to the global environmental problem of climate change was within the frames and bounds of institutional routine. We do observe some innovation not in the centre but at the periphery of power, such as coalition forming between the environment and the economy, and the combinations of energy, economy and environmental interests in local initiatives. They have to be strengthened and complemented by innovation in the governmental area if there is to be a new phase in climate change politics in Germany.

Turning to Figure 4.1, we conclude that institutional de-innovation was as important as institutional innovation in the German reaction to the UN FCCC. Because key policy was linked to industry, employment, energy and transport, so stable policy cores, anxious about their competitive position and highly resistant to additional taxation, no matter how fiscally neutral, served to block new initiatives. The science was not a stimulant, nor was any feeling of environmental empathy. Cosmetic compliance was compounded by a complacency that the economic structuring of the former eastern *Länder* would do the job.

Yet Germany postured on the European stage as the leader of the climate change response pack. Now that Germany has the CoP Secretariat, and with the European Environmental Agency promoting the EU monitoring effort, it will be much more difficult for Germany to hide behind the smokescreen of economic restructuring. The future could bring more innovation in modest tax reform, in local action across a broad group of local authority policy measures, and in the emergence of whole energy management structures, especially at the local level. All of these will involve new alignments of interests – possibly more proactively developed than elsewhere. The trigger may come from the CoP and Brussels.

REFERENCES

Alber, G. (1993) Forum Klimabündnis: Erfahrungsaustausch zum Thema lokale Klimapolitik. *Kommunale Briefe für Ökologie*, 18.

Beschluß der Bundesregierung vom 29. September 1994 zur Verminderung der CO_2-Emissionen und anderer Treibhausgase in der Bundesrepublik Deutschland auf der Grundlage des Dritten Berichts der Interministeriellen Arbeitsgruppe 'CO$_2$-Reduktion' (IMA CO$_2$-Reduktion) (1994).

Boehmer-Christiansen, S. and Skea, J. (1991) *Acid Politics: Environmental and energy policies in Britain and Germany*, London: Belhaven Press.

Bundeshaushaltsplan für das Haushaltsjahr (1994).

Bundesminister für Umwelt, Naturschutz und Reaktorsicherheit (ed.) *BMU Pressemitteilungen*
BMU Pressemitteilung 89/91.
BMU Pressemitteilung 71/92.
BMU Pressemitteilung 75/92.
BMU Pressemitteilung 23/93.
BMU Pressemitteilung 44/93.
BMU Pressemitteilung 44/94.
Bundesminister für Umwelt, Naturschutz und Reaktorsicherheit (ed.) (1990) *Umweltpolitik. Beschluß der Bundesregierung zur Reduzierung der CO_2-Emissionen in der Bundesrepublik Deutschland bis zum Jahr 2005*, Bonn.
—— (1993) *Environment Policy. Climate Protection in Germany. National Report of the Federal Republic of Germany in anticipation of Article 12 of the United Nations Framework Convention on Climate Change*, Bonn.
—— (1994) *Environment Policy. Climate Protection in Germany. First Report of the Government of the Federal Republic of Germany Pursuant to the United Nations Framework Convention on Climate Change*, Bonn, September.
Bundesverband der Deutschen Industrie e.V. (ed.) (1992) *BDI Jahresbericht 1990–1992*.
Bundesverband Junger Unternehmer (BJU), Bund für Umwelt und Naturschutz Deutschland (BUND) (ed.) (1993) *Gemeinsames Statement von BJU und BUND: Plädoyer für eine ökologisch orientierte soziale Marktwirtschaft*.
Cavender Bares, J. (1993) *Germany's Policy Towards Global Environmental Risk Management*, Cambridge, MA: Center for Science and International Affairs, Harvard University.
Commission of the European Communities (ed.) (1992) *Europeans and the Environment in 1992*, Eurobarometer 37.0.
Cutter Information Corporation (ed.) (1993) Phaseout Deadlines in OECD Countries for Ozone-Depleting Compounds, *Environment Watch: Western Europe*, 3, Special.
Deutscher Bundesrat (ed.) *Bundesrats Drucksachen*
Br.-Drs. 18/91.
Br.-Drs. 103/92.
Br.-Drs. 485/92.
Deutscher Bundestag (ed.) *Bundestags Drucksachen*
BT.-Drs. 10/243.
BT.-Drs. 11/533.
BT.-Drs. 11/678.
BT.-Drs. 11/787.
BT.-Drs. 11/4133.
BT.-Drs. 11/8166.
BT.-Drs. 12/1136.
BT.-Drs. 12/2081.
Deutscher Industrie- und Handelstag (DIHT) (ed.) (1993) *Belastbare Energiepolitik für den Standort Deutschland. Positionen zu den Schwerpunkten einer Energiekonzeption für die Bundesrepublik Deutschland*, Bonn.
Deutsches Institut für Wirtschaftsforschung (ed.) (1992) *Verkehr in Zahlen*.
—— (1993) Umweltschutz und Standortqualität in der Bundesrepublik Deutschland, *DIW Wochenbericht 16/93*, Berlin, 199–206.
—— (1994) Energiepolitik und Klimaschutz in Deutschland, *DIW Wochenbericht 9/94*, Berlin.
Dyson, K. (1982) West Germany: the search for a rationalist consensus, pp.120–141, in J. Richardson (ed.) *Policy styles in Western Europe*, London: Georg Allen and Unwin.

Etzbach, M. (1993) Interview on 21 May 1993 by Jeannine Cavender Bares.

Ganseforth, M. (1993) Personal Communication, 20 October 1993.

German Bundestag (ed.) (1991) *Protecting the Earth. Third Report of the Enquete Commission of the 11th German Bundestag 'Preventive Measures to Protect the Earth's Atmosphere'*, Bonn.

Greenpeace (ed.) (1994) *Ökosteuer – Sackgasse oder Königsweg?*, Berlin.

Hansmann, K. (1992) *Bundes-Immissionsschutzgesetz und ergänzende Vorschriften*, Nomos Verlagsgesellschaft, Baden-Baden.

ICLEI (ed.) (1993a) *Cities for Climate Protection. An International Campaign to Reduce Urban Emissions of Greenhouse Gases.*

—— (1993b) *Klima schützen heißt Städte schützen. Briefing Buch für kommunale Verantwortungsträger*, Freiburg.

Institut für Ökologische Wirtschaftsforschung (ed.) (1992a) Umweltfragen in Europa unter wachsenden Verteilungskonflikten. *IÖW Informationsdienst*, 5, 1–4.

—— (1992b) Die Notwendigkeit neuer Leitbilder – Plädoyer für eine ökologisch orientierte Wirtschaftspolitik in den neuen Bundesländern, *IÖW Informationsdienst*, 6, 1–4.

Institut für praxisorientierte Sozialforschung (IPOS) (ed.) (1992) *Einstellungen zu Fragen des Umweltschutzes.*

Ipsen, J. and Tettinger, P.J. (ed.) (1992) *Zukunftsperspektiven der kommunalen Energieversorgung*, Osnabrücker Rechtswissenschaftliche Abhandlungen, 35, Köln, Berlin, Bonn, München: Heymanns.

Jarras, L. and Obermair, G. (1994) *More jobs, less pollution. A tax policy for an improved use of production factors*, Wiesbaden.

Katzenstein, P. (1987) *Policy and Politics in West Germany: The Growth of a Semisovereign State*, Philadelphia: Temple University Press.

Klima-Bündnis der Städte (ed.) (1990) *Klima-Bündnis zum Erhalt der Erdatmosphäre. Manifest europäischer Städte zum Bündnis mit den Indianervölkern Amazoniens*, Frankfurt.

Kohler, S. (1984) Geschichte der deutschen Elektrizitätswirtschaft und ihre Auswirkungen auf die kommunale und regionale Energieversorgung, in Ökoinstitut e.V. (ed.) *Kommunale und Regionale Energieversorgungskonzepte*, Darmstadt.

Köpke, R. (1992) Rationelle Energieverwendung im kommunalen Bereich. Ansätze für ein Umdenken in der Energiepolitik am Beispiel ausgewählter Städte und Gemeinden in den Bundesländern Bayern und Nordrhein–Westfalen, in Klemmer, P. (ed.) *Beiträge zur Struktur- und Konjunkturforschung*, XXXIII, Bochum: Universitätsverlag Brockmeyer.

Landesregierung Nordrhein–Westfalen (ed.) (1992) *Klimabericht Nordrhein–Westfalen. Der Beitrag des Landes Nordrhein–Westfalen zum Schutz der Erdatmosphäre*, Düsseldorf.

Loske, R. and Hennike, P. (1993) Klimaschutz und Kohlepolitik, *Wuppertal Paper Nr. 5.*

Müller, E. (1990) Umweltreperatur oder Umweltvorsorge? Bewältigung von Querschnittsaufgaben der Verwaltung am Beispiel des Umweltschutzes, *Zeitschrift für Beamtenrecht*, 6.

—— (1993a) Interview on 18 May 1993 by Miranda Schreurs and Jeannine Cavender Bares.

—— (1993b) Personal Communication, 24 November 1993.

—— (1994) *Kommt eine Energiesteuer?* Contribution to the High Level Meeting 'Schwerpunkt der Umweltpolitik während der deutschen Präsidentschaft' am 28. Juni 1994 in Hamburg, unpublished.

Prognos AG (ed.) (1992) *Energiereport 2010. Die energiewirtschaftliche Entwicklung in Deutschland*, Stuttgart.

Sauter, W. (1993) BUND–Nordrhein–Westfalen: Stellungnahme zu Kompensation/ CO_2-Abgabe, 27 May 1993, unpublished.

Schaefer, H., Geiger, B., Jochem, E. and Ott, V. (1990) *Emissionsminderung durch rationelle Energienutzung. Bericht für die Enquete-Kommission 'Vorsorge zum Schutz der Erdatmosphäre' des Deutschen Bundestages.* München, Karlsruhe, 15. März 1990.

Schallaböck, K. O. (1993a) Verkehr, Energie und Klima – Argumente zum umsteuern, in *Energie-Dialog*, Dezember 1993, 20–23.

—— (1993b) Verkehrsentwicklung, unpublished.

Schmidt, M., Wortmann, J. and Six, R. (1992) Handlungsorientiertes kommunales Konzept zur Reduktion von klimarelevanten Spurengasen für die Stadt Heidelberg. Endbericht, in Stadt Heidelberg – Amt für Umweltschutz und Gesundheitsförderung (ed.) *Klimaschutz Heidelberg*, Heidelberg.

Schmidt-Jortzig, E. (1992) Die Neuordnung der Energieversorgung in den neuen Bundesländern – Bestandsaufnahme, Tendenzen, Probleme –, in Ipsen, J. und Tettinger, P. J. (eds) *Zukunftsperspektiven der kommunalen Energieversorgung*, Osnabrücker Rechtswissenschaftliche Abhandlungen Band 35. Heymanns, Köln, Berlin, Bonn, München.

Schön, M. M. (1993) Personal Communication, 11 October 1993.

Statement by the German Delegation, 3rd Session of the UNCED PrepCom, Geneva, 12 August–4 September 1991, WG 1, Agenda item 2.

Torrie, R. (1993) *Findings and implications from the Urban CO_2 Reduction Project*, Toronto: ICLEI.

Umweltbundesamt (ed.) (1992) *Jahresbericht 1991*, Berlin.

—— (1993a) *Jahresbericht 1992*, Berlin.

—— (1993b) *Umweltschutz und Industriestandort. – Der Einfluß umweltbezogener Standortfaktoren auf Investitionsentscheidungen* – Berichte 1/93, Erich Schmidt Verlag, Berlin.

Verband der Industriellen Energie- und Kraftwirtschaft e.V. (VIK) (ed.) (1993) *VIK-Tätigkeitsbericht 1991/92*.

Vereinigung Deutscher Elektrizitätswerke (VDEW) (ed.) (1993) Kompensationsmaßnahmen als Beitrag der Elektrizitätswirtschaft zur CO_2-Reduktion, in *Argumente*, 1. Juni 1993.

Vorholz, F. (1993) Der Weg nach Ökotopia, in *Die Zeit*, 34, 3.

Voss, G. (1990) *Die veröffentlichte Umweltpolitik. Ein sozio-ökologisches Lehrstück*, Kölner Universitätsverlag, Köln.

Weale, A., O'Riordan, T. and Dramme, L. (1992) *Pollution Control in the Round*, London: Anglo-German Foundation.

Weidner, H. (1989) Die Umweltpolitik der konservativ-liberalen Regierung. Eine vorläufige Bilanz, in *Aus Politik und Zeitgeschichte*, 47–48.

Weizsäcker, E. U. v. and Jesinghaus, J. (1992) *Ecological Tax Reform*, London: Zed Books.

Wicke, L. (1993) *Umweltökonomie*, München: Verlag Vahlen.

Würzner, E. (1993) Personal Communication, 27 October 1993.

8

STRUGGLING FOR CREDIBILITY
The United Kingdom's response
Tim O'Riordan and Elizabeth J. Rowbotham

The UK has almost a unique reputation of being a hard negotiator but a dutiful implementor of international agreements. The Climate Convention makes modest demands on a major economy, at least at the outset, so we shall see that the UK has embarked on institutional reforms that are purposeful but low key. The most important implication of the British experience is the interconnection between climate policy and other policy arenas, notably tax, industrial and transport policy. Institutional adaptation in the UK is therefore only partly climate driven. Its strength lies in the mutuality of purpose between macroeconomic strategy and climate response, plus the emergence of cost-effectiveness appraisals of policy mixes to achieve a given objective but involving a number of government departments and policy initiatives. The pace is incremental, almost imperceptible, but change both in outlook and in administrative structures has taken place and will continue. The focus will shift from climate to sustainability in its broadest sense, and for the UK to Europe more generally. But the character and political significance of the UK response can be measured, evaluated and predicted. Hence the structure and content of this chapter.

THE INFLUENCE OF POLITICAL CULTURE

To assess institutional response in the UK, one has first to understand how the political culture of the nation operates. By the term 'political culture' we refer to the institutional structures and perspectives as identified in Chapter 4, most particularly those which apply to policy formulation and policy change.

The British policy style has been examined by Jordan and Richardson (1982), Marsh (1983) and Greenaway *et al.* (1992) amongst many others. Jordan and Richardson characterize the British approach on the broad themes of consensus building and imposing will by consent. This implies a client-orientated, bargaining approach that relies on negotiation amongst elite communities of contending parties. This seems to take place in two ways. There is a relatively stable core of influential groups and organizations, with coherent structures and strong positional influence, that tends to dominate day to day politics. But on the periphery there are more open structures where there is scope for negotiation and bargaining in a more informal and exploratory manner. The first tends towards formalization and regularity; the second to innovation and fragmentation. We argue here that the UK government's response to climate change displays elements of both, though the core component remains dominant.

There are various models of British politics. Greenaway and his colleagues (1992, 46–68) summarize many, but come to no specific conclusions. Yet their review is helpful as a basis of interpretation of the dynamics of institutional adaptation. In simplified form, four models are regarded as representative:

1 *The democratic sovereignty model* centred on Parliament and public debate over policy options and their underlying ideological rationale. Issues of taxation and the privatization of major utilities are handled in this realm, mostly for symbolic and rhetorical purposes.
2 *The party government model* circulating around party politics with its dominant supporting and opposing interests. The ideology of the market place, consumer choice in, say, transport, and the contested role of state intervention in the privatized utilities are relevant here.
3 *The adversarial politics model* based on interparty squabbling over basic principles and rhetoric. This is less common in recent years with a centre–right shift in the three main parties from previously more fragmented ideological positions. Thus, for example, the ruling Conservatives and the opposing Labour parties have little to say, but essentially little to differ, over sustainable development at the national and local levels. Many key areas of 'core' conventional politics do not experience severe interparty wrangling. But when environmental themes open into the much more contested terrains of social and fiscal policy, then the squabbling really has an effect. This is possibly how matters may evolve.
4 *The cabinet government model* relies on a powerful role for policy communities circulating around cabinet ministers and advisers. Here interdepartmental tensions, prime-ministerial preferment, and political calculus play a more important role than just the merits of the positions taken. We shall see that the relatively weak position of the Environment Secretary contrasts with the growing political influence of the Secretary of State for Transport in shaping long term climate-related strategies.

None of these models provides a satisfactory explanation of policy processes in the contemporary UK.

In the study that follows we argue for a core of stability in both style and policy adjustment, and a periphery of altered perspectives and changing interorganizational alignments and relationships. Each is necessary for the other: each plays a vital role in mediating institutional adaptation in a world where environmental concern is becoming enmeshed in international obligations, economic and social policy arenas and fresh ethical positions. It is this intermixing of policy and ideological positions that creates the conditions for institutional change.

For our purposes, we draw on a different set of qualities regarding the British political culture from Jordan and Richardson though the overlaps are clear. They argued for sectorization around ministries; clientelism around ministers; consultation to gain legitimacy, especially when new policy arenas have to be established; institutionalization of compromise, especially when hard-won agreements are forged; and the creation of common views around new perspectives. All of this can be handled under the broad heading of bureaucratic accommodation.

Our analysis leads us to look more closely at the role of policy communities, also advanced by Jordan and Richardson (1982) and especially by Marsh and Rhodes (1992). This perspective is covered in detail in Chapter 4. Though nominally and constitutionally Parliament is supreme, in practice the UK Parliament as a collective debating chamber has become seriously devalued. Whipping of party allegiance means that renegade opinion is usually strictly controlled, especially in circumstances of small party majorities. If the dominant party majority is large, or where preferment to ministerial office is in party patronage, or when a government policy is under fire so that defeat would be a matter of confidence, a compliant Parliament is more likely. Consequently policy influence circulates around 'the gatekeepers', namely those who advise ministers, together with senior civil servants and client groups whose support is sought to legitimize or at least clarify policy acceptability or opposition. The roles of organizational coherence, credence of argument and positional influence all play a part here. The success of any institutional change can be measured to some extent by the degree to which the gatekeepers are forced to, or willingly, allow entry to new groups, procedures or perspectives. Obviously all this depends on how many ministerial departments are involved. In the case of response to climate change, at least four are active, namely Treasury, Trade and Industry, Transport, and Environment. Institutional response will reflect repositioning of these policy communities around these four ministries and the coordinating machinery of the Cabinet Office. We shall see how fresh approaches to cost-effectiveness analysis of environmental policy options are assisting in this innovation.

In the period studied here, Conservative administrations have reigned. These are characterized by a belief in the market, minimal state intervention except where socially or structurally justified, privatization of the economy, deregulation and freedom of consumer choice based on information, guidance and limited regulatory interference. According to Rhodes (1994, 138) the 'hollowing' of governmental responsibilities is a function of modern time, and not just of Conservative political ideology. But it has meant that decision making is becoming more dispersed, less politically accountable, and rather more incoherent. This is not a friendly administrative culture for climate change politics.

This is so because a diffuse policy arena such as climate change is so disconnected that it allows for a chaotic mass of policy initiatives both to advance it and to impede it. As a result the policy communities are not directed at all to climate change. They swirl in and out of the policy 'nebula' sometimes connecting to climate change and sometimes not. This suggests that a climate change policy is a chance process where the pattern of implementation becomes decentralized, disjointed and opportunistic. These observations are developed further in Chapter 12.

Policy unity has never been possible, so it should not be a surprise that integration of disparate but essentially interconnected policy arenas is very uncommon in the British political culture. There are various reasons for this:

- There is a clear pecking order of departmental influence. Ministers and their advisers like to maintain this by reinforcing hierarchy at every opportunity.
- The policy arena becomes a power struggle between competing departments each seeking favour and ministerial preferment. Environmental matters have always been low in the pecking order, so it is always more of an effort to gain political legitimacy.
- The departmental policy communities enjoy relatively stable positional relationships so do not take easily to cuckoos in their midst – unless realignment becomes desirable or necessary.
- The British civil servant is renowned for being a generalist, capable of constructing a clear brief but dependent on specialist advice. That advice tends to be narrowly framed as the science advisers are also departmentally confined.

These are broad generalizations. Some important distinctions could now be made to sharpen this picture in the context of the study that follows:

- Parliamentary surveillance is becoming more pointed, well researched, well documented and successfully critical of policy assumptions and policy biases. The committee reports are revealing, but the influence remains modest. So long as the backbench MP is beleaguered by pressure of party loyalty and constituency demands, it is unlikely that backbench muscle flexing will have much effect until significant policy community realignments take place.

- Interdepartmental groupings, often of an *ad hoc* nature are becoming very common. These groupings are sometimes highly specific and short-lived, but many are designed to be coordinative. Their influence is shadowy, and they still are affected by hierarchies of traditional bias, but they do act as focal points for policy reassessment and coordination.

- New administrative and regulatory agencies are forming. Some of these fit the corporate mould, outlined in the introductory chapter, but a number have the capability of having specifically transformational elements in policy direction – though always within the framework of the domestic governing ideology and departmental pecking orders. For the purposes of this study, the changing fortunes of the Energy Efficiency Office, the Energy Saving Trust and the consumer protection offices of tariff regulation to the electricity and gas industries deserve special attention.

- International pressures created by non-binding international agreements are beginning to create areas of political and moral commitment which limit the room for future manoeuvre by national governments, despite domestic political priorities to the contrary. This point is the subject of Chapters 2 and 6 so will not be elaborated here. It is only pertinent to point out that such agreements create a tension between legal require-ment, moral responsibility and counteracting political expediency. This is the turmoil that will be examined in the context of the theory that drives this whole project.

THE UK CONTRIBUTION TO GLOBAL WARMING

The UK contributes approximately 3 per cent of global radiation forcing gases. Of these, CO_2 is responsible for 87.3 per cent of the total, methane 8 per cent, nitrous oxides 4.4 per cent and other gases 2.3 per cent (HM Government, 1994b, 13). Within the dominant CO_2 sector, coal contributes over a third of the total, petroleum almost the same amount, and natural gas around a fifth. In terms of emission sources, power stations contribute a third, industry about a quarter and road transport just over a fifth. Via end users, industry and the domestic sector produce a little over a half of the CO_2 output, and transport over a fifth. These data are summarized in Tables 8.1 and 8.2. For methane, landfill waste and agriculture account for a third each and coal mining 16 per cent. By far the largest contribution of N_2O is the nylon industry (nearly two-thirds) with agriculture adding a further 17 per cent. Most of the remaining gases are industry sourced.

These data are subject to the various caveats that Susan Subak analyses in Chapter 3. The key point from her review is that the British scientific establishment is seeking to lead the way in the agreed measurement of greenhouse warming potential, as well as in the improved monitoring and evaluation of emissions. This effort, closely connected with the emergence

Table 8.1 United Kingdom carbon dioxide emissions projections (1990–2005) (MtC)

	1990	*1995*	*2000*	*2005*
Government				
Households	41	39	41	42
Industry/agriculture	56	56	58	61
Commercial/public	24	23	26	30
Transport	38	41	45	49
Total	160	159	170	183
Barker et al.				
Total				
	160	152	157.9	166.7
1995 DTI Review	158		144–152	154–165

Sources: DTI (1992, 1995a) and Barker *et al.* (1993). Recently the time series data on carbon dioxide emissions have been revised downwards to take account of new measurements regarding the carbon content of North Sea gas. This has had the effect of reducing 1990 emissions estimates from the 160 MtC used as the basis for the DTI's EP59 document to 158 MtC. The Barker team assume greater energy efficiencies in future patterns of UK economic growth, together with specific energy pricing measures; hence the discrepancy. The 1995 assessments take into account a host of factors analysed further in the text, and presented in Tables 8.3 and 8.4.

of the European Environment Agency charged with publishing quality data of comparable reliability throughout the European Union, is a good example of the significance of an international treaty in focusing attention on the political role of scientific measurement in guiding the compliance component of an international convention. Wynne (1993) provides a more detailed analysis of this theme.

Table 8.2 CO_2 emissions by source and by final energy consumer

CO_2 emissions, 1990

	By source		By final energy consumer	
	MtC	*%*	*MtC*	*%*
Power stations	54	34	–	–
Refineries	5	3	–	–
Households	22	14	42	26
Industry and agriculture	35	23	50	32
Commercial and public sector	8	5	28	18
Road transport	30	19	33	21
Other transport	4	2	5	3
Total	159		158	

Source: Department of the Environment (1993b, 14)

GLOBAL WARMING AND BRITISH PARTY POLITICS

Climate politics do not rank high on the agendas of any of the three British political parties, namely the ruling Conservatives, the opposition Labour, and the minority Liberal Democrats. For the Conservatives, the strict commitment of the UN Convention, namely return to 1990 levels by 2000 of the three key 'non-Montreal' greenhouse gases (CO_2, CH_4, N_2O), has been regarded as manageable with little in the way of climate-specific policy innovation. Official estimates suggest that CO_2 levels will initially fall due to economic recession and industrial restructuring (Department of Trade and Industry, 1995a, 93–99). There is still much disagreement as to the scale of this 'background' reduction. The most recent calculations suggest that structural changes in the economy, the virtual removal of the remaining coal industry due to privatization, the switch to combined cycle gas and gasification plants, and fiscal policies to raise the prices of gasoline, plus modest energy efficiency investments, could all lead to a reduction of greenhouse gas emissions by well over the 1990–2000 stabilization target, even to 2005. Table 8.3 summarizes the overall findings. This matter is both central to the analysis that follows and highly contentious. It will be assessed later in the chapter.

The Conservatives see the climate issue as more a matter for fiscal policy, choice in energy markets favouring gas and renewables, industrial management and long term grappling with the car economy. Labour and the Liberal Democrats, on the other hand, favour a carbon tax as a basis for revenue redirection. These points are made in policy documents published in 1994 (Labour Party, 1994; Liberal Democratic Party, 1994). Both of these parties are beginning to tinker with the very core of the tax gathering mechanisms.

Table 8.3 Latest UK projections on CO_2 equivalent greenhouse gas emissions reductions

	Emissions (Mt)			Emissions (Mt CO_2 equivalent)		
	1990	2000*	2010	1990	2000	2010
CO_2	158	150	162	158	150	162
N_2O	0.11	0.03	Stable	8.0	2.2	Stable
CH_4	5	4.4	Falling	14.6	12.9	Falling
Total				180.6	165.1	?

* Without the special climate change measures, the 2000 CO_2 emission would be 158 MtC.
Note: These projections are based on central estimates of models derived from high, medium or low energy prices, and different rates and structural characteristics of economic growth. A 'worst case' analysis of CO_2 emissions for low economic growth and high energy prices (which could be stimulated by a levy, for example) would reduce emissions by an additional 8 Mt by 2000 alone. These conclusions are speculative as the text subsequently discusses.
Sources: Department of Trade and Industry (1995a, 98–99); ENDS, No. 242 (1995, 24)

They see mileage in shifting more to consumer or expenditure taxes rather than income tax and value-added tax with the revenues from ecological taxation being channelled into employment generation, skill training, labour relocation and investment in technology and management innovation. This is potentially the tip of a fiscal iceberg. Both parties regard taxation reform as a major area for policy innovation and social acceptance. As the public sector expenditure burden grows, raising revenue yet stimulating the economy at the same time become of paramount importance.

The Liberal Democrats claim to go further. They are prepared to embrace the prospect of specific taxation on environmental bads, via an extension of the polluter pays principle. They see a move away from the highly regressive and consumer-directed VAT in favour of using the tax regime to change behaviour and to alter attitudes. Thus carbon, chlorine and nitrogen taxation in various forms would, they believe, send a precautionary signal to the consumer and the citizen in favour of a different approach to the emission of greenhouse gases and chemical pollutants. So far the Conservatives, already beleaguered by a wide range of new taxes imposed to correct a huge public sector borrowing requirement, have not been willing to join in. Should they eventually do so, climate politics will diffuse into the tax, employment and industrial sectors, possibly with health and education not too far behind.

Because CO_2 emissions are primarily the result of fossil fuel consumption which is presently essential to economic activity, reduction of CO_2 emissions raises profound political and economic issues regarding the patterns of energy supply, structure and consumption. However, owing to the pervasive use of energy in society, attempts to address these issues raise further questions concerning the price and the end use of energy. As we have noted, such issues involve industrial competitiveness, transport, building regulations, access to information, and participation in decision making. Consequently, it can be seen that the measures needed to combat global warming will need to be diverse, sequential and comprehensive and will likely have effects upon the current structure of UK society that are only indirectly related to the narrower issue of climate change.

THE BRITISH RESPONSE TO THE UN CONVENTION

As required by the UN Framework Convention and as agreed as a European Union objective, the British government published its strategy to reduce greenhouse gas emissions to 1990 levels by 2000 (HMG, 1994b). This document appeared on 25 January 1994 along with three other responses to the Rio requirements, the whole package covering sustainable development (HMG, 1994a), climate change (HMG, 1994b), biodiversity (HMG, 1994c) and forestry (HMG, 1994d). All of these reports passed through a prolonged consultative phase with a wide range of invited interests. This in itself was something of an innovation. Prior to the publication of the climate change 'blue book' informal discussions were held with key interests covering industry, trade unions, environmental groups, the media and science. These discussions outlined the broad parameters of the response in a multipolicy domain that arguably might not have emerged otherwise. This informal phase was followed by a CO_2 consultation document (Department of the Environment, 1992a). This was analysed in two formal conferences where the Department of the Environment made use of consultant facilitators to gain a cross-section of opinion in a manner that was ostensibly designed to be neutral and 'green'. Ostensibly this was an example of co-alition building. In practice various key assumptions about future emissions were already being prepared that limited this process to one of awareness raising and preparing for future coalitions of interest should the need arise.

Inevitably such a procedure was a mixed success. Intermediaries do not always pick up the nuances of opinion, rooted in particular biases. And faithful reporting is always likely to be deflected by the articulation of particular argument. Nevertheless, the fact that three phases of independent consultation took place, each carefully monitored, is in itself a matter of note. This is especially the case given the government's willingness to establish an annual review, again in conference format. This review is in line with the participatory frame arising from all the UN Conference on Environment and Development proceedings, namely to include 'stakeholders' in policy formulation and review. How successful this process has been will be evaluated at the end of the chapter.

The government accepts that the application of the precautionary principle applies to the justification of the action plan, and in so doing also expects all other major economies to pull their weight. But any future UK commitment beyond 2000 will explicitly require collaborative action throughout the developed world. It seems as though the British government is looking to this commitment as a basis for determining future climate policy, rather than a homegrown version of the precautionary principle. Boxes 8.1, 8.2 and 8.3 summarize the UK response. To assess how difficult that process may be, we need to look at the changing structure and accountability of the energy supply and demand control sectors in the UK. This is a picture that has been enormously coloured by the politics of privatization and utility regulation over the past ten years.

ENERGY POLICY IN THE UK

The primary focus of energy policy in the past decade (and before) has been on the provision of secure supplies of energy at the lowest possible cost. This strategy has incorporated two key, and often competing, objectives: (1) the realization of energy self-sufficiency within the UK; and (2) the introduction of the free market and the price mechanism into energy supply and conservation. The pursuit of a diverse and self-sufficient energy supply

BOX 8.1 THE UK CLIMATE STRATEGY: 1994 VERSION

The UK climate change response strategy is based on the objective of reducing possible 2000 emissions of CO_2, CH_4 and N_2O to 1990 levels over the remainder of the decade. The government operates on a gas by gas approach, run through domestic policies with no offsets or joint implementation schemes.

(87.3%) *Carbon dioxide* Likely future emissions are to be cut by 10 MtC to a total of 160 MtC:

- 4 MtC off domestic energy consumption via an extension of VAT to domestic fuel improvements to the Home Energy Efficiency Scheme and grant aid for energy use efficiency via the Energy Saving Trust. With the failure to increase VAT to more than 8 per cent, this figure has been revised downwards to 2.5 MtC.
- 2.5 MtC off business energy use via grant aid for energy efficiency through the Energy Saving Trust and advice and informative schemes.
- 2.5 MtC off transport via a 5 per cent annual 'escalator' on the excise duty of petrol (gasoline) for the foreseeable future. The additional rise of 3 per cent on fuel in 1995 will not change the final figure, which is regarded as an overestimate.
- 1 MtC off public sector energy consumption through the imposition of targets for central and local government and public sector bodies.

(8.1%) *Methane* An estimated 10 per cent reduction by 2000 via:

- a reduction in landfill via a landfill levy;
- encouragement of the recovery of landfill methane for energy;
- regulatory incentives for waste minimization via integrated pollution control;
- tougher guidelines on the limitation of methane emissions from coal production.

(4.4%) *Nitrous oxide* An estimated 75 per cent reduction by technological innovation in the nylon manufacturing industry.

(2.2%) *Other trace gases*

- Carbon monoxide – a reduction of 50 per cent via the phase introduction of three-way catalytic converters.
- Nitrogen oxides – a reduction of 35 per cent via NO_x removal in large power stations to meet the requirements of the EC.
- Volatile organic compounds – a reduction of 20 per cent via controls on ozone depleting substances, catalytic converters, carbon canisters on cars and further limitations on the chemical industry. All of this is primarily aimed at reducing photochemical smog rather than greenhouse gas (GHG) emissions as such, but obviously there are joint benefits.

237

Box 8.1 (continued)

- Perfluorocarbons are mostly associated with the aluminium smelting industry and technological substitution backed by regulation via integrated pollution control should reduce emissions by more than 90 per cent by 2000.

Note: All of these strategies are either connected to other policy arenas such as fiscal or industrial pollution control, or covered by other international environmental agreements such as EC directives on air pollution or the Montreal Protocol on the reduction of ozone depleting substances.

BOX 8.2 ELECTRICITY SUPPLY INDUSTRY (ESI)

Prior to 1990 the ESI was a nationalized industry and the Central Electricity Generating Board (CEGB) had an integrated monopoly on both generation and distribution. In 1990 the ESI was privatized by *The Electricity Act, 1989* and the greater part of generating capacity of the CEGB was divided between three companies: National Power, PowerGen and Nuclear Electric. National Power and PowerGen became privatized companies (although the government still retains a large ownership share) while Nuclear Electric remained a nationalized company, at least to 1996, when the most profitable reactors will be sold to the private sector (Department of Trade and Industry, 1995b).

Distribution of electricity to consumers is handled by the regional electricity companies (RECs). The RECs currently enjoy a monopoly of supply to customers using less than 1 MW of electricity. This monopoly was reduced to 100 kW in 1994 and will be removed altogether in 1998. Customers which fall outside the monopoly are free to choose their own electricity supplier and the RECs must make their networks available for use by third party suppliers (Department of Trade and Industry, 1995b, 30).

Transmission of electricity is the responsibility of the National Grid Company which is ultimately owned by the RECs. In addition to transmitting electricity the National Grid Company is also responsible for operating a wholesale electricity market (the 'pool') which is designed to match supply and demand for electricity at any given time (Department of Trade and Industry, 1995b, 30).

BOX 8.3 COMBINED HEAT AND POWER (CHP) PLANTS

CHPs are the most efficient and lowest emitters of greenhouse gases of the gas-fired electricity generating plants that can be built. In addition to producing electricity, they also utilize the heat that is produced. As a result, these systems lose only between 20 and 40 per cent of the total energy content of the fuel compared with the 60 to 80 per cent lost as waste heat by traditional generation methods (Greenpeace, 1993, 69). Furthermore, gas CHP could save more SO_2 and twice as much CO_2 per extra unit of gas burnt as a coal-fired station at little or no extra capital cost.

At the present time there are approximately 2000 MW of CHP plants operating in the UK at approximately 600 sites. This represents approximately 2–3 per cent of total UK electricity supply (HMG, 1990).

One of the largest CHP power stations in the world is located at Teesside in the UK opened in April 1993. The station, which is fully computerized, employs 65 persons compared with the 650 needed to operate a coal-fired plant of similar size. It emits one-half of the CO_2 of a coal plant, its sulphur emissions are negligible and it produces one-ninth of the nitrous oxide waste of a coal plant (*Financial Times*, 22 June 1993).

---

policy stems largely from the oil crisis of the early 1970s and the negative macro- and microeconomic effects it had on the UK and other states in Western Europe (Caminus, 1993, 38.). The notion of competition and market choice, on the other hand, is a fundamental tenet of the Conservative philosophy. The major changes to UK energy policy in the period 1985–1995 lay in the privatization of the oil, gas, coal and electricity sectors, the savage reduction of the coal sector to less than a third of the 1980 operating capacity and the ring-fencing of the economics of nuclear power (McGowan, 1993). This was more a matter of ideology and macroeconomic strategy than energy politics. It is no wonder that the Department of Energy was transferred to the Department of Trade and Industry in 1992. The control over possible abuse of monopoly power has been put in place via independent regulatory agencies, notably the Office of Gas Supply (Ofgas) and the Office of Electricity Regulation (Offer). Also various acts designed to encourage greater interfuel competition and ownership were passed during the latter part of this period. A study by the Science Policy Research Unit (1993, iv) concluded:

> The government has failed to articulate a transparent energy-environment policy based on a 'set of harmonious principles' to which it aspired in its White Paper (HMG 1990). It has fallen back on a 'clutter of expedients' which it sought to avoid.

The economic importance of energy efficiency has long been recognized in the UK though not reflected in policy. In its first report on the Energy Efficiency Office, the House of Commons Energy Committee outlined the importance of the efficient use of energy to the UK. It stated, *inter alia*, that energy efficiency was important to energy self-sufficiency; that reduction of energy waste in the domestic sector through increased efficiency may be more cost-effective than through social security or transfer payments; and that industrial performance could be enhanced more cost-effectively through the promotion of energy efficiency than through other direct aid to industry (House of Commons Energy Committee, 1985, vii.). The Committee further concluded that the price of energy played an important role in promoting energy efficiency but cautioned against using energy prices as a form of revenue generation to reduce public sector borrowing (House of Commons Energy Committee, 1985, vii). Despite these statements supporting increased and improved energy efficiency, little has actually been accomplished. The most recent edition of UK Energy Statistics (Department of Trade and Industry, 1993b) reveals that energy use per unit of economic output increased by over 3 per cent since 1989, rising every year. This has been put down to declining real prices, a rise in vehicle transport, increased use of gas, and greater electricity consumption. In recession the UK is, in relative terms, still wasting its energy.

Figure 8.1 shows that the real price of energy from all fuels has declined since the early 1980s by approximately 25 per cent (Environmental Data Services Ltd, ENDS, No. 232, 1994, 22). The sharpest fall is in electricity because of factory coal prices following the failure of the miners' strike and the opening of the international coal market to the utilities. These data reveal a considerable consumers' surplus in energy, and hence the attraction of taxes on fuel. The cornerstone of the UK strategy is as much a mopping up of spare consumer energy cash as it is a specific fuel tax policy.

On top of this is a policy of downward pressure on fuel prices caused by intervention of the electricity and gas regulators in their fuel markets, and the promised removal of the 10 per cent surcharge on electricity prices placed in 1989 to finance the decommissioning and fuel disposal costs of the nuclear industry. Further promised deregulation of the gas industry could cut gas prices by an additional 8–13 per cent by 1996, and electricity prices could go down by 5–7 per cent because of moves by Offer to reduce the transmission cost mark-up, and by another 10 per cent when the nuclear levy is removed.

The policy of reducing further fuel prices is ostensibly at odds with measures to increase energy use efficiency, except that conceivably cash could come from tax income otherwise levied on energy to direct investment in energy conservation. We shall see that government policy makes this possible.

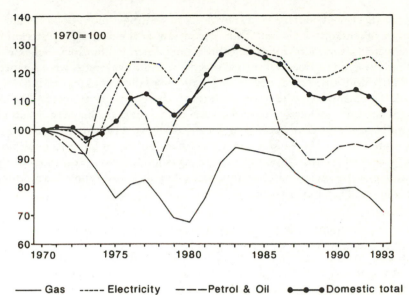

Figure 8.1 Changes in real fuel prices in the UK 1970–1993. Although there is much frustration, there has been a fairly steady downward trend since 1983. The policy of both the fuel regulators, and the government generally, is to continue to force real prices down. In principle this could leave room for some form of energy carbon tax gathering regime to mop up the consumer surplus. Such a move would be very difficult to make in the current 'anti-tax' British political climate

To reinforce the general sluggishness with which the government is responding to energy efficiency options, a private member's bill to promote energy savings in homes was blocked in the report stage by a series of unanticipated government amendments. The bill would have required local authorities to prepare plans to improve the energy efficiency of all homes in their areas, including energy audits of private homes. The government claimed the measures were unnecessary, expensive and too dictatorial. In the spring of 1995 the same bill was passed into statute. This is regarded as a significant victory for the energy conservation lobby, the astute politicizing of the Environment Minister in charge of the Energy Efficiency Office (Robert Jones), and the sustained campaign by a fascinating coalition of interests, including the insulation industry, social care groups and environmental organizations, who saw this as an opportunity to stimulate energy conservation technology, create local jobs and improve the health of the disadvantaged. This is a new policy community, an example of the socio-environmental coalition that is gathering strength, and that is here to stay.

The only significant change of policy lay in the transfer of the EEO to the DoE in 1992. This separated the cabinet ministers promoting the causes of supply (DTI) and conservation (DoE), so generating a more open debate in favour of energy efficiency. This separation enabled the DoE to argue convincingly for more funds for the Home Energy Efficiency Scheme in 1993, 1994 and in 1995. As it turned out these additional funds were financed from the VAT taken on domestic fuel – a form of directional taxation.

We comment below that the failure of the government to require the Ofgas and Offer regulators to levy a special 'tax' on gas and electricity in the cause of new efficiency investments, caused the government to inject more cash into the EEO (£25 million in 1995–1996). This is an imperfect solution, but a sign of a commitment to recycling some of the VAT money into 'fiscally neutral', 'sustainability' investments. This is embryonic institutional innovation of some significance.

THE POLITICS OF THE UK RESPONSE

The UK ratified the UN FCCC in December 1993. As noted earlier, the government has published a strategy that reviews the measures which it proposes to achieve its commitments as outlined in Box 8.4.

Proposals for carbon dioxide reduction

Much of the actual achievement in CO_2 emissions reduction depends on the accuracy of the data from which the calculations are made. This is a more contentious area than many realize, and provides a key analysis of the sleight of statistical hand under which the UK government could arguably be operating. Susan Subak discussed this issue in some detail in Chapter 3.

Since 1970 CO_2 emissions have fallen by 23 MtC, from a peak of 182 MtC to 159 MtC (see Figure 8.3). This fall is due primarily to the industrial restructuring that took place during this period, with a 47 per cent reduction in CO_2 emissions from the industrial sector (Maddison and Pearce, 1994, 9).

In 1992 the government published Energy Paper 59. This report identified the total amount of CO_2 emissions by the UK in 1990 as 160 MtC. In addition, it outlined three possible emission projections for the year 2000 reflecting a growth in CO_2 emissions of between 2 and 12 per cent (Department of Trade and Industry, 1992, 6–8). The lowest emission scenario placed CO_2 emissions at 157 MtC while the highest emission scenario placed them at 179 MtC.

In March 1995 the Department of Trade and Industry published dramatically revised statistics of UK energy demand, its fuel mix, and as a result its CO_2 equivalent emissions to 2020. These data are presented in Tables 8.3 and 8.4. The analysts looked at various combinations of energy prices, industrial structure and patterns of economic growth, as well as the character of electricity production by fuel throughout the period. On central estimates of growth and energy prices, CO_2 emissions are due to fall to

BOX 8.4 SOURCES OF SAVING ON CARBON EMISSIONS IN THE UK

Sector	Carbon saved (Mt)	
	1994 estimates	1995 estimates
Energy consumption in the home	4	2
VAT on domestic energy use		
Energy Saving Trust		
Energy efficiency advice		
Ecolabelling		
EC SAVE Programme		
Improved building regulations		
Energy consumption by business	2.5	2
Making a corporate commitment		
Best Practice Programme		
Energy Efficiency Offices		
Energy Management Assistance Scheme		
Energy Saving Trust		
Energy Design Advice Scheme		
EC SAVE Programme		
Improved building regulations		
Energy consumption in the public sector	1	1
Targets for public sector bodies		
Transport	2.5	2
Road fuel duty increases		
Total	10	7

Sources: HMG (1994b, 16); DTI (1995a, 100–101)

150 MtC in 2000 compared with 158 MtC in 1980, and the total CO_2 equivalent emissions from 180.6 MtC in 1990 to 165.1 MtC by 2000 (Table 8.3). Towards 2020, the CO_2 component could be as low as 171 MtC compared with the 158 MtC baseline with no new measures, though it could also be as high as 197 MtC.

The reason for this unexpected set of forecasts lies in the huge expected switch away from coal by 2020. Table 8.4 indicates that this could be responsible for only 2 gigawatts of electricity production compared with 40 GW today. Investment in highly efficient combined cycle gas technology and new forms of integrated gas conversion technology using oil-based derivatives could fill the gap, with considerable CO_2 savings. This of course is speculation. The necessary investment programmes are not yet in place, though the switch to gas is very noticeable – from zero in 1990 to 16 GW by 2000, much of which is already built or under construction.

Table 8.4 Contribution of energy sources to the electricity supply of the UK (GW)

	Coal	Coal FGD	Oil	Oil FGD	IGCC	CCGT	Nuclear	Renew-ables
1990	40	0	12	0	0	0	11	1
2000	17	10	8	0	0	16–19	12	3
2010	6	6	6	2	3–4	21–33	7	3–4
2020	2	1	3	0	12–30	23–37	1	4–5

FGD flue gas desulphurization
IGCC integrated gas combustion cycle
CCGT combined cycle gas technology

Note: The range is based on assumptions about price and economic performance. Note the decline in coal-fired power, worth over 10 MtC by 2000 compared with earlier estimates, and the huge range of uncertainty over integrated gasification plants (using oil or orimulsion) and combined cycle gas technology. These uncertainties relate to the relative price, technological advances and the investment strategies of the privatized companies. Yet it is on this range of assumptions that the UK is basing its negotiating position for the CoPs beyond 1997.

Source: Department of Trade and Industry (1995a, 130–131)

Value-added tax

In March 1993 the Chancellor of the Exchequer (Finance Minister) introduced the prospect of a levy of the value-added tax (VAT) to domestic fuel and power. A VAT levy of 8 per cent was imposed as of April 1994 designed to increase to the full rate (currently 17.5 per cent) from April 1995. The government estimated that imposition of the VAT would lead to CO_2 savings from households of 1.5 MtC (HMG, 1994a, 19), and that the full levy in 1995–1996 would increase Treasury revenue by £2.8 billion. Ostensibly the government introduced the measure to match up to VAT levies imposed in all other EU Member States, to mop up some of the fall in real domestic energy costs, and to equalize the VAT take on fuel and insulation and other energy efficiency materials (which had anomalously been subject to VAT). But the introduction of VAT on home energy use was widely criticized because of the negative effect it would have on those in the UK who already exist in a state of fuel poverty. VAT is a regressive tax, and VAT on all fuel does not constitute anything near a carbon tax. The poor spend five times as much of their income on energy compared with the wealthy. Crawford *et al.* (1993) show that the lowest income decile spends 15.5 per cent of income on domestic fuel and 0.5 per cent on petrol, compared with 3.9 per cent and 3.3 per cent for the highest decile. Table 8.5 also reveals that the poor will suffer disproportionately more as a result of the VAT imposition, and will have to restrict their energy use even more severely. Pensioners, and those living in colder Scotland and Wales, are especially hard hit. Since the government had announced that it had no plans to raise

Table 8.5 The short run effects of imposing VAT at 17.5 per cent on households of different income

Income quintile	Additional tax paid per week	Additional tax as % of spending	% change in energy consumption
1 Poorest	£1.56	2.0%	−9.2%
2	£1.83	1.3%	−8.3%
3	£2.11	0.9%	−6.2%
4	£2.18	0.7%	−4.2%
5	£2.63	0.6%	−1.1%
All households	£2.06	1.1%	−5.8%

Source: Crawford *et al.* (1993, Table 4.1)

VAT before its election in 1992, and given that Scotland and Wales elect very few Conservative MPs, one can readily imagine the political furore.

Whether or not, and to what degree, the VAT levy on domestic fuels will be successful in reducing CO_2 emissions will depend to a large extent on two factors. The first factor is that of the short run demand elasticity for energy. For example, Maddison estimates, based on a demand elasticity for energy of −0.4 per cent, that the possible effect of the VAT levy may be a reduction of 2.9 MtC not the 1.5 MtC as suggested by the government (see Maddison and Pearce, 1994). This estimation, however, may be skewed and the effects of the tax delayed if the full response to price changes tends to be sluggish. Consequently, the savings in CO_2 emissions could be less than the government estimates, and much below the Maddison and Pearce predictions. This is particularly the case in the light of reimbursement to low income households.

The second factor relates to the compensation response. This issue is complicated by the fact that state benefits are automatically operated by the retail price index (RPI). This in turn includes an energy element. But the RPI does not properly reflect the entire burden of energy spending on the poor. So the Chancellor proposed to give single pensioners £1.00 per week and married couples £1.40 per week in addition to uprating their benefits. He also provided £1.50 on top of the £6 per week cold weather compensation payments, and added an extra £35 million to the Home Energy Efficiency Scheme to help insulate the homes of pensioners. The whole relief package will cost £1.3 billion in 1996–1997, or over half of the total revenue originally designed to be raised.

As mentioned, the decision to raise the VAT on domestic fuel was originally couched as an environmental measure. The strength of popular resentment against the proposed tax caused ministers to shy away from the CO_2-related issue (as well as to halve their original forecast of CO_2 reduction)

and to concentrate on the necessity of reducing the public sector borrowing requirement.

In November 1994 the government faced defeat from its own back-benchers over the additional 8 per cent VAT rise, scheduled to be imposed in April 1995. Consequently the second tranche, worth over £1 billion to the Treasury, was abandoned in favour of a package of measures, including an additional penny on a litre of petrol. The net effect is to reduce likely CO_2 emissions by 1–1.5 MtC by 2000, a gap in the projections that will not be made up by the decision to increase the EEO Home Energy Efficiency Scheme by an additional £30 million in 1995–1996. In addition the VAT take on energy efficiency materials was cut to 8 per cent in line with the fuel levy. In the furore over the VAT debate, the UN FCCC commitment was never noticeably mentioned. This is the danger of piggy-backing climate policies onto tax strategies that are ill thought through.

A carbon/energy tax as proposed by the EC could have generated similar revenue with much more CO_2 reducing effect. According to studies by Barker and his associates (1993) a carbon/energy tax on its own of $10 per barrel of oil equivalent by 2000 would reduce CO_2 emissions in the UK by 13.7 MtC compared with the 1.5 MtC predicted for VAT on fuel. So we have to ask why was a carbon tax ignored and VAT introduced? Here the introductory theory helps. The UK was never keen on the EC carbon/energy tax because of its influence on industrial competitiveness, the link to energy generally and hence to the newly privatized sectors, and because the policy communities around Treasury, Trade and Industry and Transport, to say nothing of the anti-European coalitions around the Foreign Office and Cabinet Office, fought the imposition of such a tax. This is the policy process at work, defining the agenda and resisting a Euro-dominated UK strategy.

Above all, any EC-imposed tax would be an anathema to the ruling Conservatives. As pointed out in Chapters 5 and 6, the EU does not have competence over tax policies, but it could have competence in international environmental agreements. The issue could be resolved by more domesti-cally based tax rises built on a strategy for making the economy better and less environmentally demanding. Barker (1994) for example estimates that a 10 per cent fuel escalator and a carbon/energy tax would meet a target of 20 per cent CO_2 reduction by 2005, yet increase employment by at least 750 000 and decrease energy waste. These are hardly the statistics that suggest a negative effect on industrial competitiveness, as claimed by the CBI and DTI . Yet the eco-tax strategy still has no effective policy com-munity. Environment communities remain weak; indeed none of the major environmental groups has a qualified tax economist intellectually equipped to grapple with such matters. The time is very ripe for entry into the policy arena.

At the periphery of the policy arenas – in interdepartmental committees, in environmental group lobbying, in academic analysis – there was excitement around a carbon tax, and possibly (but less enthusiastically) the EC carbon/energy tax. However, it was also accepted that, to be effective, any carbon tax had to be progressive and its revenue directed into energy saving measures, public transport and offsetting the added financial burden on the poor. These were not policy measures that the Treasury was prepared to stomach. The VAT route appeared to be the easy option. The resulting political débâcle showed all too clearly that strategic policy coordination is still awaited in British political and administrative circles.

Transport fuel duties

Also in the March 1993 budget the Chancellor of the Exchequer increased the tax levy on transport fuel by 10 per cent. An additional levy of 3 per cent per annum in real terms was promised for future years. The expected saving of this measure in CO_2 emissions is 1.5 MtC by the year 2000. In November 1993 the Chancellor raised this levy to 5 per cent per annum, reduced the level of the rebate on large company cars, imposed a 3 per cent surcharge on the insurance premium on cars and promised the prospect of motorway charges beyond 1997. The increase in fuel levy, hardly noticed in the public debate so far, should raise around £2 billion additional revenue by 1996–1997. This extra measure is designed to add another 1 MtC to the CO_2 reduction strategy. In effect the transport sector will grow to only 31–33 MtC by 2000 instead of 36.5 MtC as forecast, from a 1990 base of 29.3 MtC.

Again, however, as with the VAT levy on home fuels, the realization of these savings depends to a large extent on the demand elasticity for transport fuel. If the demand is highly inelastic, that is if there are no perceived acceptable transport alternatives to private car use, then the effect of the rise in petrol rates will likely not have the expected effect. Given government plans to privatize British Rail it is possible that the switch from private to public transport implicit in the government savings projections will not be realized, and as a result, the savings in CO_2 emissions will not occur.

This apparently bold move must be put in its price context. Road fuel prices have declined by 35 per cent since 1980, and may fall further still as Russian oil floods recession-hit markets. Arguably once again the Treasury has seized on a soft consumer surplus to raise revenue from a form of 'sin tax' on cars. Conceivably there is even more revenue potential from the road vehicle in future years. According to the Barker *et al.* (1993) model, a 17.6 per cent escalator (in line with VAT) could reduce CO_2 emissions by 7.6 MtC by 2000 and 25.9 MtC by 2005 yet raise £73 billion in revenue.

The Royal Commission on Environmental Pollution (1994, 111–114) believed that the 5 per cent escalator was insufficient to meet the longer term

UN FCCC objective. It estimated that there would be a shortfall of 8.3 MtC by 2020, if its target of a 20 per cent reduction in CO_2 emissions from the road transport sector was to be met. So the Commission argued for a doubling of the real price of fuel by that date, or an escalator of 9 per cent per year at least to 2005 (see Figure 8.2).

The government has yet to respond to this recommendation. The signs are that it will react cautiously but favourably, for such is the shift in the climate of public opinion towards pricing the car closer to its full social cost. However, the government will have to rid itself of the stigma. That means some interesting footwork over the recycling of revenues. Here is one area where institutional innovation could be very dramatic.

The fuel price escalator is a real piece of policy innovation. For the first time a Chancellor has sought statutorily to bind the hands of his successors. The petrol price rise will continue annually and indefinitely, so long as ministers hold their nerve. This should have the effect of encouraging more

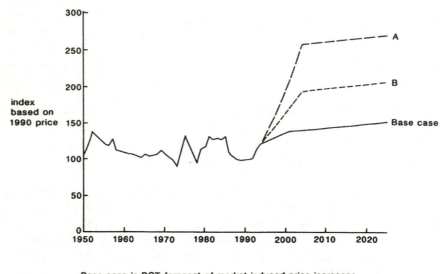

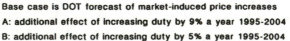

Base case is DOT forecast of market-induced price increases

A: additional effect of increasing duty by 9% a year 1995-2004

B: additional effect of increasing duty by 5% a year 1995-2004

Figure 8.2 The effect of increasing the duty on petrol by 9 per cent per year. Even this would only cut projected traffic emissions by around 15 per cent by 2010. The willingness to pay for the costs of car travel is very great, given the inconvenience and cost of moving middle to long distances by any other means. To reduce CO_2 emissions from the transport sector to any sizeable extent would require persistent price increases signalled over a long period, heralding an era of fuel-efficient vehicles, and a less car-dependent life style

Source: Royal Commission on Environmental Pollution (1994, 112)

fuel-efficient cars and in causing, at the margin, fewer vehicle journeys. The test will come in the political determination to continue the increase, year in and year out, and the manner in which the additional revenue will be spent.

Here is an example where the market force interest of the government coincides with the determination to raise revenue by indirect taxation. The two motives coincide. Although the scale of the levy will be too small significantly to affect transport use in the foreseeable future, an innovative policy measure has been introduced, under the fury of the VAT on fuel debate. Its future influence will be a matter of great interest. The policy periphery was not active in the decision, for in that arena the interest was more in favour of removal of company car tax breaks. This is a case of government ideology favouring market forces in a single-minded manner.

Building regulation

Although it is now possible to design and build an energy-efficient home for practically the same cost as a traditional house of the same size (ACE, 1993; Boardman, 1992) existing British building standards are similar to those in Sweden in the 1930s. As a result, experts believe that as much as 90 per cent of the energy currently used to heat British buildings could be saved. Given present rates of regeneration, it would take 2700 years to replace all existing homes in the UK (Boardman, 1992).

The Commons Environment Committee (1993) was critical of the apparent prospect of relaxation of building codes as part of the deregulation initiative proposed by government, and of the narrow scope for investment in the Home Energy Efficiency Scheme. The government rejected the latter criticism and claimed that its already proposed new building regulations will reduce CO_2 emissions by between 0.25 and 0.5 MtC by the year 2000 (Department of the Environment, 1993a) (see Box 8.5). Yet the government appears to have no plans for the domestic sector beyond 2000. The Building Research Establishment told the Committee that there was no policy framework in which to plan for emission reduction targets beyond 2000.

Given the fact that an energy-efficient home is as cost effective to build, if not more so, than a traditional home there should be no reason why the government is delaying the implementation of more stringent building codes. Although it is an example of state intervention in the market place such considerations should surely be outweighed by social equity concerns. However, such a move would represent a shift in favour of more target-driven regulation, not welcomed by a building industry experiencing very tender re-emergence after a three-year slump, and well known for its generous financing of Conservative political funds.

It is worth noting that estimates by various analysts indicate that a policy of deliberate house insulation backed by tougher building codes could result

---

BOX 8.5 HOME EFFICIENCY

In terms of ranking renovation efforts for their energy saving effectiveness, loft insulation ranks first, followed by cavity wall insulation and double glazing. Between the years 1974 and 1989 only 20 per cent of all properties which had loft insulation installed had done so solely through private initiative. The remaining 80 per cent have installed loft insulation as the result of regulatory requirement, with the aid of partial government grants or as a result of landlord action. The rate in insulation has been virtually static since 1986 and since February 1988, eligibility for grants for loft installation has been restricted to low income households (Boardman, 1992).

The benefits of cavity wall insulation are generally undervalued by householders. In the past 20 years only 1.5 million properties, or 20 per cent, have installed cavity wall insulation (Department of the Environment, 1993c, 3). It has never received positive government encouragement in the form of grants although, according to the Department of the Environment, more heat is lost through the walls than any other part of the house (Department of the Environment, 1993c, 5).

Double glazing is likely the most popular measure undertaken by homeowners with over 45 per cent now having at least some double glazing installed. Grants have been available for double glazing installation to reduce noise but not to improve energy efficiency (Boardman, 1992).

---

in a 25 per cent energy saving in the domestic sector over 20 years, or up to 4 MtC by 2000. This would not only be an important contribution, it would also create jobs amongst unemployed groups who could, with modest retraining, form energy saving action teams. Various trade association groups argue that 650 000 jobs throughout the Community as a whole could be created via a modest acceleration of home insulation. Pilkington Glass plc estimates that 20 000 jobs over ten years could be created to convert some 15 million dwellings to 1980 insulation standards (Eurisol, 1993). Given that Local Agenda 21s are also in the offing, where the application of the global Agenda 21 is expected to apply at the local government level, here is a case of an institutional innovation for the taking. But such a change would cut across sectorization, across non-intervention and across the client groupings that buttress ministerial hierarchies. This is British institutional design at its most unresponsive.

The Energy Saving Trust

The Energy Saving Trust (EST) was established in November 1992 as an independent body comprising the DoE, the Scottish Office, the 12 RECs,

Scottish Power and Scottish Hydro-Electric. Its aim is to increase the cash available for tackling the barriers to energy efficiency, primarily amongst domestic and small business consumers. The EST will initially look at three schemes: financing the improvement of heating in low income households; promotion through incentives, and advice for those replacing their heating so as to encourage the maximum cost-effective improvement in the use of gas (mostly through the purchase of condensing boilers); and stimulation of investment in small scale or residential application in combined heat and power installations. Future schemes include the development of local energy advice centres for domestic and small business consumers, and the promotion of low energy light bulbs.

Since 1992, the £200 rebate provided by the EST on gas condensing boilers created 5200 successful conversions and a lot of publicity. The number of companies providing the boilers has doubled from four to eight, and the price has dropped by 10 per cent. So the rebate has been reduced to £150. Yet, as we shall see, the scheme was curtailed by Ofgas in April 1994. In 1993 the Trust subsidized compact fluorescent lamps by £5. In eight weeks over 750 000 were sold, more than in the whole of 1992 (Owen, 1994b, 14). Yet the scheme did not encourage manufacturers to drop their price, partly because of the uncertainties over its continuance.

Financing for the EST was originally designed to come from British Gas and the RECs. These companies were in turn supposed to recover the cost of their contribution through the levy of an 'E factor' which is to be added to the price of gas and electricity. The amount of the levy was expected to be approximately 1.5 per cent. Funding for the first year was only £6 million though the target was planned to be £1.5 billion by 2000. The Commons Environment Committee received evidence that the electricity industry was only likely to provide £100 million by 1998. The gas industry originally confessed to an amount equivalent to £450–500 million by 2000. So the EST would still be short by over £700 million.

The dilemma of cash shortfall for the EST was made much more serious by the change of Director-General of Ofgas. The previous DG, Sir James McKinnon, conceived the scheme and was prepared to support it with a gas consumer levy. But his successor, Claire Spottiswoode, argued that such a levy constituted a tax. As such, she believed it was beyond the powers of Ofgas, and refused to support it. In the process she implied Sir James was acting illegally, an allegation she subsequently was forced to withdraw before the Commons Environment Committee.

This unresolved difficulty raises two issues. One is the legality of the levy and the other is the independence of the regulatory office. On the first point the evidence is sufficiently ambiguous that Ms Spottiswoode may be acting within her rights. She claims such a levy would be discriminatory whereas the EST argues that poor households, though being helped by the investment, would save on fuel bills more than the levy would cost them. On the independence matter, it is clear that the government is in an awkward position. It has set up these utility offices to be tough and separate, yet wants them to pursue government-inspired pricing measures. The legality issue could only be resolved by new legislation; the independence issue is less simple to sort out.

In the event, the government backed off. It abandoned any attempt to change the law, partly because it realized that any new gas levy, no matter how well intended, would be a prime target for further opposition criticism and backbench Conservative unrest. So it has added £25 million to the coffers of the Energy Efficiency Office instead, financed from the taxpayer, but in effect from the VAT domestic fuel tax. This is hardly compensation, and in a sense acts to serve the same purpose as the EST levy. The policy innovation failed because of old-fashioned clientelism: the government could not generate the political will to create a new interventionist, non-market measure in the face of a highly brittle political coalition within its own party. Established industry and ideological pressure groups circulating around DTI and the Treasury could easily see off the tax. No doubt the more favourable, but highly uncertain, energy–GHG projections helped their cause. Trying to piggy-back climate politics onto other policy arenas may be essential, but it also carries its dangers.

The Director-General of Offer, Stephen Littlechild, has sanctioned £100 million from the RECs for energy efficiency projects to 1998. The EST will merely have a supervisory role in this expenditure. Yet Professor Littlechild is unwilling to go any further, on the grounds that this would 'raise issues more appropriately dealt with through general fiscal policy' (quoted in Elliott, 1994, 39).

The EST, therefore, faces a severe crisis. Its aim is to raise £400 million annually by 1998–1999. At present it is earning some £80 million and if the DGs of Offer and Ofgas retain their strongly expressed views, the EST will be well short of projected cash. Its future will symbolize the government's commitment to its CO_2 strategy. EST is very much the kind of institution that is currently in political favour. It combines a private money raising initiative with a public purpose, and couples a consumer levy to a consumer gain without the need for intervention. The government faces a sterner political task of presenting this as a reallocative device, not a tax. Inevitably it will be labelled as a tax, unless the revenue is clearly targeted.

Meanwhile only ten of some 2000 possible small CHP schemes have been funded by the EST, with little sign of any significant support for this sector of industry. The condensing boiler subsidy of £200 per unit has been scrapped owing to lack of funds. Yet the target of 70 000 replacement boilers, up to 25 per cent more efficient than conventional boilers, was due to save 0.8 MtC by 2000 alone. Clearly this will no longer be the case.

The policy arenas are at cross purposes. Energy prices are falling just at a time when capture of these consumer savings could be transferred into targeted energy saving schemes. This is not going to happen: even the VAT increases may be offset by falling fuel prices. The CO_2 strategy is not going to be won without more effort.

The ambivalent role of the privatized utility regulator in the UK can be explained in four ways:

- The structure was never properly thought out. In effect non-accountable administrations have enormous discretionary powers to determine their own rules and set their own idea of the public and private interest.
- There is no experience of this kind of office or independent agency in the UK. Experience is being learnt through mistakes, crises, experiments and reactions to criticism.
- The supply promotion and demand reduction roles are not regarded as equivalent or necessarily complementary. There is no justification in law for the Director-General to match up these strategies.
- Ministers did not realize just how non-accountable these offices were to become. Policy control is fairly restricted.

Despite this prognosis, it is too early to estimate just how significant will be the role of the EST in helping the UK meet its UN FCCC objectives generally and whether it will meet the ambitious target of 4.0 MtC by 2000. To do this:

- it must obtain far more money than it has at its disposal now;
- it must intervene beyond the industrial market and further into the domestic market than is already the case;
- it must go beyond grant aid into R&D for energy saving technology into the next century;
- it must interlink with the EEO and sponsor energy saving targeted programmes for industry based on cost-effective investments determined on a strategic basis.

If it is to win at the climate–energy policy issues, the EST will have to cross departmental boundaries, be proactive and interventionist. It may yet do so, for it is a key player in any policy mix over the next decade. Its future is shaping up to be an example of institutional change that will test the theory outlined in Chapter 4. For example, it could be a catalyst to the opening up of the electricity and gas supply–demand restriction opportunities that many observers see as the mix necessary for fully efficient energy management services. A survey by Owen (1994a, 3) indicates that the gas and electricity utilities remain unsure of how much they can combine both fuel mix strategies and customer 'whole management' services, based on a single package as a single payment. This would have to be connected to changes in outlook and advice from the various consumer councils and other fuel use advisory bodies. Owen (1994b, 13–15) believes that the EST could be the catalyst for the revolution in energy management services, coupled to competition within an envelope of special financial measures, that could guarantee a major and persistent contribution to the UK climate warming strategy over the coming decades. She is probably right. Despite its initial setbacks, the EST remains in pole position to play this crucial role.

Appliance standards

The adoption of labelling and standards for household appliances are expected by the government to contribute to savings of 0.5–0.7 MtC by the year 2000 (ENDS, Nos 218 and 222, 1993). These standards are expected to be introduced on an EC-wide basis.

The government is insistent that energy efficiency labelling and minimum standards for electrical appliances be harmonized at the EC level because it is of the belief that overly stringent environmental requirements may make its products uncompetitive within the UK, within the internal market of the EC and internationally. The preference of the UK government would be the negotiation of voluntary agreements on minimum standards of energy efficiency with standards prescribing regulations being the least preferred option. In the 1990 White Paper on environmental strategy, the government announced its intentions to achieve these agreements within the EC (HMG, 1990, para. 5.31). To date, no voluntary agreements have been reached as the UK government is allegedly obstructing the establishment of EC-wide regulations.

The reasons for this are obscure. They are tangled up with ecolabelling of energy efficiency in appliances. Here the government's fundamental belief supports information and exhortation. The EC regulations would benefit German and to a lesser extent Dutch manufacturers who already meet the tougher standards. In an open market, British suppliers would initially be the losers. Core policy and institutional frames operate around the goals of the Department of Trade and Industry to reduce any wider response to the Convention. If this policy arena is to move forward it will depend on the legal competence of the Commission leadership to set appliance efficiency standards by majority vote. At present, that competence is too ambiguous to permit aggressive action. As 2000 approaches this may change.

The transport 'crunch'

One of the most obvious candidates for the inclusion of climate change considerations in general policy decisions is transport. As mentioned earlier, transport is the largest growing source of CO_2 emissions and presently accounts for approximately 30 per cent of CO_2 emissions with one-half being attributable to private car use and another one-quarter being associated with the industrial, commercial and public sectors. Although the government, in the CO_2 strategy, advocated the adoption of several measures aimed at reducing the amount of emissions from the use of private transport, only a few specific measures were put into effect to achieve this reduction. These include: an increase in the price of petrol (discussed above), and a decrease in the tax advantages enjoyed by the corporate car fleet. However, tax allowances of 33 per cent for the first 10 000 business miles and 67 per cent for the next 10 000 or more business miles erode the rising petrol tax for business travellers – 40 per cent of all vehicle miles in the UK.

With respect to the reduction of CO_2 emissions, other measures discussed, but not yet acted upon, include: motorway tolls; road pricing; congestion charges; and taxing of less fuel-efficient vehicles. The effectiveness of these measures, implemented or proposed, will depend largely upon whether or not there is a change in the long-established roads and vehicle bias in transport policy. What is of interest here is the convergence of three ideologies. One is the need to involve the private sector where the public sector normally monopolizes the scene. The second is revenue generation from socially unacceptable behaviour such as road congestion and car-generated air pollution. The third is the desire to cut public expenditures, and an easy target is the unpopular road building programme estimated originally at £23 billion by 2010. Recent proposals aimed at reducing the public sector borrowing requirement suggest that this programme will be cut by over 100 out of 400 originally proposed schemes. As always, where motivations coincide new policy is possible. But the proposals on a congestion tax depend on reliable tagging technology, which in turn requires proactive investment from an interventionist government. As of yet, insufficient funds are being made available.

The car is not just an emitter of CO_2. It also releases noxious amounts of carbon monoxide, hydro-carbons, nitrous oxides and small particulants. The last, sometimes referred to as PM10s, can singly or in combination with other toxic pollutants aggravate respiratory ailments. The Royal Commission on Environmental Pollution examined the health effects of car-generated air pollution and concluded that up to 10 000 deaths per year could be attributed to this source (RCEP, 1994, 31). Public anxiety, coupled to extensive pressure group lobbying, has forced the government to link the health and global warming issues into a single policy arena. This is an example of issue networking that may well result in the health benefits of controlling car use having greater sway than the gains from CO_2 emissions reduction. Because the health and environmental departments are cooperating via cost-effectiveness studies, the likelihood of greater policy coordination is greater in that framework than just through combining policy communities and issue networks. In 1995 the Environment Secretary enabled local authorities to control car traffic on health grounds by such devices as restricting land use decisions that generate traffic, and by controlling polluting vehicles.

The great transport tussle demonstrates many of the themes outlined at the outset of the chapter. Cars are regarded as both a desirable and a crucial indicator of economic activity. Over 400 000 jobs hinge on the successful car industry in direct terms and three times that in associated services. The public itself is ambivalent. A recent poll showed that 43 per cent of a representative sample wanted more spending on roads, with 46 per cent wanting more on public transport (ENDS, No. 224, 1993). Yet given a choice of new investment, a majority (56 per cent) favoured public transport over roads.

The dilemma here is that the 'car culture' grips the individual as consumer more than as global citizen. People want cars yet recognize their polluting nastiness. Locked into the car economy, people do not really wish to escape, and governments of all political colours support the industry. So road building will continue, though probably at a lower rate than planned a few years ago, and the car will increasingly be seen as a source of tax revenue with less political protest except from rural communities. Crucially, however, the present administration is unlikely to invest any more money in public transport. This is core institutional inertia very much at work. The Department of the Environment is still the junior partner against the might of the Departments of Trade and Industry and Transport, both still locked into the car economy, despite the many pronouncements by the Transport Secretary in favour of more dedicated bicycle routes, and the capping of future rail fares below the future rate of inflation.

One interesting debate on the institutional periphery is the growing interest in restricting land use planning decisions that generate car use at the expense of less mobility elsewhere. The Planning Policy Guidance Note No. 13 published by the Department of the Environment in 1994 reveals a willingness by the Department to help planning authorities and planning inquiry inspectors (on appeal) to resist out of town schemes such as supermarkets or industrial parks where insufficient public transport is provided. Essentially, however, this is doublespeak. While such guidance is extremely important for planning authorities to resist applications, and for planning inspectors to recommend refusal, nevertheless it has to be said that a large number of the out of town supermarket moves have been completed. The recession and tax rises of mid-1994 are deflating the consumer boom and causing loss of profit in the oversupplied retail sector. Just because some public transport is provided this will not necessarily significantly stop the increased use of the car for shopping, leisure and commuting. This particular battle has a long way to run.

The car is the fastest growing contributor to CO_2 in the UK, and it could be the major reason why the UK may not meet its longer term GHG emissions reduction targets beyond 2000. Here is the clash of principles outlined at the outset. The policy communities encircling the major economic ministries will not countenance any decline in the production of cars or any savage reduction in road building. Sectorization of departmental responsibilities puts the CO_2 and other environmental aspects of the private car at a low priority compared with its economic and social advantages. Reduction in use cannot easily be handled by price alone, because car travel is increasingly necessary due to cut-backs in rail and bus investment. Cycle lanes require local authority spending that is starved by Treasury-inspired Department of the Environment yardsticks. There is no feasible alternative to steadily rising car ownership and use given the current institutional mix and policy response. Despite the exhortations of critiques by the Royal Commission on Environmental Pollution and by parliamentary committees, to say nothing of pressure group opinion, there is no obvious shift in view from the government. So much is this the case that leading environmental groups have reduced their campaigns against the car as such, and have concentrated instead on the environmental disbenefits of more road building.

Here is a case where our theory holds: centrist institutional inertia reflecting and buttressing lack of governmental will means that the UN FCCC will not seriously alter transport politics in the UK before 2000. Beyond that, the role of the EC coupled to the political strength of the Conference of the Parties may force a change – and when that comes it will surely be in the realm of pricing.

Meanwhile a modest institutional reform is in place. A Greener Transport Forum meets regularly to discuss the coordination of strategy and lobbying. This involves the environmental NGOs, the two automobile associations, the motor trade association and the British Road Federation. This is not, as yet, a negotiating forum, but at least it is an arrangement through which interested parties can talk (Rawcliffe, 1995).

Public awareness and corporate commitment

Since 1983 UK governments have promoted awareness of energy efficiency and its associated benefits. These information programmes are a key component of the government's response strategy to climate change. As the thrusts of the programmes focus on the voluntary adoption of energy efficiency measures by both individuals and industry these programmes complement the general philosophy of non-market intervention. The most widely recognized of the information campaigns are those promoted by the Energy Efficiency Office (EEO). These include the Best Practice Programme; the Making a Corporate Commitment Campaign; the Home Efficiency Scheme; the Helping the Earth Campaign; and the Energy Management Assistance Scheme (for details see Box 8.6).

Falling real energy prices during a recession are hardly conducive to encourage energy conservation investment within industry. The EEO has found that over 90 per cent of business contracted in its Making a Corporate Commitment Campaign would have invested in energy savings anyway, and

BOX 8.6 THE ENERGY EFFICIENCY CAMPAIGNS

The Helping the Earth Campaign

The objective of the Helping the Earth Campaign is to increase awareness among householders of the link between global warming and energy efficiency (Department of the Environment, 1993a). The plan was to spend £10 million over the following three years on advertisements to encourage householders to reduce energy consumption in their homes.

The Best Practice Programme

The Best Practice Programme was designed to provide advice on energy efficiency to business and industry. Over 1000 publications have appeared under the Best Practice Programme where the adoption of CHP technology is encouraged.

The Making a Corporate Commitment Campaign

The Making a Corporate Commitment Campaign was aimed at securing top management commitment to energy efficiency in both the business and public sectors. The central feature of the campaign is the signing of a non-legally binding Declaration of Commitment by the Chief Executive which sets out the guiding principles behind an effective energy efficiency programme and shows that a company is making a corporate commitment to responsible energy management and the promotion of energy efficiency throughout its operations. Over 1300 businesses have joined this programme.

The Home Energy Efficiency Scheme

The Home Energy Efficiency Scheme was to provide grants to low income households (Department of the Environment, 1993a) and was run by the fuel-poverty charity Neighbourhood Energy Action. The grants were initially given to cover up to 90 per cent of the purchase and installation of the most basic energy efficiency practices such as loft insulation. At current rates, this scheme would have taken 27 years to cover every householder entitled to a grant (Boardman, 1992). In 1988, the loft insulation programme was reduced by one-third (Friends of the Earth, 1990, 43).

The Energy Management Assistance Scheme

The Energy Management Assistance Scheme (EMAS) provides help with the cost of consultancy for managing energy efficiency in small and medium-sized business. Only companies with fewer than 500 employees worldwide are eligible for a grant.

that only 42 per cent of companies have made any capital investment to meet environmental objectives (ENDS, No. 221, 1993, 26). Only 3 per cent of companies who have signed up to the Campaign actually have published targets. Yet a 0.5 MtC reduction by 2000 hinges on this scheme.

Similarly the EEO Energy Management Assistance Scheme (EMAS) is supposed to save 0.8 MtC by 2000. By 1993 the target for saving had been reduced from £200 million per year to £20 million per year because of the effect of recession in small business (ENDS, No. 232, 1993, 23). Those businesses which have signed up are unlikely to save much energy

As mentioned earlier, the shift of the EEO from DTI to DoE signalled a new era of higher political profile for energy efficiency. Friends of the Earth have criticized the EEO for being ineffective in its information dissemination role, and wasteful of precious funds (ENDS, No. 206, 1992). One test for the EEO will be the interconnection between the British Standards Institution's BS 7750 environmental management scheme, the proposed EC eco-audit directive for most industry, and the scope for energy saving by sound management. As noted above, little progress is being made at present. There will have to be a shift in the UK industrial culture to make BS 7750 a truly effective vehicle for the post-2000 CO_2 removal strategy.

Other greenhouse gases

Box 8.1 shows that about 8 per cent of UK greenhouse gas emissions come from methane, 4.4 per cent from nitrous oxides and 2.2 per cent from other trace gases. Methane emissions have remained fairly stable since 1980, except for a dip during the 1984–1985 miners' strike. The privatization of coal will further reduce mine-related methane emissions only if appropriate regulations and incentives are in place. Otherwise the government is relying on reductions in landfill sites and better methane control of existing sites for its methane removal. These are optimistic expectations, as are those proposals assuming that BS 7750 will also contribute to a reduction in industrial emissions. As for N_2O, the strategy relies almost totally on improved performance from the nylon manufacturing industry. This essentially involves two chemical plants owned by duPont. By 1996 new technology could remove nearly all the 80 000 tonnes of N_2O currently being emitted, equivalent to 6 MtC on a global warming basis. But the introduction of catalytic converters will increase N_2O emissions for 'clean cars' putting further pressure on restricting car use. These conclusions appear in the data presented in Table 8.3.

The group of fluorinated gases that have very high radiation forcing qualities are now very much in the frame. These include the perfluorocarbons such as carbon tetrafluoride (CF_4) and hexafluoroethane (C_2F_4) and sulphur hexafluoroethane (SF_6). These are used as fire retardants, while SF_6 is increasingly used as an insulating agent in the electricity industry. At present their total contribution is only 1 per cent of the UK emissions, but their use is growing at over 2 per cent annually.

Here is where new policy initiatives involving industry and voluntary compliance may form. The eco-audits will force industry to search for substitutes and to change practices. This is another example of a policy measure that is met by new initiatives established for a more general social and industrial purpose.

CONCLUSION

This chapter, in assessing the UK's policy response to the UN Framework Climate Change Convention, has highlighted the following themes.

The inadequacy of the target

The commitment in the UN FCCC, or at least in its subsequent interpretation by the EC, to stabilization of greenhouse gases to 1990 levels by 2000, barely affects the climate warming scenarios. If nothing more is done the predicted rise in temperature to 2.5 °C warmer by 2100 will be delayed by only 4–5 years. The IPCC continues to call for a 60 per cent reduction in emissions by 2040 to meet the UN FCCC objectives, yet at best present measures will amount to 7 per cent.

The reason for this is obvious. Climate change avoidance is regarded as a long term objective with distant benefits. Indeed, rarely in any political discussion are the benefits of any reducing measure spelt out. Only the costs and the socio-economic consequences are stressed. In the political mind, greenhouse gas reduction is a disadvantage to core values of economic growth and redistribution out of an ever expanding economic pie. This is a major reason why the Department of the Environment and its policy moves are so marginalized in the UK. The crunch will come when real economic growth prospects run up against future environmental regulations mandated by international agreements requiring joint action. The British public is not at present prepared for that day.

The supremacy of core institutional positions and world views

As indicated in Chapter 4, institutional patterns are shaken up by a set of external forces, such as changing public perceptions of scientific discoveries, shifting interest group alliances, and policy measures adapted for a variety of purposes. In the UK the scope for institutional innovation depends crucially on the political culture, and that in turn is influenced by policy community sectorization, ministerial pecking order, party political ideology and attitudes of secrecy and limited accountability.

The UN FCCC has forced the government to focus on a greenhouse gas reduction strategy which certainly would not otherwise have been the case. Constituencies have been mobilized which would not otherwise have talked to each other. There is an opening up of interdepartmental communication that did not previously exist in order to produce the CO_2 document and the UK's proto-Agenda 21. There are procedural moves but they do not appear to have influenced position.

Any fresh moves can be put down more to convergence with ideology than with UN FCCC response primacy. In summary, the only identifiable departures of policy are:

- the persistently rising petrol duty at 5 per cent per year;
- the energy efficiency subsidy schemes of the Energy Saving Trust;
- financial upgrading of the promotional schemes of the Energy Efficiency Office;
- small changes in company car taxation;
- VAT imposition on domestic fuel (mostly for revenue raising);
- PPG 13 on limiting out of town car-travel generating developments.

This is a modest package. The cost is low, the impact will contain cost-effective zones and the structural change will be all but zero. Most of the promises are based on the rules of information provision, voluntary compliance and general exhortation to civic duty. Non-intervention, a core institutional parameter, remains a primary policy factor in limiting action. Box 8.7 summarizes the present position.

The crunch to come

As matters stand the UK is making only a modest contribution to its own objectives as laid down both by its own policies and by its membership in the European Community. It has seemingly blocked both the EC carbon energy tax proposals and stringent energy efficiency standards in appliances. It has failed to exercise tough building codes, and is still starving most of its energy efficiency programmes of money. With respect to the car, the most important growth source of CO_2, it is doing too little to limit a doubling of usage within 50 years. As for air travel, the only hope appears to rest with improved technology (see Figure 8.3). Surely this is a sector that will face the same degree of contradictory criticism in years to come as the car culture is receiving today. Again the British public is unprepared for this debate.

Where further institutional change in the UK will emerge, it should lie in the greater penetration of the traditional policy cores by aggressive realignments at the periphery. Look for developments in the following areas:

BOX 8.7 KEY INSTITUTIONAL INNOVATIONS IN THE UK RESPONSE

- **The gasoline tax escalator of 5 per cent per year for as long as ministers hold their nerve**
  - **1.5 MtC cut annually**
  - **a signal for fuel-efficient vehicles.**

- **Value-added tax on domestic fuel at 8 per cent**
  - **1.0 MtC cut annually**
  - **rebates for poor.**

- **Energy Saving Trust funded by special levies on gas and electricity plus 'make-up' government funding**
  - **0.5 MtC annually**
  - **financed by gasoline and VAT levy.**

- **Whole energy management in the private energy utilities**
  - **sponsored by deregulation**
  - **promoted by Energy Saving Trust.**

- **Local Agenda 21 initiatives**
  - **housing insulation schemes**
  - **bicycle routes**
  - **land use planning controls**
  - **cogeneration of wastes.**

- pricing of car travel, notably in congestion taxation;
- tighter controls over land use planning linked to local Agenda 21 programmes;
- greater incentives to reduce the journey to work, cut down on mobility generally and encourage the use of the bicycle; this will mean more financial encouragement to telecommuting and teleconferencing, more provision of local leisure facilities, greater restriction on out of town developments, and the specific provision of cycle access routes as part of local Agenda 21s;
- greater use of the Energy Saving Trust into a wider array of cost saving measures coupled to the vital shifts in 'whole' energy management, cross-sectoral fuel supply, and single billing;
- job creation schemes in domestic energy efficiency programmes, and other energy efficiency measures;
- subsidizing R&D on innovative technology coupled to more flexible use of BS 7750 eco-audits and industrial capital allowances so as to reduce energy and materials throughput in industry;
- widening the basis of environmental education to stimulate a real sense of civic responsibility in all individuals and create a shared belief that control of greenhouse gases is a good thing in itself;

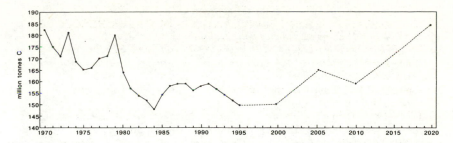

Figure 8.3 CO_2 emissions in the UK since 1970 expressed at MtC. Note the general trend downwards due to industrial restructuring and the reduction in coal burnt in electricity generation. Forecasts show a steady rise as the fuel switch in favour of gas stabilizes and transport-related emissions grow

Source: DTI reports

- generating comprehensive and publicly debated policy mixes aimed at linking climate policy to economic strategy, sound welfare, public health, employment and education via interconnected issue networks;
- steadily opening up the floodgates of targeted revenue expenditure on the basis of indirect taxation to build public support for wholesale taxation reform.

What emerges is not so much institutional innovation as institutional diffusion. The translation of climate policy into fiscal and energy policies, the emergence of citizens' environmental initiatives as a policy dimension embracing social cohesion as well as quality of life values, and the incorporation of energy conservation strategies and social welfare policies combine to mean that the climate message is becoming incorporated into wider policy arenas. It is possible that the climate message may then get lost as an identifiable, scientifically biased, moral theme. The UK response will increasingly depend on other countries also taking action, and successful metamorphosis of social and economic dimensions into a more multidimensional approach to the Rio agenda. The problem is that, at present, the necessary realignments of interest coalitions and policy communities are not prepared for this revolution. If global warming is to be curtailed, significantly new dimensions of institutional innovation will be required, few of which need be specifically climate driven. The UK cannot afford to bask in speculative forecasts of energy consumption reduction and chance connections of policy agendas. The sustainability dynamic deserves something more focused and programmed than that. For this to happen will be the ultimate institutional innovation.

263

REFERENCES

Anderson, V. (1992) *Energy Efficiency Policies*, London: Routledge.

Association for the Conservation of Energy (1993) *Climate change: our national programme for CO_2 emissions – A response to the government's discussion document*, London: Association for the Conservation of Energy.

Bannister, D. (1992) Transport, pp. 120–156 in P. Cloke (ed.) *Policy and Change in Thatcher's Britain*, Oxford: Pergamon Press.

Barker, T. (1994) Taxing pollution instead of employment: greenhouse gas abatement through fiscal policy in the UK, Discussion Papers on Energy-Environment-Economy Modelling, Cambridge: University of Cambridge, Department of Economics.

Barker, T., Baylis, S. and Masden, P. (1993) A UK carbon-energy tax: macroeconomic effects, *Energy Policy* 21(3), 296–308.

Birtles, W. (1994) Environmental information and audit, pp. 25–48 in A.S. Boyle (ed.) *Environmental Regulation and Economic Growth*, Oxford: Clarendon Press.

Boardman, B. (1992) *Energy and Environment: Paying for Energy Efficiency*, Brighton: National Society for Clean Air and Environmental Protection.

Boehmer-Christiansen, S. and Skea, J. (1991) *Acid Politics: Environmental and Energy Policies in Britain and Germany*, London: Belhaven Press.

Caminus Energy Limited (1993) *Taking a Long Term View: Preparing for the 21st Century*, London: Caminus Energy.

Clean Air (1993) Climate Change UK National Programme For Carbon Dioxide Emissions: Comments by NSCA on Department of Environment Discussion Document, *Clean Air* 23, 5–11.

Cornot-Gandolphe, S. (1993) The environment and the role of gas, *Energy and Environment* 4, 110.

Council for the Protection of Rural England (1993) *Comments on The National Plan for Carbon Dioxide Reductions*, London: CPRE.

Crawford, I. *et al.* (1993) VAT on domestic energy, *Institute of Fiscal Studies Commentary* No. 39, London: IFS.

Department of the Environment (1992a) *Climate Change: Our National Programme for CO_2 Emissions – A Discussion Document*, London: DoE.

Department of the Environment (1992b) *The UK Environment*, London: HMSO.

Department of the Environment (1993a) *Climate Change: Our National Programme for CO_2 Emissions – A Discussion Document – Addendum*, London: DoE.

Department of the Environment (1993b) *Climate Change: Our National Programme for CO_2 Emissions – Report of Conference*, London: DoE.

Department of the Environment (1993c) *Helping the Earth Begins at Home*, London: DoE.

Department of the Environment (1993d) *Digest of Environmental Protection and Water Statistics (1992)*, London: HMSO.

Department of Trade and Industry (1992) *Energy Related Carbon Emissions in Emissions in Possible Future Scenarios for the United Kingdom (Energy Paper 59)*, London: HMSO.

Department of Trade and Industry (1993a) *The Prospects for Coal: Conclusions of the Government's Coal Review (White Paper on Coal)*, Cmmd. 2235, London: HMSO.

Department of Trade and Industry (1993b) *Energy Statistics*, London: DTI.

Department of Trade and Industry (1995a) *Energy Projections for the UK*, Energy Paper 65, London: HMSO.

Department of Trade and Industry (1995b) *The Prospects for Nuclear Power in the UK: Conclusions of the Governments's Nuclear Review*, London: HMSO.

Elliott, D. (1994) A moment to saver, *New Statesman and Society*, 25 March, 39–40.

Eurisol (1993) The implications of a carbon tax, Mimeo, Brussels.

Friends of the Earth (1990) *How Green is Britain?*, London: Hutchinson Radius.

Friends of the Earth (1992) *Friends of the Earth's Response to the Department of Environment's Discussion Document on Climate Change (Our National Programme for CO$_2$ Emissions)*, London: FoE.

Giles, C. and Ridge, M. (1993) The impact on households of the 1993 Budget and the Council Tax, *Fiscal Studies* 14, 1–20.

Gill, J.P. (1993) *Climate Change: The Contribution of Science to UK Policy*, Edinburgh: Safe Energy Research.

Greenaway, J., Smith, J. and Street, J. (1992) *Deciding Factors in British Politics*, London: Routledge.

Greenpeace International (1993) *Towards a Fossil Free Energy Future*, Boston: Stockholm Environment Institute.

Grubb, M. (1991) *Energy Policies and the Greenhouse Effect*, Aldershot: Dartmouth Publishing.

HM Government (1990) *This Common Inheritance: Britain's Environmental Strategy*, Cm. 1200, London: HMSO.

HM Government (1991) *This Common Inheritance: The First Year Report*, Cm. 1655, London: HMSO.

HM Government (1992) *This Common Inheritance: The Second Year Report*, Cm. 2068, London: HMSO.

HM Government (1993) *Open government*, Cm. 2290, London: HMSO.

HM Government (1994a) *Sustainable Development: The UK Strategy*, Cm. 2426, London: HMSO.

HM Government (1994b) *Climate Change: The UK Programme*, Cm. 2427, London: HMSO.

HM Government (1994c) *Biodiversity, The UK Action Plan*, Cm. 2428, London: HMSO.

HM Government (1994d) *Sustainable Forestry: The UK Programme*, Cm. 2429, London: HMSO.

Houghton, J.T., Jenkins, G.J. and Ephraums, J.J. (eds) (1990) *Climate Change – The IPCC Scientific Assessment*, Cambridge: Cambridge University Press.

Houghton, J.T., Jenkins, G.J. and Ephraums, J.J. (eds) (1992) *Climate Change 1992: The Supplementary Report to the IPCC Scientific Assessment*, Cambridge: Cambridge University Press.

House of Commons Energy Committee (1985) *The Energy Efficiency Office*, London: HMSO.

House of Commons Energy Committee (1989) *Energy Policy Implications of the Greenhouse Effect*, HC 192-I, London: HMSO.

House of Commons Environment Committee (1993) *Energy Efficiency in Buildings*, HC 648, London: HMSO.

House of Commons Trade and Industry Committee (1993) *British Energy Policy and the Market for Coal*, HC 237, London: HMSO.

House of Lords Select Committee on the European Communities (1993) *Structure of the Single Market for Energy*, HL Paper 56, London: HMSO.

Jäger, J. (1992) Developing Policies in response to climate change, pp. 45–62 in *Klima und Strukturwandel*, Bonn: Economica Verlag.

Jordan, G. (1990) Policy community realism versus the new institutional ambiguity, *Policy Studies* 32, 470–484.

Jordan, G. and Richardson, J. (1982) The British policy style or the logic of negotiation?, pp. 80–110 in J. Richardson (ed.) *Policy Styles in Western Europe*, London: George Allen and Unwin.

Labour Party Policy Commission of the Environment (1994) *In Trust for Tomorrow*, London: Labour Party.

Liberal Democratic Party (1994) *Paying for Pollution*, London: Liberal Democratic Party.

Liberal Democrats (1993) *Taxing Pollution Not People*, London: Liberal Democratic Party.

Lowe, P. and Flynn, A. (1989) Environmental politics and policy in the 1980s, pp. 71–90 in J. Mohan (ed.) *The Political Geography of Contemporary Britain*, London: Macmillan.

Lukes, J. (1974) *Power: A Radical View*, London: Macmillan.

Maddison, D. and Pearce, D. W. (1994) *The United Kingdon and Global Warming Policy*, GEC 94-271, Norwich and London: Centre for Economic Social Research on the Global Environment.

Marsh, D. (ed.) (1983) *Pressure Politics: Interest Groups in Britain*, London: Junction Books.

Marsh, D. and Rhodes, R.A.W. (eds) (1992) *Policy Networks in British Government*, Oxford: Clarendon Press.

McGowan, F. (1993) Energy policy in the UK to1992, pp. 1–14 in S. Thomas (ed.) *Energy Policy: An Agenda for the 1990s*, Brighton: Science Policy Research Unit.

Mintzer, I.M. (ed.) (1992) *Confronting Climate Change: Risks, Implications and Responses*, Cambridge: Cambridge University Press.

Newberg, D. (1993) The impact of EC environmental policy on British Coal, *Oxford Review of Economic Papers*, 9(4), 1–28.

O'Riordan, T. (1992) The environment, pp. 247–324 in P. Cloke (ed.) *Policy and Change in Thatcher's Britain*, Oxford: Pergamon Press.

O'Riordan, T., Kemp, R.V. and Purdue, H.M. (1988) *Sizewell B: An Anatomy of the Inquiry*, London: Macmillan.

Owen, G. (1994a) *Energy services market: will competition be left to chance?*, London: Energy Saving Trust.

Owen, G. (1994b) *From energy supply to energy services: the role of the Energy Saving Trust in transforming the energy market*, London: Energy Saving Trust.

Rawcliffe, P. (1995) Making inroads: transport policy and the British environmental movement, *Environment* 37(3), 16–20, 29–36.

Rhodes, R.A.W. (1994) The hollowing out of the State: the changing nature of the public service in Britain, *The Political Quarterly* 65(2), 138–151.

Royal Commission on Environmental Pollution (1971) *First Report*, Cmnd. 4585, London: HMSO.

Royal Commission on Environmental Polution (1994) *Transport and the Environment*, Cmnd. 2674, London: HMSO.

Sands, P. (1994) Access to environmental justice in the European Community: principles, practice and proposals, *Review of European Community International Environmental Law* 3(4), 204–214.

Schneider, S.H. (1992), Global climate change: ecosystems effects, pp. 142–148 in *Interdisciplinary Science Reviews* 17.

Science Policy Research Unit (1993) *Energy Policy: An Agenda for the 1990s*, Brighton: University of Sussex.

Secretary of State for the Environment (1993) Energy policy and the environment: annual address to the Parliamentary Group for Energy Studies, *Energy Focus* 10, 38.

United Nations Framework Convention on Climate Change, UN Doc. A/AC.237/18(Part II)/Add/1/Corr.1.

Virley, S. (1993) The effect of fuel price increases on road transport CO_2 emissions, *Transport Policy* 1(1), 43–48.

Woolf, T. (1993) Developing integrated resource planning policies for the European Community, *Review of European Community and International Environmental Law* 1, 118–124.

World Energy Council (1993) *Energy For Tomorrow's World*, New York: St Martin's Press.

Wuebbles, D.J. and Edmonds, Jae (1991) *Primer on Greenhouse Gases*, Chelsea, Michigan: Lewis.

Wynne, B. (1993) Implementation of greenhouse gas reductions in the European Community, *Global Environmental Change* 3, 101–112.

9

NORWEGIAN CLIMATE POLICY
Environmental idealism and economic realism
Anne Kristin Sydnes

INTRODUCTION

The tension between Norway's leadership ambitions in international climate politics on the one hand and its role as petroleum exporter on the other forms the political framework for this chapter. Norway has become internationally known as a high profile environmental negotiator, with a self-declared goal to act as a driving force in international climate policy making. In 1989 the Labour Party government put forward a proposal for stabilizing Norwegian CO_2 emissions by the year 2000. However, it became clear that a majority in the Storting (the Norwegian Parliament) supported a more ambitious Norwegian target of stabilizing CO_2 emissions at the 1989 level by the year 2000. An ideological attitude for a strong and lasting Norwegian political commitment to the cause of sustainable development appears to have had a considerable effect on the climate–political debate at the time the CO_2 goal was first adopted.

However, if we take into consideration Norwegian petroleum economic interests and other important features of the Norwegian economy, we find little to indicate a role for Norway as a driving force in climate policy. According to official estimates, emissions of CO_2 are expected to rise by approximately 16 per cent by the year 2000 compared with 1989, taking the effects of the CO_2 tax into account; 65 per cent of the increase is expected to come from growth in gas production and transport.

As Norwegian climate policy has developed over time, some critics maintain that it has increasingly tended to favour Norwegian economic interests. Gunnar Bolstad (1993) of the Norwegian Society for Conservation of Nature is among those who emphatically claim that the political will to carry out Norwegian climate policy in its 'original' form has been seriously weakened. The International Energy Agency (IEA) (1992) maintains that the means employed hitherto are inadequate, and that CO_2 taxes must be employed far more actively to realize the target (Aftenposten, 1994).

To what extent has a change really taken place in Norwegian climate policy? Have Norwegian economic interests increasingly succeeded in getting their views accepted? Has the effect of 'Brundtlandism' been maintained undiminished, or has the ideological platform been redefined in favour of specific economic interests? What are the driving forces in the decision making processes of Norwegian climate policy – and what counter-forces have made themselves heard in these processes?

In trying to find an answer we shall consider two arguments in the Norwegian climate debate: a change has taken place in climate policy goals – the original environmental 'idealism' has declined – while the 'fossil fuel–industrial complex' has strengthened its position and influence on climate policy. Or alternatively, material interests cannot explain Norwegian climate policy. The aim of the present analysis is to discuss the relevance of these assertions, to trace developments in Norwegian climate policy empirically and to see what driving forces have proved the stronger in various phases of policy formation. A key question is how the organization of the decision making processes and the pattern of participation in these processes have influenced the development of Norwegian climate policy. The analysis in this chapter will primarily focus on the formative period from spring 1989 to autumn 1992, but the broader aspects of Norwegian follow-up of the Rio conference will also be discussed.

OFFICIAL CLIMATE POLICY AND GREEN CRITICISM

The main elements of Norwegian climate policy and the Norwegian negotiation position within the Intergovernmental Negotiating Committee (INC) were in place by September 1991. While the relative attention paid to the various issues has changed somewhat over time, the major components have remained unchanged.

A preliminary target was agreed upon by the Storting to stabilize CO_2 emissions at the 1989 level by the year 2000. However, the Norwegian government stressed the need for a comprehensive approach, including the reduction of all greenhouse gas emissions, in dealing with climate change. The principle of cost-effectiveness should guide practical measures and solutions. The original position favoured economic instruments; CO_2 taxes are an important ingredient in Norwegian climate policy on the national level. Norway also favours international cooperation on joint implementation and a consensus on one or more protocols to include quantitative targets for reductions of greenhouse gas emissions. Developed countries should cover incremental costs related to developing countries' policies to reduce growth in greenhouse gas emissions. Financial resources transferred in this regard should be new and additional to current development assistance budgets.

269

In 1992, Norway ranked first among the aid donors (1.09 per cent of GNP), and had the highest equivalent national contribution to the Global Environment Facility. Aid for environmental purposes in addition to ODA was NOK (Norwegian Kroner) 138 million, including pledges to the GEF.

The Framework Convention on Climate Change was ratified by Norway on 9 July 1993. At that time ratification hardly represented an important achievement for Norway, because of the limited obligations to be fulfilled by the Parties. The Norwegian ratification process normally takes a year so the procedure was essentially uncontroversial.

Nevertheless, 'green' criticism of Norway's climate policy increased in 1993. Representatives of Norwegian NGOs commented very sarcastically on Norway's role in the negotiations leading up to the first CoP in Berlin.

> Of course, as any of Mrs Brundtland's officials will assert, Norway is already doing her part. It was the first country with a CO_2 commitment, and its CO_2 tax is light years beyond anybody else's. True. It is also true that Norway is one of the world's largest petroleum exporters, that the role of oil production was doubled from 1986 to 1991, and that the government is considering opening more fields for production.
>
> (Moe, 1993)

Environmental NGOs have also voiced much scepticism regarding the Norwegian attitude favouring joint implementation. They fear that joint implementation will be used by the authorities as an excuse for not implementing the national CO_2 target.

The government is also under strong pressure from industry and labour organizations equally alarmed that an active climate policy will harm economic activities in different sectors and increase unemployment. Industrial interests groups have been active in arguing against radical climate policies, in particular against the CO_2 tax scheme. While some voices in petroleum industrial circles have been raised against scientific legitimacy of global warming, most industry groups seem to take the challenge of climate change seriously. Several climate policy studies have been undertaken by industry itself, and efforts have been made to participate constructively in the debate. Still, the argument of international competitiveness is used actively against unilateral action like the present CO_2 scheme. Also, representatives from industry strongly support the official government proposals for joint implementation.

Economic interests at stake

The most significant negative impacts on the Norwegian economy of an efficient climate regime are expected to be indirect effects of changes in the petroleum markets. So far, however, the major climate–political controversy in Norway has been linked to the national target of stabilizing CO_2 emissions. Since Norwegian energy consumption is based largely on hydro-electric power, the country's potential for reducing CO_2 emissions through changes in its energy consumption patterns is limited. Meeting the 2000 stabilization objective will require the introduction of measures in the indus-try and transportation sectors which will meet strong resistance.

Also crucial are the predicted increases in CO_2 emissions from petroleum activities on the Norwegian continental shelf, estimated to rise by approxi-mately 60 per cent from 1989 to the year 2000. The very substantial economic interests attached to the production and export of oil and gas make it diffi-cult to imagine any appreciable restrictions on these activities, at least in the coming decade. In 1992, nearly a third of total Norwegian export income came from petroleum export – as against only 0.5 per cent 20 years earlier. Petroleum export income as a share of total exports is expected to increase to 35 per cent by 1996. Income from petroleum activities as a share of revenue accounted for in the state budget has increased from 0.2 per cent in 1973, peaking at 20.7 per cent in 1984, reaching the level of 9.6 per cent in 1992, and has been estimated to fluctuate for some years between 9 and 12 per cent (The Norwegian Ministry of Finance, 1994).

In any country attempting to establish an ambitious climate policy, conflicts of interest are bound to arise. Several factors indicate that, in Norway's case, contests between opposing interests can prove especially severe. According to various calculations, including analyses done by ECON (the Centre for Economic Analysis), income from Norwegian petroleum exports will be adversely affected to a severe degree if tough international climate agreements are concluded. ECON's calculations show that losses in petroleum income for Norway may well lie somewhere between NOK 10 and 50 billion a year. Included here are expected losses in oil income of between NOK 10 billion and 40 billion per year and losses in gas income of up to NOK 10 billion per year (Haugland and Roland, 1991). These calcu-lations are based upon a number of preconditions which admittedly are highly uncertain, but which still must be assumed to be reasonably realistic in light of prospective climate negotiations. Another figure being used in the debate is an expected 15 per cent reduction of the value of Norwegian oil reserves if CO_2 taxes are implemented globally (Dagens Næringsliv, 1994). In the longer run, these issues are likely to become a far more important source of interest conflict than domestic CO_2 stabilization.

The structure of Norway's energy consumption, its dependence on petroleum exports, and the fact that the country has already introduced high energy taxes compared with other countries, all these factors are expected to make it costly and politically difficult for Norway to live up to her goal of stabilizing greenhouse emissions at the 1989 level by 2000. The losses in GNP by 2000 have been estimated to 2.2 per cent in a scenario where Norway implements policy measures unilaterally, and 2.7 per cent if other countries also introduce measures to stabilize CO_2 emissions (NOU, 1992, chapter 8).

These measures would also have an influence upon regional policy in Norway. Both geographical and demographical factors add to the problem of stabilization. Long distances and a dispersed population make for a high demand for transport services. Implementing new taxes on transport is politically very difficult since people in the rural areas with an income below the national average will be strongly affected. It is important not to forget that distributional justice is an important factor also in climate change policies. The social democratic government of Norway will have an even harder time than other governments to implement measures which can be shown to be distributionally unfair.

Norwegian energy consumption

Figure 9.1 shows the composition of Norwegian energy consumption in 1990. We note that while oil consumption corresponded to 9 million tonnes of oil equivalent (Mtoe), and coal 1 Mtoe, hydro-power consumption corresponded to 24 Mtoe – almost three-quarters of total energy consumption.

Industry was the largest end user of energy in 1991, with a 40 per cent share. The transport sector followed, with approximately 20 per cent.

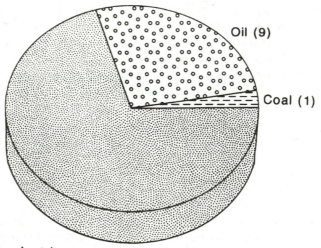

Oil (9)

Coal (1)

Hydroelectric power (24)

Figure 9.1 Energy consumption in Norway in 1990, distributed among energy carriers (million tonnes of oil equivalent)

Source: BP Statistical Review of World Energy, 1991

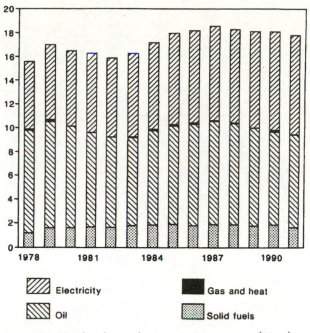

Figure 9.2 Total end use of various energy carriers (Mtoe) 1978–1991

Between 1978 and 1991 only minor changes have taken place between these two sectors, with a 2–3 per cent increase in transport and a corresponding reduction by industry. During this period a new trend emerged, with energy carriers moving from oil to electricity (Figure 9.2).

The transport sector is dominated by oil (Figure 9.3). From 1960 to 1989 energy consumption for transport increased at the rate of 4 per cent a year.

CO_2 emissions

Norway's contribution to global CO_2 emissions are marginal: 0.2 per cent of global emissions. In 1993, Norwegian emissions of CO_2 were approximately 35.7 million tonnes. Estimates for 1994 indicate emissions on the level of 37.2 million tonnes. Road traffic contributed about 24 per cent of the emissions, while coastal traffic and fishing accounted for 10 per cent. Oil and gas production accounted for 23 per cent, while emissions from industrial processes constituted 19 per cent of the total. Emissions of the various greenhouse gases in 1989, 1990, 1993 and projections for the year 2000 are presented in Table 9.1.

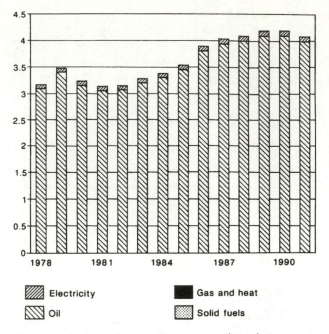

Figure 9.3 End use of energy for transport (Mtoe) 1978–1991

CO_2 emissions increased substantially from 1960 to 1980. Throughout the 1980s emissions were relatively stable. In this period there was a large increase in the consumption of natural gas by the oil industry, while the use of fuel oil was considerably reduced. CO_2 emissions were reduced by about 4 per cent from 1989 to 1991, but from 1991 to 1993 emissions increased, partly due to increased oil and gas production and pipeline transport.

Table 9.1 Emissions of the various greenhouse gases, 1989, 1990, 1993 and projections for 2000 (million tonnes CO_2 equivalents)

	1989	1990	1993[1]	2000	Changes 1989–2000
Total	50.1	50.2	48.1	52.9	+6%
Carbon dioxide	35.2	35.6	35.5	39.5	+12%
Methane	6.6	6.7	6.8	6.4	−2%
Nitrous oxide	4.3	4.2	3.8	4.4	+4%
PFCs	2.0	2.1	1.5	1.4	−26%
Sulphur hexafluoride	2.0	1.7	0.5	0.6	−71%
HFCs	0.0	0.0	0.0	0.6	−

[1] Preliminary figures.

Source: Norway's National Communication under the Framework Convention on Climate Change, Ministry of the Environment, Norway, September 1994

The overall composition of energy consumption renders energy-related emissions of CO_2 per unit of GNP relatively low for Norway as compared with other countries (0.49 tonnes of carbon per USD 1000, calculated in 1985 prices, compared with 0.96 tonnes as the OECD average in 1993). Energy consumption per capita in Norway is amongst the highest in the world.

If, then, Norway's contribution to the greenhouse effect is marginal in a global context (approximately 0.2 per cent of global emissions), why is it so vital for Norway to reduce its emissions? The political problem facing Norwegian politicians has nothing to do with the magnitude of total Norwegian emissions, but concerns energy-related emissions per capita. Despite the dominant role of hydro-power in Norwegian energy supply, Norwegian CO_2 emissions per capita in 1990 were roughly the same as the average for the EU area.

FROM DECLARATIONS TO INTEREST STRUGGLE AND POLITICAL NEGOTIATIONS

We now return to the various stages in the development of Norwegian climate policy, and the content of that policy in different phases from winter/spring 1989 until the Rio conference in 1992 and the follow-up of the conference. How did conflicting interests influence decisions made in this period?

The analytical approach employed here is a traditional one in political science. The actors identified are assumed to represent certain interests, and participation in the decision making processes will be used as an indicator of influence (Sydnes, 1985; Engesland and Sydnes, 1990). The analysis will be limited to the participation of the relevant ministries, as we expect them to reflect the different interests at stake in climate policy negotiations. We also assume that the idea of the segmented state is valid. Ministerial and other state actors together with private and non-governmental organizations form vertical alliances, or policy communities, which correspond to the different policy areas covered by the various ministries.

The Norwegian participation in the IPCC process was of significant importance to the research and the understanding of the climate problem that emerged on the national level. Although the Ministry of the Environment was alone in representing Norwegian authorities at the start of the IPCC process, Norwegian representation was far broader by the time the first assessment was concluded in Sundsvall in Sweden in 1990. There, representatives from the Ministries of the Environment, Foreign Affairs and Finance participated. Norway was amongst the first countries to include officials from Finance in the delegation, though it was still Environment that played the leading role in the IPCC process on behalf of the Norwegian authorities.

Enthusiastic target setting

Phase 1 in the formation of Norwegian climate policy – enthusiastic target setting – was in no manner typical of the Norwegian climate debate. Official decision makers were generally much more moderate as regards the level of ambition than individual MPs tended to be. Elements of this enthusiasm can nevertheless be identified in parallel with the climate–science research phase reviewed in Figure 1.6. Indeed, this is typical of the early phase of the so-called issue attention cycle mentioned both in Chapter 1 and in Chapter 4.

Enthusiasm reached a peak during the Storting debate on White Paper 46 (The Norwegian Ministry of Finance, 1988–89) – the report on Norwegian follow-up of the World Commission on Environment and Development (the Brundtland Commission). In the White Paper, the Brundtland (Labour) government proposed that Norway's CO_2 emissions should be stabilized by the year 2000. This was a goal far less ambitious than the one which the Storting was finally to adopt.

The Centre Party proposed that emissions should be halved by the year 2000. The Socialist Left Party wanted stabilization at the 1987 level by 1993, thereafter a halving of emissions by 2025. The Christian People's Party wanted to stabilize emissions by 1995. The neo-liberal Party of Progress was alone in opposing any stabilization proposal (see below). Despite this euphoria, the proposal put forward by the Conservatives obtained most support: stabilization of emissions at the 1989 level by 2000.

The Conservatives thereby won a 'green' victory over Labour. A number of Labour MPs – led by Prime Minister Gro Harlem Brundtland – declared that agreement on crucial environment issues was a good thing, but that it would not be easy to reach the target. The Prime Minister's scepticism was related to the major political problems likely to occur with the implementation of practical measures. In the winter of 1989, a government-appointed committee had tabled the first report on economic measures to reduce certain types of emissions. This report emphasized the magnitude of emissions from road transport and the need for higher petrol prices, if a reduction of these emissions was to be made.

In the course of the Storting debate, several politicians emphasized the need for action in the transport sector in order to stabilize total emissions of CO_2. The Prime Minister claimed that many nations could cut their CO_2 emissions without doing anything in the transport sector, concentrating instead on the large emissions emanating from industry and power generation. Said Mrs Brundtland: 'We lack that possibility in Norway, so it is necessary for us to take most from transport, a sector especially important for us in our geographically extensive country, sparsely populated as it is in parts' (quoted in Bolstad, 1993, 17).

No formal vote was taken in the Storting; it was the government's interpretation of the debate afterwards that established the target. Active support for the proposal was given by the Office of the Prime Minister as well as the Ministry of the Environment. But, the Ministries of Finance, Industry and Energy have proved sceptical, as expressed in connection with work in the so-called Green Tax Commission and internal discussions during preparation for the Berlin CoP. Amongst the more sceptical decision makers in the administration were some who underestimated the political clout behind this 'resolution'. They probably expected the Storting to discover quite soon that this alternative would simply be too expensive. The repeated confirmation of this target by both the Storting and the government came as a surprise.

However, the Standing Committee on Local Government and the Environment has claimed that the goal is only provisional, subject to assessment in light of continuing climate policy deliberations. Moreover, the former Conservative government had in its budget a modification to the target from the Ministry of the Environment to the effect that it was preliminary, and would be 'assessed in light of the reports as they came in, the technological development, international negotiations and agreements'.

The national research phase – the Interministerial Climate Report 1989–1991

An Interministerial Working Group published a comprehensive report on climate policy in March 1991 (The Norwegian Ministry of the Environment, 1991), primarily as part of the efforts to consolidate a national consensus on the natural science basis for climate negotiations. It also dealt with strategies and measures to reduce emissions of greenhouse gases. Furthermore, it contained a separate chapter on the economic effects of limiting CO_2 emissions, as well as a chapter on climate measures within each of the key economic sectors. This was a major integrative document. In the context of this book, it reflects one of the triggers identified in Figure 4.1, namely the attempted unification of natural science analysis and socio-economic appraisal.

The Report reiterated the main conclusions from IPCC. Other main conclusions may be summarized as follows:

- With regard to climate changes and the impacts for Norway, the most negative effects will probably emerge as a result of the consequences for world trade and the effects in other countries. If global climate change leads to destabilization of trade and cooperation patterns worldwide, this would probably have consequences for Norway as well.
- Viewed in isolation, Norway is unlikely to be affected by climate change as gravely as many other nations.
- As far as Norwegian fisheries are concerned, there is reason to expect that a warmer sea will lead to improved conditions north of the 62nd parallel, while important fish species in the North Sea may decline.

Ocean resources will probably benefit from a warmer climate, though an aggravation of problems related to algae and diseases may be expected.

- Climate change is not expected to have any significant additional health effects in Norway.
- Some sectors of industry, such as building construction, may gain from milder winters with less snowfall.
- An effective international climate strategy should seek the most cost-effective solutions regardless of national boundaries, sectors and climate gases.
- To avoid any undesired changes in terms of trade, harmonization of measures will be an important item in the negotiations.
- Efforts to establish an effective global climate agreement must be given very high priority. Parallel work must be done in order to achieve cost-effective solutions in which Norway in collaboration with its main trading partners as well as competitor countries may achieve the goals set for stabilizing climate gases.
- Industrialized countries bear a particular responsibility for solving global environmental problems and must take the lead with measures to limit their emissions of climate gases. As an initial step, these countries should stabilize their emissions of CO_2 at the 1989 level by the year 2000.
- Industrialized countries have a duty to assist developing countries with additional financial resources and transfers of energy-efficient technology.
- In drawing up an agreement, parties must aim at strengthening the international system for handling problems of this magnitude. In this context, there may be a need for new institutional mechanisms and more binding commitments pursuant to The Hague Declaration.

Disagreement erupted within the interministerial group on the issue of the effects a climate agreement might have for Norwegian oil revenues. During these discussions, the major disagreements between the various ministries crystallized very markedly. Despite the family-like atmosphere in interministerial relations in Norway, the climate issue did not produce the kind of consensus looked for by theorists examining the effectiveness of international environmental regimes as reviewed in Chapter 1.

Yet the work on the Interministerial Climate Report was an important part of the educational process, strongly supported by the Prime Minister's Office and the Ministries of the Environment and Foreign Affairs. In this round, however, consensus building reached farthest in the natural science areas. The chief protagonist was the Ministry of the Environment, strongly inspired by the IPCC process.

The climate group comprised high level ministerial representatives from the Prime Minister's Office, the Ministries of the Environment, Foreign Affairs, Fisheries, Transport, Industry, Finance, Oil and Energy, and Agriculture. The Secretariat for the group was housed in the Environment

Ministry offices, and Environment seemed to play a key role in the process. The Office of the Prime Minister also played a prominent part in this work. The Ministry of Finance entered the process in earnest only in connection with the environment tax reports, and these came later.

International climate negotiations: INC – February 1991–1992 – national and international tug-of-war

A key element in the Norwegian negotiating strategy was the proposal for a clearing-house mechanism and joint implementation of climate efforts. This must be viewed in connection with the special characteristics of Norway's energy consumption patterns and the high marginal costs Norway will be burdened with in putting into effect domestic stabilization targets. According to an OECD environmental performance review of Norway,

> in many cases, larger CO_2 reductions per unit of expenditure could be obtained by financing energy efficiency measures abroad rather than in Norway. The concept of a clearing-house mechanism to facilitate joint measures was one of many Norwegian proposals for institutions to help solve a difficult and costly global problem.
>
> (OECD, 1994, 119–120)

As noted in Chapter 2 the Climate Convention establishes firm principles about a comprehensive approach, including joint implementation and reasonable burden sharing – all crucial principles in the Norwegian strategy of negotiation. As to distribution of any burden, the Convention firmly states that special consideration must be paid to nations dependent upon export of fossil fuels. On the issue of burden sharing, Norway was especially prominent when it came to considerations for developing countries, and the Norwegian delegation actively promoted the idea of additional resource transfers from north to south to finance climate measures in the south.

On the other hand, the Norwegian proposal for a common stabilization target for the OECD area achieved no breakthrough in the Climate Convention. Norway wished to join in the pursuit of such a common goal for a larger area, *inter alia*, in order to be able to discuss with other OECD countries joint implementation of cost-effective measures within the region.

In cooperation with the Global Environment Facility, Poland and Mexico, Norway is currently running two pilot projects intended to demonstrate the potential for joint implementation of measures to mitigate climate change. The project in Mexico focuses on improving energy efficiency by introducing compact fluorescent lamps in two major cities. The project is estimated to result in net annual emission reductions of 101 000 tonnes of CO_2 and 220 tonnes of methane. The project in Poland covers the conversion of a number of coal-fired boilers to gas, and is linked to several World

Bank loans to the energy/environment sector in Poland. The estimated emission reductions are 6700 tonnes of CO_2 per year.

INTERMINISTERIAL COORDINATION

These are early examples of informal joint implementation. The Berlin CoP made it clear that such schemes cannot – at least not in the near future – be credited against Norwegian greenhouse emissions inventories. Norway linked these schemes to technology transfer, scientific and technological capacity building and additional development aid. These are innovations in policy, and couple the Ministries of Environment, Industry and Foreign Affairs in a common endeavour.

Norway's participation in the international climate negotiations was prepared by an interministerial group with representatives from the Ministries of the Environment, Foreign Affairs, Finance, Transport, Industry and Energy (the Ministries of Oil and Energy and of Industry had merged) and Agriculture. Officials from the Prime Minister's Office also took part. The Ministry of the Environment has led and housed the Secretariat for this group (The Climate Secretariat). In addition to the ministerial circle, CICERO – Centre for International Climate and Energy Research at Oslo University – participated in discussions of certain issues.

The Norwegian delegation consisted of nine representatives from four ministries: Environment, Foreign Affairs, Finance, and Industry and Energy plus CICERO. A representative from the Ministry of the Environment chaired the delegation, while Foreign Affairs had the deputy leader. The Environment Ministry had gradually achieved a strong political position for international climate issues. It is also possible that Environment was given leadership because it was able to produce an individual with a formal ranking higher than the representative from Foreign Affairs, namely its Secretary General.

It was apparently the wish of Finance and other sector ministries that Foreign Affairs should lead the negotiations. Foreign Affairs also played a prominent part in the negotiations. In addition to the vital functions as representative and spokesperson for the Norwegian delegation in a number of contexts, Foreign Affairs exercised an important diplomatic function internally in the Norwegian delegation, by building consensus.

The Secretary General of the Ministry of the Environment regarded training, education and awareness raising as vital parts of the interministerial process. He placed great importance upon including the various sectoral ministries to ensure a broad political understanding of climate problems. Still, the ministries played individual roles with varying degrees of involvement. Agriculture, although much engaged in connection with the Interministerial Climate Report, did not participate in the climate negotiations. Finance held a high profile in the internal bargaining processes in the

Norwegian climate delegation. By the time the climate negotiations really got going, it had decided to enter the scene, because it felt that there was an urgent need for the kinds of competence and analytical approaches it could offer when it came to identifying measures and solutions.

The crucial role of the Ministry of Finance became evident in connection with the drawing up of the Norwegian positions at the start of the Third INC meeting in Nairobi in September 1991. At this meeting, Norway was able to table a proposal covering all the main elements of a climate agreement. Norway had worked out a framework for the whole agreement as early as June that year. No other nation had progressed so far in preparations for the Nairobi meeting.

Environment's function in the negotiation delegation was dual. The Ministry's representatives had, together with those from Foreign Affairs, greater knowledge about the diplomatic game. Environment played a vital technical role. Moreover, Environment was important in work concerning the institutional parts of the climate agreement, reporting systems within the framework of the agreement and issues of verification. As specific proposals for solutions appeared on the agenda, however, Finance came to play an increasingly important role. The proposals for mechanisms and solutions put forward by the Norwegian delegation bore the stamp of active commitment by the Ministry of Finance.

There is reason to assume that had Environment been more dominant in the Norwegian delegation, more weight would have been placed upon traditional environmental policy measures. Above all national commitments might have been more centrally placed. Established environment agreements such as the Montreal Protocol and measures agreed upon there would have been more of an inspiration.

AFTER RIO: THE TUG-OF-WAR CONTINUES

There have been no significant changes in the Norwegian negotiating position within the INC since Rio. However, work with the national plan of action has proved that Norwegian ambitions as a leader in international climate politics have become increasingly more difficult to fulfil. In its report to the Storting on the Rio conference, the Norwegian government announced that it would develop a national plan of action – as a follow-up of the Convention and the Norwegian CO_2 target (The Norwegian Ministry of the Environment, 1993a). In the event, the actual report was delayed by one and a half years from when it was first promised in late 1993. Part of the reason for this was that the work was decentralized, with the Ministries of Foreign Affairs, the Environment, Finance, Transport, Industry and Energy, and Agriculture each writing separate parts on their respective fields of responsibility. The differences in points of view caused bigger problems than originally expected.

The Ministry of the Environment acted as coordinator. It also has the main responsibility for Norwegian participation in the Commission for Sustainable Development (CSD), and has taken a lead responsibility in the Commission's work on consumption and sustainable development. A 'high level group on climate and acid rain' was established after the Norwegian Rio Report had been presented in January 1993. It worked as a steering group for work on the action plan. All drafts of parts of the action plan were handled interministerially, with all policy-related questions taken up for assessment by the high level group.

From the Norwegian authorities' point of view it is important to ensure that the climate policy measures which Norway actually has introduced are made visible through these kind of action plans. There also seems to be a desire to stress that Norway's future climate policy behaviour will be influenced by the policies adopted by other countries – not least that it will be difficult for Norway to continue to apply the CO_2 tax in its present form if other countries fail to introduce corresponding measures.

The Minister of the Environment reported to the Storting in April 1994 (Berntsen, 1994) that the costs of Norwegian climate measures must be compared with costs of similar measures in other countries to avoid the economic burden on Norway becoming greater than that carried by Norway's most important competitors (Ministry of the Environment, 1994–1995, 10). This could be interpreted to mean that Norwegian authorities are willing to accept that the costs of climate policy measures will be higher, but not unreasonably higher than in other industrialized countries. However, the interministerial process handling these issues uncovered different opinions about the matter.

During the work on the action plan, officials explored the possibilities for climate policy collaboration between industry and public authorities. Norwegian industry has shown an interest in implementing its own measures to contribute to the stabilization of CO_2 emissions. This reflects a trend found generally in these case studies, namely that industry is willing to play a part in 'good housekeeping' so long as this is voluntary, or at least flexible in the method of achievement.

The Norwegian action plan was finally presented in June 1995, after a considerable delay due to conflicting interests related to the Norwegian stabilization target. The development of voluntary agreements with industry represents the new element in the Norwegian climate strategy. Positive experiences from other countries have inspired Norwegian authorities. Considerable efforts must be expected to be put into developing the cooperation between industry and government in this field in the years to come (Ministry of the Environment, 1994–1995).

The action plan does not break new ground in defining new policy instruments. However, it presents a thorough analysis of the impacts on the Norwegian economy of the implementation of different climate policy instruments. This analysis is the common ground for the discussion of Norway's role in international climate efforts: what should and what could be the role of Norway in the development of the Climate Convention, and what measures can be implemented in Norway independently of the international process? The action plan seems to represent an important step forward in building the interministerial consensus with regard to Norwegian climate policy. Nevertheless, in defining the necessary practical measures, there still seems to be a long way to go. It is crucial in this context, though, that the dominating role of the discussion on the Norwegian national target is being scaled down. This may gradually pave the way for practical solutions.

DEVELOPMENT OF ENERGY TAXES

Norway has implemented an ambitious CO_2 taxation scheme. The CO_2 tax level in Norway is equivalent to 50 dollars per tonne of CO_2 or 20 dollars per barrel of oil. These rates are the highest compared with similar taxes introduced or proposed in other countries. The CO_2 taxes are seen by Norwegian authorities as a central measure within a future protocol under the Climate Convention, and the Norwegians look to other OECD countries to follow suit.

The specific CO_2 tax introduced in January 1991 involved a tax of Norwegian Kroner (NOK) 0.60 per litre of gasoline, in addition to the taxes motivated by road use and fiscal purposes, NOK 0.30 tax on fuel oil and diesel in addition to an already established, fiscally motivated tax of NOK 0.32 per litre and a sulphur tax, NOK 0.60 per standard cubic metre of natural gas, and NOK 0.60 per litre for other petroleum products used on the Norwegian continental shelf. A CO_2 tax on coal of NOK 300 per tonne was introduced in 1992. At the same time the fiscal tax on fuel oil was reduced bringing the total tax on fuel oils down from NOK 0.62 to 0.47 per litre plus the sulphur element.

In January 1993, the CO_2 tax on coal was increased to NOK 400 per tonne. The CO_2 tax on gasoline was maintained at NOK 0.80 per litre, and the tax on fuel oil was raised to NOK 0.40 per litre. The fiscal tax on fuel oil was removed, and this led to a reduction in total taxes on fuel oil, from NOK 0.47 to 0.40 per litre plus the sulphur element (The Norwegian Ministry of the Environment, 1993b).

The status for the CO_2 tax in August 1993 was as follows:

- The petrol tax is levied on all domestic utilization with negligible exceptions. The level is NOK 0.80 per litre, corresponding to approximately $50 per tonne CO_2 or $20 per bbl crude.
- The tax on fuel oil and autodiesel applies to household as well as industrial use and is NOK 0.40 per litre, corresponding to approximately

$25 per tonne CO_2. Exempted are the paper, pulp and fishmeal industries, which pay NOK 0.20 per litre. Fishing vessels and domestic transport of goods by sea are also exempted. The latter group accounts for approximately 10 per cent of CO_2 emissions in Norway. Corresponding exemption applies also to air transport.

- Coal tax is limited to coal used for heating purposes, with an exemption for the cement industry. A major exemption concerns emissions from industrial processing, mainly where coal is used as a reduction material in metal smelters. These account all together for 18 per cent of national CO_2 emissions.

All together, the CO_2 tax applies to approximately 60 per cent of total CO_2 emissions in Norway. Table 9.2 presents the tax rates for petroleum products, gas, coal and coke in 1994. Table 9.3 summarizes the changes in CO_2 taxes on petroleum products, gas, coal and coke for 1991–1994.

Concerning the effects of the CO_2 tax, the Ministry of the Environment is of the opinion that it is too early to draw definite conclusions; however, the CO_2 tax is considered to have limited increases in emission which would otherwise have taken place. More time will be demanded to assess the more specific effects on the environment as well as on the economic structure. In the opinion of the Ministry of the Environment, the Norwegian CO_2 tax must be regarded as an initial attempt to develop relevant means of tackling the global greenhouse challenge. The climate changes will take place over a long period of time and thus require means that can be allowed to have an effect over a lengthy period of time.

Table 9.2 Tax rates for petroleum products (NOK/l), gas (NOK/m^3) and coal and coke (NOK/kg), 1994

	Basic tax	CO_2 tax	SO_2 tax	Total	CO_2 tax per kg CO_2 emitted
Unleaded petrol	3.12	0.82		3.94	0.35
Leaded petrol	3.78	0.82		4.60	0.35
Autodiesel	2.45	0.41	0.07	2.93	0.16
Mineral oil	0	0.41	0.07[1]	0.48	0.16
Diesel, North Sea	0	0.82	0	0.82	0.31
Gas, North Sea	0	0.82		0.82	0.35
Pit coal	0	0.41		0.41	0.17
Coal coke	0	0.41		0.41	0.13
Petroleum coke	0	0.41		0.41	0.11

[1] The tax rate is 0.07 NOK per 0.25 per cent SO_2 content (1 US$ equals about 7 NOK).

Source: Norway's National Communication under the Framework Convention on Climate Change, Ministry of the Environment, Norway, September 1994

Table 9.3 Change in CO_2 taxes on petroleum products (NOK/l), gas (NOK/m^3) and coal and coke (NOK/kg), 1991–1994

	From January 1991	*From January 1992*	*From July 1992*	*From January 1993*	*From January 1994*
Petrol	0.60	0.80			0.82
Mineral oil[1]	0.30			0.40	0.41
Gas, North Sea	0.60	0.80			0.82
Oil, North Sea	0.60	0.80			0.82
Coal and coke			0.30	0.40	0.41

[1] From 1 January 1991 the CO_2 tax was levied in addition to an already established fiscal duty of 0.32 NOK/litre. The fiscal duty was reduced to 0.17 NOK from 1 July 1992 and to zero from 1 January 1993. Total taxes on light fuel diesel oils were reduced from 0.69 NOK/litre in 1991 to 0.48 NOK/litre in 1994 (including sulphur tax).

Source: Norway's National Communication under the Framework Convention on Climate Change, Ministry of the Environment, Norway, September 1994

The national tug-of-war over a CO_2 tax was not long in emerging. Even though in most cases no clear connection can yet be proved between the imposition of this tax and weakened competitive ability, spokespeople for industry have sharply attacked these measures. In response to this the Revised National Budget for 1992 produced a broad evaluation of the implications for the competitiveness of various industries. It was claimed that this environment tax renders the Norwegian economy more efficient since it corrects an externality and moreover procures revenues that may be used to reduce other taxes.

Resistance to the Norwegian CO_2 tax has been closely related to the fact that few countries have introduced such a tax, and among those who have, Norway has introduced the highest level so far – choosing not to await international agreements in this area. It thus appears quite easy to use the argument 'reduced competitiveness' for those opposed to such a tax. As of March 1994, the oil industry was again campaigning against the CO_2 tax, arguing that the Norwegian tax ought to be harmonized with the tax level in competing countries.

In June 1994, the Secretary General of the Norwegian Petroleum Institute suggested that Norwegian authorities ought to encourage other countries **not** to introduce CO_2 taxes, since this would eventually lead to a transfer of oil revenues from producing to consuming countries. He also referred to estimates showing that the value of Norwegian oil reserves will be reduced by 15 per cent if CO_2 taxes are introduced on a global scale. There is little doubt that the Norwegian government is under strong pressure from the oil industry to 'ease the burdens' on the activities in this sector.

However, the present CO_2 tax will not ensure stabilization of Norwegian CO_2 emissions. Based on the assumptions made in the regular economic review by the government, CO_2 emissions are expected to increase by 16 per cent by 2000. This indicates that the present CO_2 scheme is far from adequate. The Minister of the Environment has stressed that significantly stronger measures will probably have to be used, and that CO_2 taxes will continue to be a major tool to regulate emissions from energy consumption and production. He hoped that the government will expand the tax to include CO_2 emissions which are not yet covered by the scheme. He also said that the government will implement stronger measures if necessary. Both economic and administrative measures will be considered.

Norway officially disapproved of the combined energy/carbon tax proposed in the European Union, since coal as an energy source benefits from such an energy tax, while hydro-power is the loser. A pure CO_2 tax would be preferable, in the Norwegian view.

The methanol factory at Tjelbergodden: a climate policy test case?

On 4 February 1992 the Storting decided to build a methanol factory at Tjelbergodden in Nordmøre based on natural gas from the Heidrun Field on Haltenbanken. This issue has been the subject of much debate in the media. For many, it has become a climate policy test case, since methanol production is expected to contribute to an increase in the Norwegian CO_2 emissions. Environmental organizations regarded the decision to build a methanol factory as a symbol of the undermining of the Norwegian CO_2 objective (Bolstad, 1993, 8).

The Heidrun Field is first and foremost an oilfield, but the oil is not available for production without first extracting the gas. Gas from Heidrun is not very profitable, as export to European customers would prove too costly from this field alone. To be profitable, export-directed production would have to take place in connection with development and production from other gas fields in the same area. Since gas may not be flared on exploration sites, development of the field alone was therefore conditional upon other utilization of the gas.

The oil companies wished the gas to be brought ashore midway up the west coast of Norway, and favoured the establishment of a gas power station and a methanol factory to utilize the gas from Heidrun. Calculations, however, showed that a gas power station would mean a 5–6 per cent increase in Norwegian CO_2 emissions, and the methanol factory would lead to an increase of 1.5 per cent. The environmental movement then claimed that such increases would make it impossible for Norway to stabilize its emissions, since any addition would require major reductions in emissions in sectors where this has already proved difficult.

Other considerations also created doubts. Financial calculations presented in the press showed that the methanol factory might have to be subsidized by consumers. The state would provide most of the financing for building the pipeline, and the cheap gas deliveries could also be in contravention of GATT rules. Marketing prospects for methanol were regarded by many economists to be poor (Bolstad, 1993, 48).

This risk failed to deter the oil companies: it was the state which first and foremost bore the risk of development. The then managing director of Statoil's petrochemical activities said: 'In any case the state bears the brunt of the risk of all shelf activities. The methanol factory at Nordmøre is no exception' (Bolstad, 1993, 50).

The alliances in the struggle about the Heidrun development, according to Bolstad (1993), appeared as follows: the Oil and Energy Ministry, the Trade Union Confederation, Confederation of Norwegian Industries, Statoil, Norsk Hydro, local politicians in mid-Norway, the Conservative Party, the Progress Party, and the majority of Labour supported the plan. Opposition came from all the environmental organizations, the Ministry of the Environment, former prime minister Kåre Willoch (Conservative), the Labour youth organization, the Socialist Left Party and a wing within the Labour Party. The Centre Party, the Christian People's Party and the Ministry of Finance found themselves in an intriguing intermediate position.

The Ministry of the Environment was actively involved in the decision making processes of the Heidrun development during the time when this issue was 'politicized'. The Ministry tried to have the issue regarded in conjunction with the general work on energy saving. The then Environment Minister Kristin Hille Valla claimed that energy efficiency should be given priority before development of new power projects:

> The CO_2 goal means increased emissions from any gas power station being compensated with corresponding reductions in the use of for instance oil and petrol. . . . A gas power station would make it difficult for us to achieve our own CO_2 goal. The improved application of already developed power and energy saving measures ought to be given priority before developing new power projects.
>
> (Bolstad, 1993, 54)

On 14 May 1991 the Storting adopted the resolution that the Heidrun Field should be developed. In November of the same year the government turned down the building of a gas power station nearby. Environmental organizations noted that the pipeline for taking the gas to Tjelbergodden would be expanded to a capacity five times the needs of the methanol factory. This was interpreted as meaning that the plans for a gas power station had only been postponed and not dropped. When in October 1993 Statkraft and Statoil presented plans for the joint development of a new gas power station, the capacity of the pipeline to Tjelbergodden was crucial in the evaluation and subsequent political debate.

Supporters of the methanol factory have voiced the opinion that in terms of environmental policy, the factory might just as well be located on Tjelbergodden as for instance in the Netherlands, which could be a possible alternative. Concerning the gas power station, one argument holds that gas from Norway could well replace nuclear power in Finland. If Finland should wish to reduce its power production from nuclear stations, this would have to be replaced by either coal or gas. In a climate context, gas is by far the more acceptable alternative.

ECONOMIC AND ADMINISTRATIVE INSTRUMENTS IN NORWEGIAN CLIMATE POLICY

Norwegian authorities have so far favoured the use of economic and other market-oriented instruments in building up a climate strategy. These priorities are based on the nature of the climate problem itself and the need for cost-efficient measures to limit the assumedly high costs for Norway in reducing greenhouse gas emissions.

Administrative instruments in climate policy, such as direct regulations, public persuasion and research policies are also considered to be important greenhouse policies. However, other countries put far more emphasis on these instruments than Norway. It is interesting to note that the last OECD environmental performance review urges Norway to pay more attention to direct regulation through administrative instruments. The OECD finds it virtually impossible for Norway to reach the CO_2 target by the use of economic instruments alone: 'With the possible exception of the hydro carbons sector, however, the CO_2 tax has limited incentive effect. . . . As a result, Norway may find it difficult to reach its stabilisation target without additional CO_2 reduction measures' (OECD, 1993).

Norwegian authorities have decided to invite industry to participate in voluntary agreements to help limit CO_2 emissions. Efforts related to information and education to improve energy efficiency will also be expanded. Special arrangements with economic support for the introduction of new technology will be continued, and the government will encourage the further development of regional energy efficiency centres.

Direct regulation may become a more important instrument in Norwegian climate policy in the future. However, there are differing views between the ministries on how to prioritize between economic instruments and direct regulation. Finance, Industry and Energy and Foreign Affairs are the main proponents of economic instruments while Environment puts more emphasis on administrative measures. So far, Environment has had few allies in this discussion. But if the government is not able to expand the use of economic instruments to meet the CO_2 target, proponents of supplementing these strategies with more direct means may have an easier task.

ORGANIZATION OF THE PROCESSES – PARTICIPATION AND INFLUENCE

The negative impacts on petroleum markets and the Norwegian economy of an efficient international climate regime may prove to be dramatic. It could, then, be claimed that Norway in its climate policy has actually put climate issues before its own economic interests – at least if we compare Norway with other countries correspondingly dependent on petroleum exports. However, both national and subnational economic interests have become increasingly visible compared with environmental considerations. This is expressed primarily through the pattern of participation in the climate processes and in the development of the general approach to climate problems. Actors representing economic interests have become increasingly active in the decision making processes over time.

Though Norway still adheres – officially – to the goals originally set, climate policies have become increasingly characterized by a more traditional economic approach, illustrated by the importance placed upon the principles of cost-effectiveness and the proposal for joint implementation. Is this because Norwegian economic interests, after a tough struggle with environmental interests, actually carry more weight in climate policy than before? Perhaps it is more correct to regard the changes as a result of a learning process, where authorities in the different ministries have concluded that it is appropriate, also from an environmental point of view, to allow economic principles to form the basis for climate policy.

A prominent characteristic of decision making processes in Norwegian climate policy has been *broad participation*. A number of different actors at governmental and non-governmental level have participated. This was the case with the Interministerial Climate Report of 1991 in which all affected ministries participated. The authorities also chose to invite several external research institutions to contribute to the analytical work. The Report resulted in a common understanding of climate change which became the basis for Norway's participation in the international climate negotiations.

Interministerial collaboration and coordination has continued to be a vital characteristic of the policy process. The influence of the various ministries has fluctuated over time, dependent upon the particular phase of the process. The crucial point here is the relationship between the various ministries and economic/political interest groups in Norwegian society. The links between the environmental movement and the Ministry of the Environment are reflected in both the aggressiveness and willingness of the Ministry to take the lead in Norwegian climate policy.

The role of powerful policy communities around the Ministry of Industry and Energy (which includes the former Ministry of Petroleum and Energy) and the interests of the oil industry and other industrial economic interests have been reflected in test cases like the establishment of a methanol factory at Tjelbergodden in Norway and the debate on the CO_2 tax.

The Ministry of Finance has all the time striven to contribute to international arrangements that are as effective and as flexible as possible. Finance has also been concerned that the costs of an international climate regime should not be disproportionately high for Norway. The Ministry has always been of the opinion that general economic measures would be the most suitable means. There has been a sceptical attitude to the more traditional approach involving proportional reductions of emissions within each country. This approach, which has also been applied to the sulphur and NO_X negotiations, is a vital matter of principle for the Ministry of Finance.

Representatives of all the ministries interviewed for this study have stressed the advantages of the interministerial process in developing a mutual appreciation of each others' positions. Natural sciences has been taught by the Ministry of the Environment. The 'gang of four', the Ministries of Finance, Industry, Oil and Energy, and Transportation (now the 'gang of three' since Industry and Oil and Energy merged), has been on the offensive in identifying practical means and measures to reduce emissions. The Ministry of Foreign Affairs has played the diplomat's role, building up consensus internally in the Norwegian camp, to provide a sound platform to build on internationally. This is an important aspect of institutional learning, just as important as specific organizational relationships or structural linkages.

However, there are still great difficulties in defining the level of ambition. The Ministry of Finance has been concerned that Norway will have only marginal influence on the direct and indirect income effects produced by a climate regime. It was thus essential to concentrate on the choice of mechanisms built into the regime, trying to keep the costs involved as low as possible. Cost-effectiveness and joint implementation are crucial points here.

The Ministry of Industry and Energy has been concerned about how a climate regime will influence the distribution of the oil revenues. Some favour improved producer cooperation capable of reducing oil supply so as to push up prices. This would mean that oil consumption could be reduced without sacrificing producer income.

Traditionally, there has been a lack of cooperation between the Ministries of the Environment and Finance, but in the late 1980s this situation began to change. The Office of the Prime Minister established in a circular that Finance and the other ministries concerned were to be involved right from the preliminary stages in environmental issues, through the process of dealing with each issue and in the ensuing work. The Ministry of the Environment for its part cooperated, but the Finance Ministry was far less active in involving Environment, even when tax and national income matters were significant for environmental policy making.

From the point of view of the Ministry of the Environment, it could be advantageous to have several bilateral processes – perhaps preceding the multilateral ones. This would provide the environmental authorities with better control over the processes, through forming alliances with other like-minded actors.

The fact that the Ministry of the Environment exercised secretariat functions and hence overall responsibility for interministerial coordination has different effects. It helped to achieve a breakthrough for its views with regard to some issues, but it also meant that the Ministry adopted a more neutral stance, and hence did not feel so free to fight for its viewpoints.

Another relevant factor concerned the academic background of the individuals participating. Generally speaking, the Ministries of Finance and Transportation employed officials with a background in economics, while Environment officials usually had technical or natural science backgrounds. While the economists usually take a top-down approach, those with a technical background often adopt a bottom-up perspective. Exchange of officials between the ministries has become more frequent over the last few years, and together with changes in recruitment policies this may improve the potential for cooperation. However, such a development will not necessarily strengthen the influence of the perspective traditionally advocated by the Ministry of the Environment.

Norway's role as a driving force – the various ministries' assessments

At an early stage, the government made it clear that Norway wished to act as a leader in climate negotiations. Lengthy discussions were conducted to establish the precise content of such a role, with considerable disagreement between the various ministries.

The Ministry of the Environment favoured enacting national climate response measures to serve as a good example to other nations. Representatives of this ministry have maintained that Norway should achieve its CO_2 goal, even if this were not to prove cost-effective in an international perspective. It appears, however, that Environment has gradually been persuaded to support the idea of joint implementation. However, if it should turn out that joint implementation cannot be established within a reasonable period, demands from the environmental movement for a Norwegian option will gain strength. This in turn would allow the Ministry to bring pressure to bear on the government.

The Ministries of Industry and Energy and Finance argued that national solutions to climate problems would not be effective. The problem demands a global approach: joint implementation of climate measures is therefore preferable to national goals, not only for national economic reasons, but also to create an efficient climate regime. These ministries maintained that if the CO_2 goal is executed through non-cost-effective measures, this would weaken Norway's driving force role rather than strengthen it. A domestic climate policy must be defined as part of a larger cost-effective system, not something preceding the establishment of such a mechanism.

THE ROLE OF POLITICAL CULTURE, IDEOLOGY AND COGNITIVE FACTORS

The Norwegian experience must be analysed within the broader framework of Norwegian foreign policy. Without the cross-party consensus-oriented tradition associated with Norwegian foreign policy, the minority government of the Labour Party would have had greater difficulties in being so active on the international scene. This consensus-oriented tradition of Norwegian policy making also corresponds with Katzenstein's (1985) image of the corporatist state: corporatism is described as continuous negotiation and cooperation between consensus seeking elite cultures. The Norwegian bargaining process was structured to accept the IPCC analysis and the broad purpose of the UN FCCC. Apart from the formal right wing Progress Party, all the Storting's political parties were in favour of positive action in the face of international obligation. This is part of Norway's image of moral leadership in international affairs.

Norway's climate policy must also be viewed in conjunction with external as well as internal factors. Progress in other nations' understanding and handling of climate change issues was an important external condition for Norwegian policy making in this field. Figure 9.4 presents a rough sketch of the material, ideological and cognitive factors that have emerged in the tug-of-war between central decision makers in Norwegian climate policy and the balancing process between the various interests.

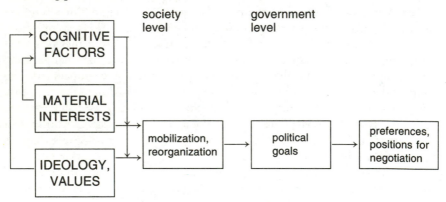

Figure 9.4 Factors influencing the shaping of climate policy

The interest conflicts sketched in the figure – between climate policy goals on the one hand and other Norwegian economic interests on the other hand – constitute an increasingly difficult dilemma for decision makers in Norway. The closer decisions are likely to have a specific influence on the Norwegian people, the greater the influence of key economic elites – for instance, prior to the decision on building the methanol plant at Tjelbergodden.

Norwegian climate policy is increasingly under pressure from business and industry, who see their interests threatened by CO_2 taxes in particular and Norway's climate policy ambitions in general. For these interests the role of ideology and cognitive factors is less important. We observe in the Norwegian case not only a challenge to the identification and implementation of policy. We also see a challenge to the traditional consensus–corporatist styles of policy making in Norway. In this respect as in the other case studies, climate change politics impinge on changes in policy style that are occurring for other reasons. The key for the visionary leader is how to adapt to these new styles, by incorporating climate change into a variety of policy arenas and interest coalitions, yet making it stick.

To do this, the Norwegians have looked to widely held attitudes as regards nature and environmental protection, norms for international collaboration and norms for economic policy and the role of the state in society. These factors pull in the direction of an active international environmental policy. Broad environmental mobilization is expressed through active participation in NGOs, media focus on local, national and international/global issues, broad agreement across party lines on crucial environmental goals and a relatively high degree of institutionalization of environmental considerations in public administration. Though conflicts and disagreement may exist as to when and how environmental challenges should best be met, there is still a widespread awareness of environmental issues. Furthermore, one reason why the precautionary principle is so strong in Norway may be connected to the general risk aversion typical of a welfare state.

Norway's wish for a strong climate regime must be viewed in the light of two ideological factors in Norwegian politics. First the 'small state' tradition in foreign policy, which seeks to solve international conflicts and challenges through broad multilateral collaboration among states. Bilateral solutions are often less satisfactory for small countries who wield limited power, as they tend to favour the larger powers.

The other factor is the attention given to the responsibility of the state in shaping consumer behaviour in favour of a wider public interest. In Norway there is considerable latitude with regard to seeking to influence consumer choices through taxation and fees. This ties in with a radically different attitude as to the role of an active state and a more general acceptance of applying such economic measures in Norwegian society compared with what we observe in many other western countries.

Cognitive processes apply primarily to the interpretation of climate change as a problem – in the natural science, political or economic sense. We have seen how this viewpoint varied from ministry to ministry with limited success in developing a unified view. This is, however, healthy for Norwegian policy making. The important institutional innovation lies with the *process* of interministerial analysis and bargaining. Through that process, cognitions did change. Though subtle and difficult to observe, this was a significant institutional response.

CONCLUSIONS

Norwegian authorities are faced with a difficult climate policy dilemma. On the one hand the government has long intended to act as a leader in international climate policy making. On the other, the dominating role of hydro, the high profile of transport, and the role of Norway as a major exporter of oil and gas create barriers to a radical CO_2 emission strategy. Despite the official target of stabilizing CO_2 emissions at the 1989 level by the year 2000, emissions are expected to rise by 12–16 per cent by the turn of the century.

It may be impossible for Norway to adopt domestic policy measures which can guarantee an implementation of the national CO_2 target. The Norwegian ambition of being a leader in international climate politics still seems to have domestic support. Norway's relative influence will, however, also depend on what amount of resources the Norwegian government is willing to put into preparations for and participation in the climate negotiations in the years to come. It will certainly also depend on the performance of other countries and their wish, and ability, to take the lead.

With regard to Norway's score in the green 'beauty contest', it is worth while noting that the Norwegian institutional response to the challenge of climate change has been quite efficient. The broad participation of the various actors and interests involved must be seen as an expression of the importance attached to this issue by Norwegian authorities. Much emphasis has been put on building a consensus across traditional institutional cleavages and interministerial disagreements. Participation has allowed a thorough discussion of potential solutions. Representatives from all the ministries concerned have stressed that interministerial cooperation has been a vital learning process. This experience must also be seen in light of the general consensus-oriented policy style in foreign policy matters. To the extent that climate policy is seen as a foreign policy matter, the need for a broad consensus nationally has been accepted by most actors, including the major political parties.

The pattern of participation indicates that environmental considerations do not rule the ground alone. Economic interests play an increasingly important role. The much acclaimed learning process has so far produced little when it comes to practical policy coordination. Oil and transport policies are still on different tracks, largely unaffected by the climate debate, driven by domestic interests and powerful lobbies. When it comes to implementing policy measures at the national level, the Norwegian policy style is very similar to that of most industrialized democracies. The various segments of state and society fight for their interests, and in this game long term interests very often lose – be they environmental concerns or economic ones.

Measures introduced in the transport sector will be crucial with regard to reaching the CO_2 target. Politically it is very difficult to handle this sector, mainly because of the long distances and a highly decentralized settlement pattern. A major problem here is that Norwegian taxes on transport fuels are already among the highest in the world. To cut any further in this sector, more efforts will have to be put into overall transport planning and its environmental component.

Politicians seem to be more ambivalent today than they were when the national CO_2 target was adopted in 1989. They have learned that climate policy measures are more difficult to implement than they expected. They also seem to experience more fragile public support and a decline in environmental concern. These factors go a long way in explaining the current ambiguity in Norwegian climate strategy.

At present, Norwegian authorities seem to be faced with the following options:

- concede that stabilization, not to mention the reduction of CO_2 emissions in Norway, is politically impossible, and thereby risk a severe blow to its reputation as an environmental frontrunner;
- impose draconian measures on the transport, petroleum and industrial sectors, and risk resulting economic losses and increased uneven distribution of wealth;
- seek credible partners for joint implementation and present a scheme for international authorization, either to future Conferences of Parties or to the European Union.

Unless Norwegian authorities choose to follow up their activist approach by well-argued and supported action at the international level, there is little left of what was presented as a radical climate policy several years ago. There is no doubt that stabilizing or reducing Norwegian CO_2 emissions will be very difficult if not implemented jointly with other countries. Since there is yet no scheme for international authorization of carbon credits, Norwegian efforts to help establish such a system will be an indicator of the dedication of Norwegian climate policy. Look out for further bilateral 'offset' deals along the lines of those begun in Poland and Mexico. Another possibility is that Norwegian industry, within the voluntary agreements for CO_2 reduction, will establish joint implementation projects with partners in other countries on a non-governmental basis. Norwegian authorities probably also have a good case in arguing that since Norwegian natural gas exports may serve the purpose of replacing coal in importing countries, Norway should not bear the burden alone of the CO_2 emissions resulting from the production and exports of the gas.

Nevertheless, in the absence of an operative global regime for joint implementation, the pressure on domestic implementation will not go away. If Norway is going to act in accordance with the environmental concerns which guided the World Commission on Environment and Development, controversial choices will also have to be made, between short term economic interests and consideration for the global environment.

REFERENCES

Aftenposten, 7 March 1994.

Berntsen, T. (1994) *Environmental Policy Report to the Storting*, Oslo: Ministry of the Environment.

Bolstad, G. (1993) *Inn i drivhuset – Hva er galt med norsk miljøpolitikk?* (Into the greenhouse – What is wrong with Norwegian environmental policy?), Oslo: Cappelen.

Central Bureau of Statistics (1991) *Electricity Statistics 1991*, No. C34, Oslo.

Dagens Næringsliv, 1 June 1994.

Engesland, B. and Sydnes, A.K. (1990) Competing Interests and Risk Aversion: Norway and the Call for International Energy Cooperation, in *Naive Newcomer or Shrewd Salesman? Norway – A Major Oil and Gas Exporter* (Bergesen, H.O. and Sydnes, A.K. eds), Oslo: The Fridtjof Nansen Institute.

Haugland, T. and Roland, K. (1991) *Klimapolitikk og de internasjonale energimarkedene* (Climate policy and the international energy markets), ECON Report No. 34/91, Oslo.

International Energy Agency, *Energy Balances of OECD Countries 1990–1991*, Paris.

International Energy Agency (1992) *Energy Prices and Taxes, Fourth Quarter 1992*, Paris: OECD/IEA.

Katzenstein, P. (1985) *Small States in World Markets: Industrial Policy in Europe*, Ithaca, NY: Cornell University Press.

Moe, C. (1993) *ECO Newsletter, Climate News*, Geneva.

Norway's National Communication under the Framework Convention on Climate Change, September 1994, Ministry of the Environment, Oslo.

OECD/IEA (1992) *Energy Prices and Taxes, Fourth Quarter, 1992*, Paris: OECD/IEA.

OECD (1993) *Environmental Performance Review: Norway*, Paris: OECD.

OECD (1994) *Climate Change Policy Initiatives, 1994 update*, Paris: OECD.

Sydnes, A.K. (1985) *Norges stillingtagen til Nord-Syd-konflikten i oljemarkedet – Petroleumsøkonomiske interesser og utenrikspolitiske orienteringer* (Norway's position on the north–south conflict in the oil market: petroleum-economic interests versus foreign policy considerations), R:004-1985, Oslo: The Fridtjof Nansen Institute.

The Norwegian Ministry of the Environment (1991) *The Greenhouse Effect, Implications and Measures, Report from the Interministerial Climate Group*, Oslo.

The Norwegian Ministry of the Environment (1992) *The Norwegian National Report to UNCED*, Oslo.

The Norwegian Ministry of the Environment (1993a) *St.meld.nr. 13 (1992–93): Om FN-konferansen om miljø og utvikling i Rio de Janeiro* (White Paper No. 13 to Parliament (1992–93) on the UN Conference on Environment and Development in Rio de Janeiro, Oslo.

The Norwegian Ministry of the Environment (1993b) *Carbon taxes; Norwegian Experience*, Oslo.

The Norwegian Ministry of the Environment (1994–1995) *St. meld.nr. 41, Om norsk politikk mot klimaendringer og utslipp av nitrogenoksider (NOx)* (The Norwegian action plan on climate change), Oslo.

The Norwegian Ministry of Finance (1989) *St.meld.nr. 4 (1988–89), Langtidspro-grammet 1990–1993* (White Paper No. 4 to Parliament (1988–89), Programme of Planning 1990–93), Oslo.

The Norwegian Ministry of Finance (1992) *Mot en mer kostnadseffektiv miljøpolitikk i 1990–årene – Prinsipper og forslag til bedre prising av miljøet* (Towards a more cost-efficient environmental policy of the 1990s: Principles and proposals for a better pricing of the environment), NOU: 3, Oslo.

The Norwegian Ministry of Finance (1994) *The National Budget 1994*, Oslo.

10

CLIMATE CHANGE POLITICS
IN ITALY

Alessandra Marchetti

This chapter aims to identify the essence and the evolution of Italian climate change policy in a response to the United Nations Framework Convention on Climate Change (UN FCCC). The dynamics of this change cover domestic environmental policy within a variety of national politics and policies. Existing Italian policy making institutions, trends in the policy communities forming around climate change, and the configurations of Italian greenhouse gas reduction strategies will be granted special emphasis.

NATIONAL POLITICAL CULTURE AND CLIMATE CHANGE POLICY

The Italian political culture exhibits a few peculiarities that have become well known inside and outside of Italy. Italy has a track record of implementation failure, and while the effects of non-success are generally known, many of the underlying causes have yet to be identified.

A common reason lies in the attitudes of key players. For the most part, these individuals make promises and commitments without any serious intent. Although this is common in many other countries, in Italy this feature has been co-opted into the whole political system.

A second feature is in the widespread corruption in the ranks of all the Italian political parties. As is well known, popular protest in the form of strikes and demonstrations lead to a series of political referenda in April 1993. The main purpose of the referenda was to abolish many unsavoury aspects of the present political system and legislation, and set the foundation for change. Among the eight referenda held, one concerned the reform of the Italian electoral system, in favour of proportional representation. This was designed to wipe out many of the minor political parties, who have tended to destabilize Italian politics in the eyes of the international community. The indirect result, however, was the dismemberment of the major Christian Democratic and the Socialist Parties.

The outcome of one particular referendum of key importance to this study was the removal from USLs (Local Health Units) of existing jurisdiction over environmental quality control. The referendum created a new institution to replenish the void, namely the National Agency for the Environment (ANPA). This is expected to be shaped on the model of the European Environmental Agency. This important institutional change will be evaluated later.

The overall importance of these referenda lies in setting in motion a mechanism for institutional change throughout Italy. But it is still too early to be sure of the overall effect. Italian politics remains in a state of chaos. Whatever the outcome, climatic change politics will not play a decisive role in the institutional ferment. But that ferment does provide opportunities and openings for a host of new initiatives. This intermixing of policy opportunity and inhibition due to the piggy-backing of climate change onto other policy activities is an important theme for the book as a whole. Italy provides a marvellous example of the *potential* for this integration, but equally the potential for *disintegration*.

Italian political culture is dominated by a patronage system or clientele network, where the exchange of political electoral votes for favours is a common practice. Widespread corruption was exposed in a series of scandals between 1992 and 1994. The scandals went beyond simple issues of patronage and involved large amounts of financial transactions which are still indirectly shaking the basis of the Italian economy. Many of the scandals were linked to the bidding for large public and environmental projects. These involved, among others, energy production and hazardous materials disposal. Legambiente (a leading environmental NGO) surveyed all the major environmental projects that involved illegal operations (Anzaldi *et al.*, 1993). Among the indictments was the diversion of money allocated to air quality monitoring stations in the region of Apulia. Multiple scandals have also involved ENEL (the National Electricity Board) and several chemical industries (ibid, 26–43). The financial diversions in the environmental and energy sectors and the criminal involvement of private interests hinder the normal course of climate policy implementation in Italy.

The transition period created by the referenda has deeply unsettled the politics and economy of the country. The large government deficit has induced freezes and cuts in many government programmes, among the first being the environmental programmes. This is regarded as a typical early target (Interviews with two leading environmentalists and an energy expert 1993; and Greenpeace–WWF–Legambiente, 1992) and undoubtedly complicates the process of implementation of the new-born climate policy.

Because national climate policy is currently being defined, it is still very much influenced by traditional policy styles and outlooks of key policy makers. But, while climate policy could be framed in the old Italian political system characterized by inefficiency, bureaucratic impasse and corruption, it could also become an active part of the new political culture. If the implementation of change in Italy fails, then climate change policy will also fail. However, if the new system succeeds, climate change politics will probably represent one of the first examples of the new Italian politics. The policy implementation process offers an opportunity to both policy makers and society to demonstrate their willingness to change the perceptions, the behaviour and the actual institutional framework that formerly bedevilled Italy.

HISTORY AND DEVELOPMENT OF CLIMATE POLICY IN ITALY

In reconstructing the history of climate policy in Italy, it is necessary to keep in mind the novelty of climate change as an issue. Much of its policy is still in the process of clarification (Scovazzi and Pineschi, 1991, 184). Some Italian experts even have a problem defining environmental policy (not to mention climate policy) as 'policy' intended in its full meaning (Confindustria, 1992, 1). Italian environmental policy is in any case of fairly recent vintage: after all, the Ministry of Environment was only created in 1986 (Liberatore, 1991, 5).

Since the mid-1960s the translation of environmental issues into policy has been gradual in Italy. The very first piece of legislation containing anti-smog measures, Law 615, dates back to 1966. Environmental policy development since then has been very troublesome, and it is only recently that a more comprehensive level of policy making has emerged. In comparison with other industrialized countries, the process is far from complete. Implementation and enforcement of those regulations are even more problematic and started late. Their procedures were established in the 1970s and even then were limited to only a few areas (Liberatore and Lewanski, 1990, 13). For most of the 1960s and 1970s, environmental issues went ignored by the majority of the population, the government and all the major political parties, with the exclusion of the Green Party (VERDI) and the environmental movement.

The 1980s witnessed an increased number of environmental statutes regulating pollution of the air, water, industrial and urban waste, and nuclear energy. The 1987 referendum banning the development of nuclear power in Italy, following the Chernobyl accident, further contributed to strengthen environmentalism. This boosted the expression of enhanced environmental concern and consciousness (Borrelli and Squillaciotti, 1988, 15). The creation of the Ministry of Environment definitely marked a new era in the development of Italian environmental policy.

Environmental policy in Italy has faced several problems of formulation, implementation, enforcement and review. Barriers to its development are usually a general lack of concern and sensitivity, including inadequate education, throughout political and civil society, and the many political–institutional barriers that have impeded the unification and management of environmental policy under a single institutional body. Despite the existence of the Environment Ministry, environmental policy in Italy remains under the domain of seven different ministries without counting the regional, provincial and communal governments, which also share jurisdiction in this field (Carrubba and D'Inzillo, 1991, 472).

Yet environmental policy is starting to abandon the marginal role it has occupied for years in Italian politics. Besides the increase of public concern on environmental issues (Legambiente, 1993; Istituto per L'Ambiente/Censis, 1993) and the consequent pressure on politicians, the most influential cause lies in the role played by the European Community. Owing to the necessity of implementing European directives, the government has been forced to step up the passing of environmental legislation. Environmental laws and programmes have been adopted at an unprecedented rate in the last few years. However, laws are one thing, implementation quite another.

The necessity of responding to European directives in a fairly short time has resulted in the adoption of emergency measures. Domestic environmental policy has thus become known as 'the politics of emergency' (Confindustria, 1992, 1; Bresso, 1993, 2). As a consequence much of Italian environmental policy is fragmentary and incomplete. It has often been elaborated through a 'patchwork' reflecting measures already adopted in other sectors but never integrated in a comprehensive way.

The history of climate change policy in Italy goes back only a few years. Prior to 1988 the Italian government embarked on a few initiatives to control urban pollution. With the creation of a National Energy Plan (*Energia e Innovazione*, 1988) in 1988 and its consequent enactment, Laws 9 and 10 (*Gazzetta Ufficiale Italiana*, 1991a, 1991b), Italy took its first official steps in adopting policies and legislation aimed at regulating some of the greenhouse gases. These measures were never directly intended to minimize the problem of global warming and the effects of climate change. Their objective was energy saving and consequent protection of the environment. The National Energy Plan, known as PEN, only attempted to provide guidelines for the regulation of sulphur emissions and completely ignored carbon dioxide (a 1991 Revision of PEN first mentioned CO_2, but did not consider any specifications). In any case, this revision was never approved in Parliament.

The landmark year for climate change policy in Italy is 1990 (interviews with government officials, environmentalists and experts in climate policy), when the Italian government underwent its first commitment to a strategy to reduce carbon dioxide. The decision to stabilize and reduce CO_2 emissions by the year 2000 to the levels of 1990 was taken under the aegis of the European Community as outlined in Chapter 6. The quandary for the Italian decision makers was that the decision was taken at the EC level during the Italian Presidency. Many Italian policy actors want the credit for introducing

ITALIAN INSTITUTIONS AND CLIMATE POLICY-MAKING

ISSUE OF CLIMATE CHANGE/UN FCCC SIGNATURE

RESPONSE ELICITED IN INTERESTED GROUPS

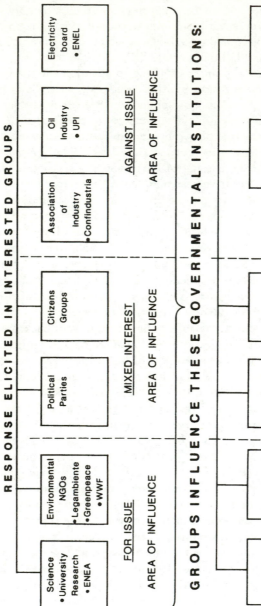

| Science • University Research • ENEA | Environmental NGOs • Legambiente • Greenpeace • WWF | Political Parties | Citizens Groups | Association of Industry • Confindustria | Oil Industry • UPI | Electricity board • ENEL |

FOR ISSUE
AREA OF INFLUENCE

MIXED INTEREST
AREA OF INFLUENCE

AGAINST ISSUE
AREA OF INFLUENCE

GROUPS INFLUENCE THESE GOVERNMENTAL INSTITUTIONS:

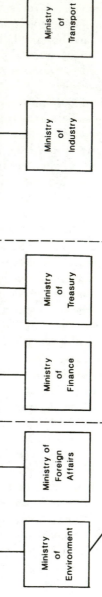

| Ministry of Environment | Ministry of Foreign Affairs | Ministry of Finance | Ministry of Treasury | Ministry of Industry | Ministry of Transport |

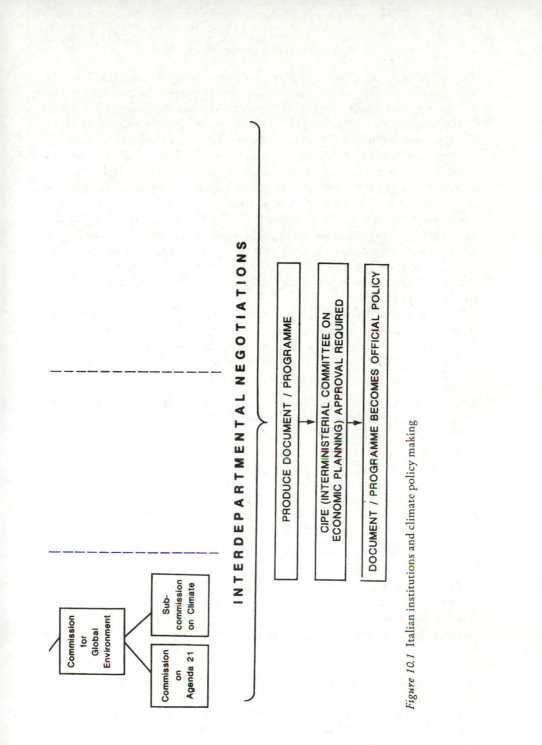

Figure 10.1 Italian institutions and climate policy making

the idea of CO_2 stabilization and of a proposed energy/carbon tax. There is no doubt that the commitment to this idea on the part of the Italian Minister of Environment at the time (Mr Ruffolo) and the influence of the EC Commissioner, also an Italian (Mr Ripa Di Meana), played a significant role in the process leading to the decision. But calling this an 'all-Italian' creation is a little far fetched. Nevertheless, from the interviews conducted on the issue, it appears many leading officials really do believe that Italy was the promoter and the main mediator of the EC decision, though other experts interviewed (who wished to remain anonymous) would not confirm these assertions. Certainly it can be assumed that without the UN FCCC Italy would have continued to ignore the problem of climate change, following its trend of confronting issues only in emergency situations. The attitude shift brought about by the UN FCCC, namely one geared towards precaution and prevention, represents a significant change for the Italian way of conducting policy.

The Italian involvement at the international level has been quite consistent, in particular in its participation to the INC (see Chapter 1). Italy has been particularly active on the issue of joint implementation, though its position is still being defined. The domestic Italian discussion on joint implementation revolves around the following points:

- Consideration of the opportunity to build up a protocol on joint implementation according to Article 17 of the UN FCCC.
- The joint implementation mechanism has to be applied to all parties to the Convention.
- A flexible approach regarding the possibility of using joint implementation to reach stabilization or reduction of greenhouse gas emissions has to be assumed.
- Joint implementation could be elaborated as a kind of voluntary financial mechanism of the UN FCCC.

In comparison with its activity in the INC, Italian participation in the IPCC has been slight in terms of both membership presence and contribution. No Italian scientist has actively contributed to the IPCC Scientific Assessment, while only three scientists assisted in the review of the 1990 Scientific Assessment (Houghton *et al.*, 1990, Appendix 3–4). The underlying reasons may be attributable to the precarious state of the Italian scientific world. However, Italy has at least paid its fair dues in comparison with other industrialized countries (IPCC 1992, 165).

In response to this, the government has expressed its intention to create an interministerial structure to provide a forum for discussion and coordination of climate change policy and sustainable development issues (interview with an official at the Ministry of Foreign Affairs). The Ministry of Foreign Affairs is among the departments that have been called on to participate, because the global scope of these issues necessitates an international dimension. The Ministry of Industry and Commerce and its aggregate industrialists association (Confindustria) are also involved. The Ministry of Environment functions as the promoter and main coordinator of climate change policy. The Ministry of Finance has to approve all the budget for the implementation of any programmes for reducing emissions and mitigating the adverse effect of climate change. This would include energy taxes and other types of environmental taxation. The Ministry of Scientific Research assists in designing a national programme on climate following the direction of the WMO and UNEP and will be providing verification of the state of scientific knowledge (Colombo, 1993).

Finally, the concerted action of different government departments, political parties, and regional and local administrations will play a role in the decision making process, implementation and enforcement of domestic climate policy. A framework of the principal institutions in charge of climate policy making in Italy is indicated in Figure 10.1. These existing and future institutions are now more closely reviewed to assess their potential role in the development of climate policy in Italy.

THE CLIMATE POLICY ACTORS

There are two distinct groups involved in the decision making process that concerns Italian climate policy. The first group is comprised of official governmental institutions embodied by departments or ministries, and their related commissions. Regional and local administrations are also a part of this group. The second clutch of interests is represented by unofficial interest groups, including industrial and oil associations, trade unions and environmental organizations.

These official and unofficial interests have a tendency to seek the support of the Ministry which will protect their self-interest. Historically, the industrial and oil associations look to the Ministry of Industry. The trade unions may find support in this same Ministry or in the Ministry of Labour. The environmentalists tend to rest their hopes on and cooperate with the Environment Ministry. It is important to note, however, that these coalitions do not always occur along the stated lines. Depending on the issue, i.e. energy/carbon tax debate, these institutions have all experienced internal splits. Owing to the multiple policy nature of climate change interdepartmental cooperation has become a prerogative for policy formulation.

The Ministry of Industry

This Ministry is better known for its involvement with energy policy. The Ministry of Industry, along with its affiliates Confindustria, ENI, ENEL and UPI, are the major opponents to environmental policies. This is the case even when certain aspects of environmental policy, such as energy saving and use of renewable resources, may actually benefit the industrial and energy sectors.

The Ministry of Industry is in charge of elaborating on the National Energy Plan (PEN). To date, it has failed to place enough emphasis on the use of renewable resources. Traditional non-renewable resources, like oil, are seen as the most convenient energy source by industry. Since the Ministry focuses on industrial interests it tends to underestimate the importance of renewable sources of energy and to hinder any move towards curbing the use of fossil fuels. The Ministry has created obstacles throughout the process leading to the Italian CO_2 stabilization plan and has succeeded in imposing modifications in the plan itself.

The Ministry of Health

Prior to the creation of the Ministry of Environment, the Ministry of Health was in charge of environmental policy in Italy. Its main purpose was to co-ordinate and set up a network of government offices that were responsible for implementing the control of air emissions and water quality. Once the Ministry of Environment was established, the Health Department was able to retain some of its power in the environmental area. This was done through the administration of the Local Health Units (Unita' Sanitaria Locale, indicated as USL), responsible for pollution control. As noted at the outset, the 1993 referenda took away this competence from the USLs (*Gazzetta Ufficiale Italiana*, 1993). The credit for organizing this outcome, belongs to the Italian branch of Friends of the Earth (Nardi and Protopapa, 1991, 28). A decisive 82.5 per cent of the population determined that the USLs should no longer retain power over this environmental matter. This response was attributed to a widespread belief that the USLs seriously mismanaged what to many Italians was their effective environmental concern, namely environmental health where they lived (Mantini, 1993, 56). The causes for this failure were attributed to the lack of financial resources devoted to monitoring programmes, insufficient staff employed for this task, and the general neglect of environmental protection and prevention.

The Ministry of Foreign Affairs

This Ministry is responsible for handling all Italian relations with the international world. It is therefore responsible for international environmental agreements to which Italy is a signatory. Since an international agreement dealing with the environment normally requires a degree of implementation at the national level, this Ministry has become an important actor in Italian environmental policy.

A significant role in environmental matters was played in Rio, when the Ministry of Foreign Affairs represented Italy at UNCED, and was subsequently recognized as the general coordinator of the Italian response to UNCED. It created an informal commission which designates various tasks and responsibilities for implementation at national level to different groups. The representatives of various departments, NGOs, private enterprises and industries, regional and local governments, and state-owned institutions are all part of this commission.

The Ministry of Environment

The Ministry of Environment (MoE) was created in 1986, and absorbed all the responsibilities previously held by other ministries such as Health, Transport, Industry, Labour and the Merchant Navy (Pinchera, 1991, 71). However, this change did not create much independence for the MoE. The policy making process, as it is today, requires that the MoE submit every policy proposal to the entire range of Italian ministries. The MoE has to negotiate with each Ministry in order to reach consensus. This often results in situations of impasse that cause MoE proposals to become unworkable.

When international environmental policy issues are raised, the MoE depends exclusively on the decisions made by the Foreign Affairs Ministry. These procedures, whether formal or informal, make this institution a rather weak one in the context of Italian politics. The MoE has no powers of final decision, yet since 1986 its importance has been increasing. This is due to the steady affirmation of environmental issues in the social and political arena for the reasons already noted.

The MoE was assigned the major task of elaborating on the national plan in response to Rio, after UNCED in 1992. This response was delivered to the UN Commission for Sustainable Development (CSD) in December 1993. The national plan explains Italy's past, present and future policies as they are relevant to the follow-up to the Rio commitments. These commitments involved Agenda 21, the United Nations Convention on Biodiversity and the UN FCCC. The MoE created a working group (an informal commission) to elaborate the response to the UN FCCC. This working group is known as the Commission on Climate.

Regions, Provinces and Communes

In the mid-1970s the Italian government decided to start a process of administrative decentralization. The regional, provincial and municipal governments were given a higher degree of autonomy. The regions were put in charge of their own energy policy. They began to share jurisdiction and influence over certain aspects of environmental policy. This was done with the other decentralized institutions, the provinces and the communes. The provinces had greater responsibility over the administration of forests. The communes increased their responsibilities over the management of urban traffic and set standards for emissions of central heating systems. As noted in other cases, this situation did not solve the problem of competence. Decentralization has caused a great deal of rivalry among these public administrations and created a situation where these administrations find justification for their inefficiency.

Unless this unclear status is improved, it is likely that it will extend to future policy implementation, such as climate change policy. This in turn will create further delays in policy implementation and effectiveness. From interviews with local and regional officials, it appears that the majority of local and regional authorities are waiting for the national government to define a policy of climate change and pass related legislation before they engage in any local initiatives. The implementation of a local Agenda 21 seems to be remote in the minds of these administrators. The reasons for this attitude partly lie in the lack of local constituencies' awareness of such issues.

Ministry for the Problems of Urban Areas

This recently established institution, created by Law 208/91, was given the task of handling problematic urban areas. The Ministry has issued a decree, following an EC directive, to reduce traffic flows in urban areas on high pollution days. The decree (*Gazzetta Ufficiale Italiana*, 1992) has been enforced in several cities, though all vehicles with catalytic converters are exempt. The Ministry is attempting to manage the Italian urban transportation problem by creating incentives to switch from private to public transport.

Interministerial committees

When dealing with certain issues that are complex and which involve different sectors and interest groups, the government has the power to create an *ad hoc* committee as well as a permanent interministerial committee. The formulation of environmental policy, which affects many sectors, often calls for a committee that includes the representatives from various ministries.

A recently instituted committee under the aegis of the Ministry of Foreign Affairs is coordinating governmental departments in the implementation of the Italian response to Agenda 21. Owing to the radical shift in the Italian government after the March 1994 elections, and consequent replacement of the Prime Minister who was in favour of it, this institutional innovation was weakened. This undermined the implementation process of climate policy in Italy.

ENEL, ENI and ENEA

These enterprises belong to a group of recently privatized or partially privatized government enterprises. However, the privatization process is far from completed and the government retains influence over their administrative proceedings.

ENEL, the Italian National Electricity Board, was privatized (Law 359, 8 August 1992) to reduce the chronic government public sector spending deficit. This action was in agreement with the EC decision to start a policy of liberalization of the electricity market, intended to break the monopolistic regime of energy prices in many Member States. However, the privatizing process is very slow, so it is too early to predict the possible changes that may directly or indirectly affect energy and climate policy.

ENI, the National Hydrocarbon Board, has practically determined the type of energy sources and quantities to be imported into the country since 1954 (Pinchera, 1991, 75). ENI's aim in the last decade has been to improve and extend national production to lower the dependency on imports.

When nuclear energy was still an option (prior to the 1987 referendum), the acronym ENEA stood for Nuclear Energy and Alternative Energy. ENEA was, in fact, the public company in charge of research and development of all energy sources. It was also responsible for promoting and providing consultation involving the rational use of energy. Today, ENEA has maintained its role in R&D in the field of energy. It has de-emphasized its nuclear energy research programmes and is expanding its interest in alternative (renewable) resources and their promotion. After various changes, due to partial privatization, the acronym ENEA has been changed to Board for New Technologies, Energy and Environment (*Energia e Materie Prime*, 1992b). ENEA has various programmes dealing with environmental issues such as emissions monitoring and inventories, climate change, environmental management and consulting. It is capable of producing qualified experts who contribute to the international community of scientists and researchers. ENEA consultants are always present at the government's table when decisions regarding energy and environment issues are raised. Their influence on the energy- and climate-related decision making process is very significant.

The non-governmental or 'unofficial' actors

The main non-governmental actors that have an influential role in shaping environmental policy in a positive way are identified in the 'green' movement in Italy. They are represented by the Italian Green Party (known as VERDI) and by various environmental organizations. The three main organizations are Greenpeace-Italy, WWF-Italy and Legambiente. Although the 'unofficial' actors are outnumbered by the governmental institutions involved in policy making, their influence is noteworthy. Their strength lies in their degree of influence on policy makers as lobbyists, in mobilizing the public and in shaping public perceptions. It can be said that these Italian 'unofficial actors' in the politics of climate have progressively gained a position of omnipresence at the bargaining table.

Confindustria

The General Confederation of Italian Industry, Confindustria, represents the interests of industrialists. During the 1970s and early 1980s, this association assumed an unfriendly position towards environmental issues, causing strong opposition from environmental groups. However, the increasing attention towards environmental issues in the social and political agenda of the late 1980s triggered a sensibility within Confindustria as well. The basis of this newly revised position has been the concept of sustainable development involving industrial production. Confindustria has created something of a new environmental dimension (Confindustria, 1992, 1), yet, on certain issues such as the use of nuclear power, its position has remained intractable. The confederation believes the resumption of nuclear energy would be the solution to the problem of climate change, and that renewable energy is too expensive (interview with Confindustria representative).

UPI

The Italian Oil Board, known as UPI, is an organization comprised of private oil companies active in the areas of oil refinement, production and distribution. UPI embodies a very powerful lobby in the energy debate and strongly protects its interests from the intervention of fiscal policies on oil prices. It criticizes the government's adoption of heavy taxation on oil, which has distorted demand for oil in favour of natural gas. UPI advocates the EC tax harmonization proposal for it recognizes that this initiative is favourable to the objective of reducing overall energy taxation in Italy. However, not surprisingly, it was strongly opposed to the energy/carbon tax.

Here we see a fairly classic mix of interested parties – governmental, quasi-governmental, industrial and voluntary. Unlike the position in Germany, these parties have not yet hammered out a common position over climate policy. The issue is still too young, and the political situation too tender, for much in the way of institutional innovation.

The post-Rio institutional framework

The post-Rio institutional framework is regarded here as the actual official response to the UN FCCC by the Italian government, the embodiment of

310

climate policy in Italy. It includes new institutional bodies like agencies and commissions that were created in Italy as a consequence of signing the UN FCCC. Programmes and documents, i.e. the Italian Agenda 21, elaborated upon for the implementation of this response are also examined in this context, because they include sections that incorporate climate policy.

The National Agency for the Environment

Some foreign countries have already experienced the establishment of environmental agencies like the American EPA and the French ADEME (Croci *et al.*, 1992, 54). These agencies have been crafted to face the complexity of environmental problems. They may differ from country to country in the extent of powers they are granted. Essentially, they all share the same goal of establishing an administrative and regulatory structure that deals with the protection of the environment.

Italy differs from the United States in that it has two governmental institutions that oversee environmental issues and policies, namely MoE and ANPA. Many critics fear that the co-existence of two similar institutions may hinder the bureaucratic process, by duplicating work and creating judicial confusion. This could indeed occur given the bad track record of Italian bureaucracy. Ministries should be in charge of policy and agencies should ensure the continuity of regulation, namely monitoring, standards setting, surveillance and enforcement.

According to the law instituting the agency, its main function will be to carry out technical and scientific activities for the protection of the environment. Clearly, this definition is vague and needs further specification and instructions from the administration. Since the abolition of the MoE is not being contemplated, the assumption made here is that the agency will act as a branch of the MoE in supporting it with technical and scientific advice. According to top advisers within the MoE (personal interview 1993), the agency will thus progressively take over the role of USLs in monitoring and controlling environmental quality. This would be done through the structure provided by the USLs.

Some critics are not satisfied by the new agency's organizational structure (Metalli, 1993a, 3). The government announced that ANPA will absorb personnel from other public agencies with the danger that personnel in the new agency would carry over the previous training in the Italian bureaucratic fashion. A renewal in the standard operating procedures of these administrators is a necessary condition to overcome the problems of bureaucratic sclerosis characteristic of the Italian style of making policy. Critics are concerned that the agency is both under-resourced and not sufficiently autonomous to do its job (Della Seta, 1993, 14; Melandri, 1993). How this matter will evolve will be a crucial test of the institutional reform as laid out in Chapter 4.

An important role for the agency would be to assume constant monitoring, to implement enforcement, and keep the public informed. ANPA could also be technically active in setting standards, providing guidelines for environmental impact assessments and coordinating research in the field. Other administrations have repeatedly failed to carry out these activities. The agency could also provide an arena where industrial and environmental forces can work together more effectively, owing to its technical (allegedly, non-political) character.

The Commission for the Global Environment

The Minister of Environment created this permanent Commission by decree in 1992. Its main function is to elaborate on and manage the National Energy Plan for the Italian response to the UN FCCC. The Commission encompasses several interests and comprises nine members, acting on a voluntary basis. These include independent experts, i.e. economists, energy experts and scientists, NGO representatives, MoE and other departmental officials. The creation of the Commission is considered to be the first-institutional step in the follow-up of UNCED. The relationship of the Commission and its related working groups in the scheme of Italian institutions involved in climate policy making is indicated in Figure 10.1.

The Italian Agenda 21 Commission

This institution can be considered as a subcommission or working group coordinated by the General Director of the MoE. It was temporarily created by the Commission for the Global Environment, and shares some of its members with that Commission.

The Italian A-21 Commission is in charge of producing the country's national document in response to Agenda 21. Its draft (Ministero Dell'Ambiente, 1993a), gained official approval by the CIPE Interministerial Committee for Economic Planning, which is coordinated by the Ministry of Budget, in January 1994 (Ministero Dell'Ambiente, 1994a). Because of the CIPE stamp of approval, the 'Italian National Plan for Sustainable Development in Enactment of Agenda 21' has become a programme to which the Italian government is legally bound. ANPA is supposed to assist the government in fulfilling the requirements of the plan.

The A-21 Commission has selected its objectives on the basis of some key sectors which are also present in the EC Fifth Action Plan. Each sector considers the actions and tools necessary to achieve the main objectives. The Italian A-21 Plan has chosen only six sectors from the multitude indicated in the Agenda 21 adopted in Rio (ibid, 1994a, 3), because they better represent the Italian socio-economic characteristics and identify any emergencies that need to be dealt with. These areas involve energy, industry, agriculture, transport and tourism, waste management, and international cooperation for sustainable development.

The sectors that are more involved with climate-related issues are the energy, industry and transport sectors. Strategies for the management of the energy sector are indicated in the Italian document and reflect the ones already found in the 1988 National Energy Plan (PEN). Two of the five objectives include energy saving and the protection of the environment. The Italian A-21 Commission recognizes the need for a revision of the 1988 PEN, in light of the Rio commitments. The document exhorts among other things the promotion of energy efficiency beyond Laws 9 and 10/91 (the two main laws concerning energy saving and efficiency). It calls for a reduction of greenhouse gas emissions that goes beyond the existing legislation.

The same lack of creativity seen in the energy section is also noticed in the sections of the document concerning industry and transport. The strategies expressed in those sections are the same strategies that have been known for the past few years. The document does acknowledge a change in the attitude of Italian industry towards the problem of environment and development, but it rests too many hopes on the voluntary role of industrial compliance. Frankly Italian industry is not as responsive as portrayed in the document.

Industry was absent from the National Conference on Climate, and demonstrated no concern about what could emerge from the climate programme. Additionally, it did not participate in the drafting of the Italian A-21 document and the 1994 CO_2 Stabilization Plan (Ministero Dell'Ambiente, 1994b), both of which directly involve industry. Yet, positional influence will dictate that industry will have the last word on all these actions.

The National Climate Programme

Italy has been slow in establishing a national programme on climate as compared with other industrialized countries (Cirillo, 1993, 34). It must be acknowledged that, after signing the UN FCCC, Italy has shown a willingness to respect its commitments. It has started a process of internationalization, for Italy is recognizing the importance of meeting collective agreements. The National Climate Programme has also helped to focus the necessary scientific support for the government in its task of elaborating response strategies to climate change. This is seen as the key to the eventual implementation of climate policy in Italy.

Although the Italian A-21 implementation programme may be criticized for its lack of originality, it offers a comprehensive and programmatic element to a process that was fragmentary, episodic and crisis driven. This is the first stage of institutional response as outlined in Figure 4.1. There is better information being collected, more environmental sympathy amongst the population as a whole, some change of outlook beyond cosmetic compliance, and at least an organizing framework around which new policy communities can congregate. As in Germany, and to some extent in the UK, industry–NGO alliances are just beginning to emerge in what should be a very formative period in Italian environmental politics.

ITALIAN ENERGY AND CLIMATE POLICY

Italy is highly dependent upon foreign imports of oil and other energy sources owing to its lack of natural resources. Energy imports amount to an average of 81 per cent (Table 10.1) of the total amount of the energy budget in this country (Ministero Dell'Ambiente, 1992a, 371). This has, in turn, forced the creation of strategies which are designed to protect the country's own interest against possible foreign oil embargoes. The Italian government has achieved low energy intensity and consumption by undertaking steps such as high energy taxation, and limited power contracts for households.

Historical elements, such as late industrialization and a traditional propensity of the population to save, have also resulted in low energy consumption patterns (Gaudioso et al., 1993, 5). The slow but steadily increasing income per capita of the Italian population versus its industrialized counterparts has also contributed to keep the consumption patterns for energy products at a low level. Economic measures taken by the government have established a fiscal system allowing for price control on oil products and energy. The significantly high price of some of these resources (Table 10.2),

Table 10.1 Primary energy budget (million tonnes of oil equivalent)

	1970	1980	1989
National production	24.8	24.7	28.7
Natural gas	10.9	10.3	13.8
Oil	1.5	1.8	4.5
Solid fuels	2.0	1.1	1.3
Primary electric energy[1]	10.4	11.5	9.1
Net imports	99.0	122.5	133.2
Natural gas	0.0	11.9	23.5
Oil	88.8	97.8	88.6
Solid fuels	9.4	11.4	13.7
Electric energy	0.8	1.4	7.4
Stock changes	3.7	0.2	0.1
Total energy	120.1	147.0	161.8

[1] Hydro- and geo-electricity; also nuclear energy in 1970 and 1980.

Source: ENEA on data from Ministry of Industry

Table 10.2 Gasoline prices in US dollars/litre converted with purchasing power

	1988	1989	1990	1991	1992
Belgium	0.61	0.68	0.77	0.80	0.81
Denmark	0.67	0.71	0.66	0.66	0.63
France	0.71	0.77	0.80	0.82	0.80
Germany	0.47	0.58	0.61	0.68	0.71
Greece	0.68	0.63	0.81	0.85	0.85
Ireland	0.79	0.84	0.91	0.93	0.88
Italy	1.00	0.99	1.03	1.04	1.02
Luxembourg	0.53	0.56	0.57	0.57	0.60
Netherlands	0.72	0.78	0.83	0.88	0.91
Portugal	1.35	1.31	1.31	1.33	1.22
Spain	0.73	0.71	0.75	0.80	0.84
UK	0.64	0.68	0.74	0.76	0.77

Source: IEA Statistics, Energy Prices and Taxes, 1993

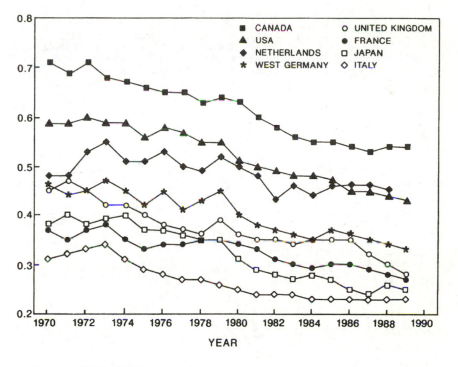

Source : CNEL, 1990

Figure 10.2 Energy intensity in some OECD countries (toe per US $1000 of GDP at 1985 prices and PPP)

in comparison with other EC countries, has in turn guided the demand for those goods. Industry has made an effort to produce more efficient items, such as automobiles with increased mileage per litre of petrol and energy-efficient domestic appliances.

Another peculiarity in Italian energy policies is of a progressive tariff on electricity for the residential sector. Together with residential contracts for power supply (mostly limited to 3 kilowatts), this tariff shapes energy consumption paths. This discourages consumers from using electric power for heating and cooking purposes. As a result, the energy intensity level in Italy is among the lowest of the OECD. The data illustrated in Figure 10.2 summarize the position.

The Italian energy efficiency is also notable in electricity intensity statistics. High prices and tariffs applied to electricity have produced a situation in which Italy is favoured in comparison with other industrialized countries (see Figure 10.3). The removal of the nuclear option in 1987 together with the oil crises of the early 1970s and 1980s provoked increased use of renewable energy in Italy. This started with the preparation of the first National Energy Plan.

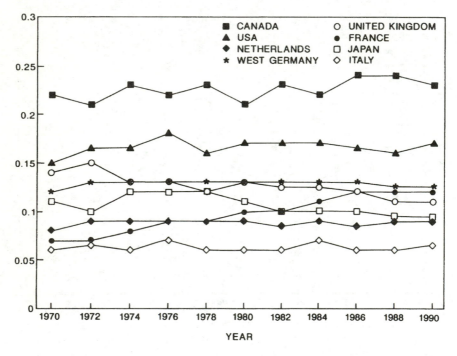

Source : ENI, ENEL, ENEA, 1990

Figure 10.3 Electricity intensity in some OECD countries (toe per US $1000)

In the opinion of an ENEA expert, and a climate scientist, Italy is already an energy-efficient country. If additional efficiency is to be achieved, in order to reduce energy consumption and consequent CO_2 emissions, this should be done in the residential and transport sectors. The industrial sector stresses that it is already efficient and should not undergo exceptional strain in trying to invest more resources into achieving efficiency. As much as this assertion may be disputable, there is the problem of foreign competition for Italian industry. As a solution to this problem many experts favour burden sharing and joint implementation, as part of a unified effort of all EC countries. We note a similar call for exemption from the industrial sectors in all the countries studied; hence the interest in a broad, Europe-wide, carbon/energy tax.

According to the policy makers interviewed, the transport sector should be the main target. This sector accounts for 28 per cent of total CO_2 emissions in Italy. Improvements could be introduced in the residential sector as well, especially regarding central heating systems and domestic appliances. However, these two sectors seem to be already heavily targeted by existing fuel taxation. Ultimately the final users are the most penalized because they have to pay higher prices. Interviews with advisers (Confindustria's researcher and a Ministry of Industry official) indicate that industry is exerting some influence on the thinking of policy makers. This clearly reinforces the status quo that energy policy in Italy still remains in the hands of the Ministry of Industry and the interest groups that support this department. As noted above, this is a common finding in all the case studies examined for this volume.

The total amount of CO_2 emissions in 1990 was estimated at 400 million tonnes, corresponding to a consumption of 163 Mtoe (Ministero Dell' Ambiente, 1992b). Total emissions were distributed among the following sectors:

- Electrical energy production 30.7 per cent
- Industry 17.8 per cent
- Residential and tertiary 19.7 per cent
- Transport 28.8 per cent
- Agriculture 3 per cent

Italy contributes approximately 2 per cent of total energy-related world CO_2 emissions (IEA, 1992). This percentage is regarded as small, according to Italian commentators (interviews with an ENEA expert, a climate scientist, and a delegate to the INC), in comparison with other industrialized countries. So there is a serious national debate on the necessity to engage in a CO_2 reduction effort independently from commitments undertaken at the international level. On the one hand, policy communities around the MoE, ENEA and Ministry of Foreign Affairs seek a serious effort (unilateral if necessary) for the adoption of 'no regret' policies. On the other hand are the

Ministry of Industry and the Ministry of Transport who together with ENEL, ENI and the Italian industrial sector tend to favour policies of 'wait and see'. These institutions directly or indirectly protect the interest of industry and big business.

Supporting the thesis of the environmentalists, together with the MoE and a large part of the scientific community in Italy, are the forecasts of future trends of greenhouse gas emissions to the year 2000 (Silvestrini, 1991). The MoE estimates that there will be increasing energy consumption due to the forecasted 2.5 per cent annual increase of GNP from 1991 to 2000. This will follow the trends seen in the period from 1987 to 1990. The trend in energy consumption would translate to 192 Mtoe in the year 2000, corresponding to 470 million tonnes of CO_2 emissions for that year against the approximately 400 million tonnes of CO_2 in 1990. This represents a 17.5 per cent increase from the 1990 level of CO_2 emissions by the year 2000. The MoE therefore calls for a series of actions that comply with the commitment taken by the government within the EC and at Rio with the UN FCCC.

The 1994 CO_2 Emissions Stabilization Plan

The Italian '1994 National Programme for the Stabilization of Carbon Dioxide Emissions at the levels of 1990 by the year 2000' is considered the first and most significant commitment for Italy within the context of the UN FCCC (Ministero Dell'Ambiente, 1994b, 1). This Programme replaces the 1992 draft by amending some of its parts, and amplifying its scope of action. The latest Plan presents a less optimistic scenario than its predecessor, and entails a modification in the CO_2 emissions inventory for the year 1990. The new goal of stabilizing CO_2 by the end of the decade corresponds to 163.5 Mtoe.

CO_2 emissions in Italy in 1990 totalled 443.9 Mt, of which 22.9 Mt were attributed to glass and cement production. However, emissions from these activities have been excluded from the stabilization target, since the reduction of CO_2 emissions in these sectors would either be too costly or reduce production in these two sectors (interview with an Italian scientist). Since reduction of glass and cement production is not contemplated, so the policy of reducing emissions in these sectors is excluded. The amount of CO_2 produced by the year 2000 is estimated to be constant at 22.9 Mt CO_2 (Table 10.3).

The energy consumption per sector and related CO_2 emissions in 1990 are broken down in Table 10.4.

The 1994 Plan forecasts indicate there will be a trend in the growth of emissions that could reach 488 Mt of CO_2 by the year 2000. However, recent trends illustrating slower economic growth show that the demand for energy

Table 10.3 CO_2 emissions in Italy – year 1990

	Mtoe	Mt CO_2
Fossil sources use	163.5	421.0
Cement and glass production		22.9
Total		443.9

Source: Ministry of Environment, 1994

has decreased. The Plan assumes that a decrease of CO_2 emissions can be expected due to ongoing recession, devaluation of the Italian currency, tendency to switch from coal and oil to natural gas use, and future effects expected from the various energy savings provisions of Law 9/91 and Law 10/91. This effect together with further interventions especially aimed at achieving the year 2000 stabilization targets would decrease CO_2 emissions in the year 2000 to a total of 430 Mt (Ministero Dell'Ambiente, 1994b, 6), as indicated in Table 10.5.

The interventions required by the Plan and already approved for the energy sector consist of:

- Technical norms and standards in energy production.
- Co-financing for energy saving and co-generation of electrical energy.
- Integrated energy management.
- Investments in new technologies.

In terms of future necessary actions to be undertaken to achieve stabilization, the Plan lists the following interventions:

Table 10.4 Energy consumption and CO_2 emissions per sector – final use

	Mtoe	Mt CO_2
Refineries	6.3	22.3
Non-energy use	8.3	6.6
Industry	55.8	131.9
Agriculture	3.7	11.5
Transport	34.4	110.6
Bunkerage	3.7	12.6
Tertiary activities	15.2	37.3
Civil/residential	36.1	88.2
Total	163.5	421

Source: Ministry of Environment, 1994, data from Ministry of Industry

Table 10.5 Stabilization of CO_2 emissions in Italy

	Mtoe	Mt CO_2
Emissions 1990	163.5	421
Trend growth 2000	188	488
No intervention	178	430
Intervention		
Objective	172	421
Stabilization		

- Incentives in the industrial and transport sectors.
- Technical norms and minimal energy efficiency standards in the transport and residential sectors.
- Technical norms and process standards in energy production.
- Voluntary agreements between public administration and private enterprises for further energy efficiency intervention in the industrial and transport sectors.
- Planning of minimal cost and demand management in the residential sector.
- Informational campaigns, ecolabelling and consumer advice in the transport and residential sectors.
- Investments in new, energy saving, technologies.

A 1992 draft itemized the sectors and the amount of emissions to be reduced. These specifications were eliminated in the final 1994 Plan, approved by CIPE. The 1993 draft was even more binding for each sector. This arrangement encountered disagreement among the industrial associations, thereby resulting in the more general format of the 1994 Plan. The 1993 draft emissions figures amount to a total reduction of 47 Mt of CO_2 and the estimated costs are as given in Table 10.6.

Table 10.6 CO_2 emissions reduction in target sectors

Sector	CO_2 emissions reduction (Mt)	Costs (billion Italian lire)
Electrical energy production	16	23 500
Industry	11	1 959
Transport	11	6 000
Civil and residential	9	3 300
Total emissions reduced	47	
Total costs		34 750

Source: Ministry of Environment, 1993 Draft Stabilization Plan

From Table 10.6, it can be seen that the effort required from private citizens is disproportionate to their contribution to the total CO_2 emissions in Italy. The transport sector appears disadvantaged in comparison with industry considering the already high fiscal component of petrol prices for transport use compared with industrial use. Because transport includes a large share of private transportation, the burden also falls disproportionately on citizens.

The energy sector, which is represented by ENEL, appears to carry a larger proportion of the expenses with a burden of 23 500 billion, of which 18 000 billion of the total had previously been allocated and approved by the government and ENEL. This was done independently from the 1993 Plan. A large part of those funds have been or are going to be paid by the government through the use of fiscal incentives. These incentives are allocated to investments in energy saving programmes.

This picture reflects the reality of the Italian political system. Industry and energy producers are very influential and able to protect their own interests, as shown by the analysis provided for Table 10.6. This is done to the disadvantage of other sectors whose positional influences around policy cores are less evident.

The answer as to whether Italy will be able to stabilize its CO_2 emissions by the year 2000 will depend on two main factors. First, the willingness to change the outlook of those favoured by the Italian economic, fiscal, juridical and political system to elude the requirements of existing environmental legislation either by engaging in powerful lobbying or by ignoring regulations. Second, in the government's determination to enforce the 1994 Plan. This will also require considerable realignment of both policy communities and standard operating procedures as outlined in Chapter 4 and summarized in Figure 4.1.

Other greenhouse gas strategies

Although the MoE has only recently started gathering data to serve as a national inventory for other greenhouse gases, this does show Italy's willingness to uphold the commitments signed with the UN FCCC. The MoE has also started to identify the first necessary measures and strategies for the containment and reduction of other greenhouse gases. The future programme to be elaborated will principally affect methane (CH_4) and nitrous oxide (N_2O). Particular sectors will be targeted, namely agriculture, waste disposal and fossil fuel combustion technologies for final industrial and residential uses.

As with CO_2, the other greenhouse gas stabilization policy will require the cooperation of all ministries involved and must overcome special interest barriers to reach a successful implementation. For example, current barriers to implementation of CH_4 strategies gravitate around the bidding for local government contracts on waste disposal. Special interests linked to this sector have succeeded in slowing the formulation of effective CH_4 reduction strategies thus far.

Environmental taxation

Within the EC context, Italy supports the idea of introducing a carbon energy tax. Existing Italian laws, i.e. Law 10, 1991, rely on the establishment of a system of incentives for energy saving and consequent emissions reduction. Furthermore, the industrial sector welcomes government guidelines and programmes based on voluntary measures. The outcome will be mainly determined by the orientation of future ruling government coalitions (the political parties that recently won the Italian elections are heavily supported by industry), the state of the economy and the social attitudes and behaviour of the Italian public. The Italian response to the requirements of the UN FCCC which is currently being developed will answer all the open questions.

Fiscal policy has long been one of many instruments utilized to gain control of Italian environmental policy. Although the use of fiscal devices for these purposes is often criticized and has given rise to controversy, it is generally acknowledged that the use of taxation becomes a simple and straightforward method once the various political debates and bureaucratic stages for its implementation are overcome (Lanza and Scabellone, 1992, 1). Environmental taxation has been increasingly applied to sectors such as energy, transport, industrial and residential waste. Its employment reflects the underlying reasons of the 'polluter pays' principle, as well as the often disputed inclination to regulate markets through the use of taxes.

The use of an environmental tax would shape the price of the pollution source. Prices in a market economy generally influence the patterns of consumer behaviour. As an example, differentiated taxes on energy products have induced a progressive switch from oil to natural gas use, as shown in Table 10.7.

Table 10.7 Primary energy consumption by fuel (percentage)

	1970	*1980*	*1989*
Natural gas	8.9	15.5	22.9
Oil	72.6	67.2	57.8
Solid fuels	9.2	8.5	9.1
Primary electrical energy[1]	8.7	7.8	5.6
Net import of electrical energy	0.7	1.0	4.6
Total energy	100.0	100.0	100.0

[1] Hydro- and geo-electricity; nuclear energy in 1970 and 1980

322

The existence of a high level of taxation and the replacement of oil with cleaner substitutes have already occurred. These were achieved independently of any EC taxation regime; thus it can be assumed that environmental taxation in the energy sector would not produce any more of an outcome than the existing one. The introduction of any future EC energy/carbon tax as a direct way to reduce energy consumption would only be relevant in Italy if considered as a revenue collection device. For this type of policy to succeed, it would have to make the new revenues available as financial resources for clean-up projects and for incentives to new technological investments leading to the employment of cleaner and alternative renewable resources.

When analysing the possible response to any future energy/carbon tax in Italy, one must take into account existing high taxation for energy products, especially products for transportation and residential use. Italian society has become accustomed to this type of government manoeuvre to increase revenues, especially for taxes on petrol, tobacco products, and one-time emergency taxes on health care and property. The country tends to be less responsive to price increases as a result (Fano and D'Ermo, 1992, 1). Driving and domestic energy consuming behaviour would not seriously be disrupted. Moreover, according to interviews with a government official and an ENEA scientist, the Ministry of Finance is not contemplating a possible change in the taxation regime should there be such a carbon energy tax. This suggests that the Italians never did anticipate that such a tax would be put in place, despite favouring the idea at Rio (Ruffolo, 1992).

LOCAL INITIATIVES FOR COMBATING THE GREENHOUSE EFFECT

Three main campaigns are mobilizing local governments to implement programmes for CO_2 emissions reduction. The promoters of these campaigns range from environmental organizations like Legambiente and Greenpeace-Italy, to government agencies such as ENEA and private international organizations (Eurosolar) that have branches in Italy.

The first campaign, which began in 1991 and is named 'Communes versus the Greenhouse Effect', is coordinated by Legambiente and has succeeded in mobilizing 54 communes (local governments) (Silvestrini *et al.*, 1992). The 54 communes, out of approximately 8100 communes in Italy, clearly represent a small percentage of the total. However, this is a positive sign that something is happening. Local governments have been chosen as the target of the campaign since they play a key role in the decentralization of energy saving programmes. These municipalities also have an important role in emissions regulation, and monitoring and enforcing programmes through their agencies (the local health units).

There are a few communes in Italy that engage in energy saving policies which are independent of the national government. This occurred directly after the 1973 oil price rise. Today, those communes remain more sensitive to energy conservation issues, dictated more by environmental quality considerations than by economic ones. Most of these city governments are participating in the urban CO_2 project (Harvey, 1993, 18), as introduced in Chapter 7.

Communes with a population of over 50 000 inhabitants are required by Law 10/1991, Article 5(5), to prepare Local Energy Plans that include energy saving procedures. Very few of these Plans have been presented, for the same reasons noted at the regional level, namely lack of funds, inadequate expertise and poor administrative competence. So the city responsibility lies mainly at the level of gathering emissions data, observing emissions trends and preparing emission scenarios to the year 2005. Energy saving plans are supposed to be included in all urban planning projects. Some communes have started various co-generation projects in collaboration with existing industries.

Some Italian cities have committed their citizens to reduce CO_2 emissions through the implementation of the following strategies:

- Switching fuels for space heating and/or boiler, and furnace efficiency improvement.
- Using renewable energy.
- Retrofitting municipal buildings.
- Increasing community outreach.
- Improving public transit and/or managing traffic demand.
- Planting shade trees.

The main problems facing successful implementation of these proposals are the necessity of increasing general public and business green consciousness, in a country where environmental education is not a priority; defying political parties' interests in public projects; educating public administrators; promoting the cause of public transport even when it is not competitive with private transport in many cities; facing the scarcity of financial resources, especially in light of Law 10 which restricts funding for communes. Moreover, owing to frequent re-elections and the typical instability of local politics, long term plans such as reforestation and emissions reductions are difficult to implement.

The two other campaigns taking shape in Italy at this time are the 'Communes for Solar Energy' (Eurosolar, 1992) and an ENEA initiative (ENEA, 1992). Both of these campaigns do not reach the extent of Legambiente's campaign, but they still deserve recognition. The Eurosolar campaign addresses the mayors of several cities offering technical and educational assistance for the development and diffusion of solar energy. ENEA's campaign takes a different approach focusing on energy saving.

Although it is not as vast in scope, each year the campaign targets one large city. The campaign concentrates its efforts towards one project at the time believing that this would obtain more significant results. ENEA cooperates directly with the administration of the chosen 'pilot-city'. It organizes every-thing from advertising the campaign to the actual implementation of the programmes. ENEA avails itself of private sponsors for the realization of its campaign. Thus far this initiative has, remarkably, involved 100 000 people.

There are other small campaigns in Italy that present a single programme. Some of these include Greenpeace's incentive campaign on fluorescent light bulbs and ENEL's campaign on electric energy saving, but these are not financed by governmental grant, and suffer from lack of official support. Note that none of these initiatives originated as a direct response to Agenda 21. Officially, no local A-21 initiative has yet been implemented in Italy.

THE ITALIANS AND GLOBAL WARMING

Climate policy can only be successful if the public perceives climate change as a critical issue for the national agenda. Climate change has to be perceived as a problem requiring prompt action and its politically unpopular policy solutions as a necessity. The most recent opinion polls (Legambiente, 1993; Istituto per L'Ambiente/Censis, 1993) show that the main environmental concerns of Italians are air pollution, deforestation and the depletion of stratospheric ozone. Results show different percentages in the rankings, with variations between 5 and 25 percentage points. Air pollution is the concern of 50 to 53 per cent of Italians. Deforestation percentages ranged from 34.9 to 59 per cent. The level of concern for the ozone hole represented 37 to 48 per cent of the public responses. The high percentage of responses indicated for the ozone hole may be attributable to recent publicity expressed in light of the new scientific evidence regarding the progressive thinning of the ozone layer in northern latitudes.

Italians appear to be more informed on CFCs and their related effects than on the wider aspects of the greenhouse effect. One of the studies (Istituto per L'Ambiente/Censis, 1993, 33) indicates that 69 per cent of the general public understand environmental problems and issues related to CFCs while only 23 per cent of the people are aware of the more general greenhouse effect. This is clearly reflected in the ranking of environmental problems, where only 12 to 25 per cent of the population show concern for the greenhouse effect. This may be due, in part, to lack of available informa-tion. The Italian government coupled with the scientific and modelling com-munities will have to prepare the public for climatic change issues if the matter is to be seriously addressed. This should be a vital early role for the National Climate Programme.

One way forward is to link climate change to the necessity of finding solutions to the problems of traffic and atmospheric pollution. Italians were

Table 10.8 Survey on the propensity to change habits for the sake of the environment

Energy	Yes (%)	No (%)
Limit energy consumption by using highly efficient domestic appliances, fluorescent light bulbs and switching from coal and oil to natural-gas-operated central heating systems	43.6	56.4
Transport	Yes (%)	No (%)
Limit use of private cars, increase use of bicycles and motorbikes, buy cars with catalytic converters	44.7	55.3

Source: Istituto per L'Ambiente/Censis (1993)

asked what behaviour they think they would modify in the future if there is an increase in these problems. The willingness of the population to change its habits in energy consumption and traffic use is suggested from the responses in Table 10.8.

Possibly a majority of Italians are not yet willing to change their behaviour for the sake of environmental problems, though approximately 44 per cent of Italians show a willingness to modify their habits if necessary. This latter evidence might facilitate the task the government faces for the implementation of climate policy.

A starting point could be relying on voluntary action while elaborating more comprehensive response strategies. The Italian government has, in fact, considered including this approach in its Agenda 21 Plan. A purposeful, voluntarist solution could alleviate requirements which might alienate many sectors and trigger an opposition to national climate policy. Once the process of attitudinal change has started, it is likely to spread to even larger sections of the population, in the same fashion social phenomena do. Here is where much more attention to Local Agenda 21 initiatives is required.

CONCLUSIONS

The Italian experience differs from those of Germany, the UK and Norway, in that there was little scientific administrative apparatus for responding to climate change when the issue became politicized in the late 1980s. Furthermore the Italian political scene was deeply distrusted for its corruption and inefficiency, leading to the squandering of precious resources and serious failure of implementation.

The political upheavals of the early 1990s came at a time when climate change politics were becoming enmeshed in the environmental policy arenas, and in wider questions of tax and industrial strategy. The formation of fairly classic administrative responses in the form of interministerial working groups, new ministers and agencies, and plans or programmes, is all the more interesting. This process is taking place as Italy struggles to regain political credibility at its centre, and is beginning to recognize the power of local autonomy and cooperation in delivering national objectives.

So we witness in Italy a pattern of institutional response to climate change that is both familiar to the other case studies yet also unique to this unusual policy. The familiarities include struggling to understand the science, to formulate it into serious policy proposals, and to grapple with economic and technological changes beyond government control. The ambivalent role of industry in this process is also familiar. Industry is slowly becoming more eco-efficient, especially when it is multinational and in the political eye. But industry in Italy hovers around a stable policy core that respects climate science, is anxious about rising costs, and fears losing competitiveness in international markets. So far the constellation of finance, industry and employment sectors, backed by the oil and gas interests, has held firm in the face of the UN FCCC process and the vociferous yapping of environmentalist groups.

So, in Italy at least, the kinds of innovation and the processes that promote these are not yet in evidence on the climate change political front. Look especially for the emergence of fresh alliances amongst enlightened industry and thoughtful environmental groups, mediated by scientists and the media, and look too for first initiatives at the local level, spurred by political aspiration and education campaigns linked to a wide array of health, environmental and neighbourhood safety issues, and to the burgeoning role of city networks in the local Agenda 21 process.

In its own context, the Italian experience of climate change response is remarkable. In the relatively short span of seven years, during a period of unprecedented upheaval, the nation has recognized its EC responsibilities, organized an emissions inventory and monitoring programme, targeted key sectors for and begun to calculate the economic, social and political costs of possible reforms. This is a modest, but still significant, achievement in institutional reorientation. The fact that there is an established Italian policy community, crossing a wide array of interests, is also something of an institutional development. Italy stands poised to graft the successes of political reform to the wider picture of social, economic and democratic initiatives that for all European countries will eventually constitute the politics of climate change. But do not hold your breath in the meantime.

REFERENCES

Anzaldi, R. *et al.* (1993) *L'Ambiente Illegale*, Roma: Legambiente e Il Manifesto.

Borrelli, G. and Squillaciotti, M. T. (1988) 'La Percezione dell'Ambiente e dei Problemi Energetici nei Sondaggi di Opionione svolti dopo Chernobyl', paper Vasa 88/13, Roma: ENEA.

Bresso, M. (1993) 'Alla Ricerca delle Politiche Ambientali Perdute', *Ambiente Italia* No. 1.

Carrubba, C. and D'Inzillo, C. (1991) 'Come Muoversi nel Labirinto della Legislazione Italiana', in G. Melandri and G. Conte (eds) *Ambiente Italia 1991*, Milano: Arnoldo Mondadori.

Cesaretti, C. (1993) 'Strumenti Economici dell'Ecosviluppo', in *Ambriente Italia '93*, Roma: Koine.

Cirillo, V. (1993) 'All' Ombra dei Progetti Altrui', *Ambiente* No. 42.

Colombo, U. (1993) Opening Statement at the National Conference on Climate, Florence.

Commissione delle Comunita' Europee (1992) 'Proposta di Direttiva del Consiglio Relativa ad un'Imposta sulle Emissioni di Biossido di Carbonio e sull'Energia', Com (92) 226 def. 30 giugno, Bruxelles: Pubblicazioni Ufficiali Comunita' Europea.

Confindustria (1992) 'La Politica Ambientale e il Rapporto Impresa-Ambiente', unpublished discussion paper, Roma: Centro Studi Confindustria.

Croci, P. *et al.* (1992) 'Le Agenzie per L'Ambiente', *Economia delle Fonti di Energia* No. 48.

Della Seta, R. (1993) 'Agenzia Nazionale per l'Ambiente', *Leganews*, supplement to *Nuova Ecologia* No. 10.

ENEA (1992) 'Campagna Pensiamocinsieme: Risparmio Energetico Interesse Comune', advertising literature, Roma: ENEA.

Energia e Innovazione (1988) 'Piano Energetico Nazionale 10 Agosto 1988', testo integrale estratto da No. 8–9, Roma.

Energia e Materie Prime (1992a) 'Aggiornamento 1991 al Piano Energetico Nazionale 1988', text of original document, No. 83.

Energia e Materie Prime (1992b) Advertisement for ENEA in No. 86.

Eurosolar (1992) 'Comuni per il Solare: Campagna Nazionale per la diffusione dell "Energia Solare"', advertising literature, Eurosolar.

Fano, P. and D'Ermo, V. (1992) 'Airborne Toxic and GHG Emissions: Italy 1991–1995', paper for ENI Workshop on Energy, Taxation and CO_2 Emissions, Milan.

Fiore, C. *et al.* (1991) *L'Arcipelago Verde*, Firenze: ISPES, Vallecchi.

Gaudioso, D. *et al.* (1993) 'Efficienza Energetica d Clima: Opzioni per la Stabilizzazione e la Riduzione delle Emissioni di Anidride Carbonica in Italis', Roma: ENEA.

Gazzetta Ufficiale Italiana (1991a) Legge 9 Gennaio 1991, No. 9 Norme per l'attuazione del Nuovo Piano Energetico Nazionale: aspetti istituzionali, centrali idroilettriche ed elettrodotti, idrocarburi e geotermia, autoproduzione e disposizioni fiscali.

Gazzetta Ufficiale Italiana (1991b) Legge 9 Gennaio 1991, No. 10 – Norme per l'attuazione del Piano Energetico Nazionale in materia di uso razionale dell'energia, di risparmio energetico e di sviluppo delle fonti rinnovabili.

Gazzetta Ufficiale Italiana (1992) 29 Settembre 1992, No. 229.

Gazzetta Ufficiale Italiana (1993) 26 Marzo 1993, No. 71.

Greenpeace–WWF–Legambiente (1992) 'Documento Greenpeace–WWF–Legambiente su Bilancio e Legge Finanziaria 1993', unpublished discussion document, Roma: Greenpeace.

Harvey, D. (1993) 'Tackling Urban CO_2 Emissions in Toronto', *Environment* 35(7), 16–21, 31–35.

Houghton, J.T. *et al.* (eds) (1990) *Climate Change, the IPCC Scientific Assessment*, Cambridge: Cambridge University Press.

International Energy Agency (1992) *Energy Statistics*, Vienna: IEA.

IPCC (1992) *Climate Change: The 1990 and 1992 IPCC Assessments*, Geneva: IPCC, WMO and UNEP.

Istituto per L'Ambiente/Censis (1993) *Osservatorio sui Comportamenti Ambientali*, Rapporto 93/04, Milano: Istituto per l'Ambiente.

Lanza, A. and Scabellone, M. (1992) 'Environmental Effects of the EEC Energy Tax Harmonization Proposals: A Quantitative Study for the Italian Economy', Milano: Fondazione Eni-Enrico Mattei.

Legambiente (1993) *Sondaggio Nazionale: La Famiglia e L'Ambiente*, Roma: Eurisko.

Levy, M. (1993) 'European Acid Rain: The Power of Tote-Board Diplomacy', pp. 75–132 in P. M. Haas, R. O. Keohane and M. A. Levy (eds) *Institutions for the Earth*, Cambridge, MA: MIT Press.

Liberatore, A. (1991) 'National Environmental Policies and the European Community: The Case of Italy', *European Environment*.

Liberatore, A. and Lewanski, R. (1990) 'The Evolution of Italian Environmental Policy', *Environment* 32(5), 16–20, 31–35.

Mantini, P. (1993) 'Referendum USL: che Fare dopo il Si', *Impresa Ambiente* No. 5, Roma: Janusa.

Melandri, G. (1993) 'Agenda 21: Ritardi e Difficolta' di Ricepimento', *Ambiente Italia* No. 2.

Metalli, P. (1993a) 'Un' Agenzia Nata Vecchia che non Sa quanto Costa l'Ambiente', *Ambiente* No. 45, Roma: Janusa.

Metalli, P. (1993b) 'Colpo di Mano di Mezza Estate', *Ambiente* No. 44, Roma: Janusa.

Ministero Dell'Ambiente (1992a) *Relazione sullo Stato dell 'Ambiente*, Roma: Istituto Poligrafico dello Stato.

Ministero Dell'Ambiente (1992b) 'Programma Nazionale di Stabilizzazione al 2000 dell'Anidride Carbonica ai Livelli del 1990' (Second Draft), Roma.

Ministero Dell'Ambiente (1993a) 'Proposta di Piano Nazionale per lo Sviluppo Sostenibile in attuazione dell'Agend XXI', (Draft), Roma.

Ministero Dell'Ambiente (1993b) 'Programma Nazionale per la Stabilizzazione delle Emissioni del'Anidride Carbonica entro il 2000 ai Livelli del 1990', (Third Draft), Roma.

Ministero Dell'Ambiente (1994a) 'Piano Nazionale per lo Sviluppo Sostenibile in Attuazione del'Agenda XXI', (Final), Roma.

Ministero Dell'Ambiente (1994b) 'Programma Nazionale per la Stabilizzazione delle Emissioni di Anidride Carbonica entro il 2000 ai Livelli del 1990', (Final), Roma.

Nardi, P. and Protopapa, M. (1991) 'Referendum USL, frattura verde?', *Ambiente* No. 25, Roma: Janusa.

Obasi, G. O. P. (1993) Opening Statement at the Italian National Conference on Climate, Florence.

Pinchera, G. (1991) 'Public Policies Affecting the Energy Sector in Italy', paper for Seminar on Environment and Energy, Bonn.

Pinchera, G. (1992) 'Limite di Emissione e Innovazione Tecnologic', pp. 102–113 in *Ambiente Italia 1992*, Firenze: Vallecchi.

Ruffolo, G. (1992) 'Carbon Tax, Technology and Global Warming: A Proposal of the Italian Delegation', Statement at UNCED, Roma: Ministry of Foreign Affairs.

Scovazzi, T. and Pineschi, L. (1991) 'Policy and Law on Global Warming in Italy', pp. 79–83 in T. Iwana (ed.) *Policies and Laws on Global Warming: International and Comparative Analysis*, Honolulu: University of Hawaii, Environmental Research Center.

Silvestrini, G. (1991) 'A Strategy to Reduce Greenhouse Gases Emissions in Italy', paper, Roma: Legambiente.

Silvestrini, G. *et al.* (1992) *Comuni Contro L'Effetto Serra*, Milano: Franco Angeli.

Vig, N. and Kraft, M. (1990) *Environmental Policy in the 1990s*, Washington, DC: Congressional Quarterly Press.

11

EXTERNAL PERSPECTIVES ON CLIMATE CHANGE

A view from the United States and the Third World

Konrad von Moltke and Atiq Rahman

THE US PERSPECTIVE *(by Konrad von Moltke)*

International relations have long focused on managing conflict and competition. The result is a virtual obsession with power and leadership, categories which are largely meaningless for international environmental management based on cooperation, transparency and accountability as its primary tools. However, old habits die hard so the question persists as to which country is 'leading' on environmental policy or which has the 'best' environmental policies. This attitude is widespread in the United States, the pre-eminent world power of the last decade of the twentieth century, and makes it particularly difficult to confront America's failing and other countries strengths in a field such as climate change.

In practice, many countries can with justification lay claim to some form of pre-eminence in environmental affairs in general and climate change policy in particular. After all, this was very much the European Community claim in the early 1990s, as noted in Chapter 6. Because climate change policy actually involves several major policy areas, a country can demonstrate 'leadership' by focusing on those areas in which it is strong and by glossing over those where there are weaknesses. The result is a world of rhetorical leaders with no followers.

Comparisons of climate change policies are difficult to undertake because so many variables must be taken into account. In practice, climate change policy touches most areas of economic and social policy. It is consequently difficult to understand specific aspects of climate change policy without an understanding of the context of economic and social policy making in the relevant country. Knowledge of the details of climate change policies in other countries is quite limited. Knowledge of the broader context in which these policies must be seen is almost non-existent.

US PERCEPTIONS OF EUROPEAN CLIMATE CHANGE POLICIES

Lacking a frame of reference, there is a tendency in the United States to view other countries' policies in lights of US policy paradigms. While understandable, this can lead to serious misunderstandings.

As the case studies presented before seek to demonstrate, in economic policy terms, the central goal of climate change measures is to achieve structural economic change by a variety of policy measures which favour activities with lesser climate impact over activities whose climate impact is high. Governments need to draw on the entire arsenal of economic policy tools to have any prospect of achieving this extremely difficult goal – including not only macroeconomic tools but also in particular components of industrial policy, education and retraining activities, and social assistance programmes. The federal government in the United States has only limited competences in many of these areas. It has virtually no authority over education. It shares its role in social policy with the states – resulting in widely differing levels of protection for workers displaced by structural changes in the economy. It eschews 'industrial policy' for ideological reasons but also because the US Constitution does not give it the necessary authority to pursue effective policies in this area.

The result has been a limited view of what constitutes economic policy, a view now shared by other countries and largely imposed on international organizations such as the World Bank, the International Monetary Fund and other international financial mechanisms. This is for the simple reason that the United States is pre-eminent on the international political and economic scene, and therefore its perceptions of what works and what does not carry particular weight. Since the US federal government has primarily macroeconomic tools – taxes, money supply and interest rates – at its disposal, it is inclined to assume that economic policy is identical with macroeconomic policy. This perception is also reflected in some 'economic' studies of climate change which seek to assess the cost of restructuring the economy with the limited arsenal of policy tools available at the federal level in the United States. The dominant economic conclusion is that the 'costs' would be prohibitive relative to the likely economic and local consequences arising from climate change in the United States. For a useful summary see Nordhaus (1991), National Academy of Sciences (1991), and Manne and Richels (1992). This conclusion is hardly surprising if purposive structural economic change is to be achieved using only the tools of macroeconomic policy, arguably the least focused and consequently the least efficient approach to the problem.

Applied to climate change policy, the US government finds it difficult to conceive of federal policies which could ensure a certain level of reductions in greenhouse gas emissions. Indeed, it is hard to visualize a package of federal measures in the United States which would be reasonably certain not only to reduce the emission of greenhouse gases but to keep them to a predetermined level. In the absence of overwhelming public concern, the two available solutions to this dilemma − novel forms of federal/state cooperation, or a change in the balance of powers between federal and state levels − are well beyond the reach of the American political consensus. In any case the separation of powers between Congress and the White House makes it difficult for, say, a Presidential tax initiative on energy or gasoline to carry approval on Capitol Hill.

Largely incapable of ensuring the achievement of a given level of greenhouse gas emissions in their own country, US policy makers have a deep scepticism about the capacity of any other country to achieve this goal, Japan is widely recognized as 'different' and can consequently make promises which seem strange to US sensibilities − although the long running saga of efforts to open the Japanese market to US products and US policy approaches has left doubts in the United States about the willingness of that country to follow through on promises. Europe is considered sufficiently similar to be judged by US standards, particularly since the United Kingdom is known to be part of Europe and visibly shares many American attitudes to economic policy orthodoxy.

For an interesting analysis of the differences in approach to policy making in the European Community, Japan and the United States, see Vernon (1993). Vernon asserts that it is the very separation of powers in the United States that gives politicians and executive officials a greater sense of initiative and flexibility in policy making. By contrast, argues Vernon, the Japanese bureaucracy is based on stability and professionalism. Its international attitude is coloured by toughness and a strong sense of teamplay. European Commission officials prefer a consultative mode and enjoy the illusion of consensus politics in preparing their bargaining positions. Hence the finding that there is some scope for collusion between the United States and the European Union.

From this perspective, US resistance to commitments on targets and timetables for the control of greenhouse gases is entirely understandable. In the view of US policy makers, greenhouse gas emissions are dependent variables arising from a number of independently adopted economic and environmental policies so it makes little sense to adjust these for climate change purposes alone, particularly since current projections show a lessening of the rates of increase in greenhouse gas emissions. These projections, however, were made before the current US economic recovery outstripped all predictions, at the same time invalidating at least some of the assumptions on which the optimistic scenarios for greenhouse gas emissions were based.

European climate change policies are seen in the United States primarily in terms of commitments to targets and timetables for the control of greenhouse gases. Only some specialists are interested in the details of measures actually being planned to meet these commitments. Unable to envisage such policies in the United States, US policy makers have always been deeply suspicious of European promises in this regard, and openly incredulous of the German commitments to 25 per cent reduction of 1990 CO_2 levels by 2005.

US perspectives on climate change are further coloured by the country's highly ambivalent attitude towards measures designed to reduce the material throughput of the economy. The United States is unique in being both a major producer and importer of oil. All other countries are either major producers or major importers. US policy with regard to oil is appropriately schizophrenic, and there is little prospect that this will change as long as underlying relationships persist. More generally, the United States is the only major industrialized country which is also a leading producer of many other important commodities. Its resistance to policies designed to increase the efficiency of resource use is rooted in the simple fact that such measures are widely perceived as harming the US producers of the relevant commodities by decreasing demand for them. In this regard, US interests – and consequently US perceptions – are dramatically different from European ones.

EUROPEAN INFLUENCE ON US CLIMATE CHANGE POLICY

Two factors have tended to limit the impact of foreign experience on US environmental policy in general and on climate change policy in particular. These are the American obsession with 'leadership' and an underlying resistance to foreign entanglements.

No country finds it easy to confront its own political inadequacies. To some extent, these inadequacies are the result of the very fabric of the political system: if it were different, the weaknesses would not be the same. Consequently the system which produces inadequate policies will find it excruciatingly difficult to recognize the problems it creates until some crisis forces it to confront them. This is particularly true of a country, such as the United States at this time, which is confronted with the apparent success of its practices, at least to judge by the number of emulators. As a result, the United States does not readily take lessons from other countries.

Perhaps more important is the abiding isolationist tendency within the nation with roots which go back to the origins of the US Constitution. At the very least it is abetted by the doctrine of separation of powers and the distribution of authority for foreign affairs within the US system of government.

In most countries, the dramatic increase in the volume and complexity of international affairs has had the effect of further strengthening the administrative branch of government. Where federal structures exist, the conduct of foreign affairs has frequently provided an opportunity to strengthen the hand of central authorities in their complex relationships with federal units.

In the United States, these phenomena are largely mitigated by the role of Congress which is in no way controlled by the Administration in power – even when the majority in Congress and the President come from the same political party. Congress tends to represent the variety of political opinion in the country and defends regional interests and those of particular states, forcing a continuous process of compromise at the federal level.

In such a situation, foreign influences are limited. Indeed the only recent instance of US policy responding to foreign interests is the success of the Israel lobby in transforming the existence of an articulate constituency within the country into a real force for political influence. This success is based essentially on converting an international issue into a domestic one. Efforts by the Canadian government to achieve a similar outcome on the issue of acid rain were significantly less successful, despite the formation of strategic alliances with key regional constituencies in the northeastern United States. It is difficult to conceive of a situation where foreign approaches to the issue of climate change could exert a major influence on US responses, certainly not without sustained efforts to transform the international dimension into a domestic one within the United States. The success of the EC in obtaining a political willingness in the United States to sign the UN FCCC in Rio was largely due to the fact that the Article 2 did *not* contain any targets. US domestic policy was frankly unready for specific commitments, even though stabilization of 1990 greenhouse gas emissions by 2000 is tacit US policy.

The US Constitution provides the Congress – primarily the Senate – with extraordinary powers in relation to foreign affairs. The fact that treaties require a two-thirds majority in the Senate has rendered ratification of agreements concluded by the Administration parlous at times. The League of Nations, the Havana Convention establishing the International Trade Organization, SALT II and, for the foreseeable future, the Law of the Sea Convention are only the more notable of the treaties left unratified by the United States.

One result has been to make the United States by far the most difficult country to negotiate with. While international negotiators can generally assume that the representatives of other countries actually reflect the position of their government, where the United States is concerned it is always necessary to consider the views of the US Congress in addition to the positions espoused by the Administration. This has sometimes strengthened the hand of US negotiators who can argue the need to convince a recalcitrant Congress to extract concessions which might otherwise not be available. On other occasions, for example in the instance of the Uruguay Round, it has simply meant that no country will act on an accord until it knows how the US Congress will respond.

One response to the problem of Senate consent has been the systematic use of 'administrative agreements', defined tautologically as instruments which do not require Senate consent, to conduct foreign policy. For example, only a single bilateral environmental treaty exists between the United States and Canada, the 1908 Boundary Waters Treaty. All other matters since then, including air pollution and the transport of hazardous wastes, have been handled through administrative agreements. While the legal effect of such agreements can be indistinguishable from those of an international treaty, they can only be negotiated in areas where no subsequent implementing legislation will be needed. This is because no guarantee can be provided that Congress, being a separate and independent branch of the US government, will actually legislate as required by an administrative agreement. In consequence standing instructions for US negotiators require them to agree only to what is already specified under US law. While this can provide quite extensive authority – as demonstrated by negotiations for the Montreal Protocol on Substances that Deplete the Stratospheric Ozone Layer – it is nevertheless fundamentally limiting. In effect, the United States is saying that it will agree internationally only to something it has previously decided to undertake domestically. This position institutionalizes the concept that US law will always 'lead' international policy. It certainly makes the United States quite resistant to international influences.

POSSIBLE CONSEQUENCES FOR CLIMATE CHANGE POLICY

Climate change policy is arguably the most complex task ever undertaken internationally. It is all but inconceivable that one country – even if it is the United States – will be capable of generating all appropriate responses through its domestic policy process alone, with international agreement designed to confirm and translate the results. There are clearly areas in which US policy lags and will continue to lag, because of the peculiar constellation of US interests, as in energy policy, or because of the distribution of domestic authority, as in many areas of structural change policy. If climate change emerges as a top priority on the political agenda – which it is not at the present time and which it will take some dramatic environmental evidence of climate change to achieve – complex international agreements covering environmental, economic and social policy will be needed. In some areas, US domestic policy will need to follow on international agreement rather than lead it.

Given the peculiar constellation of US Constitutional Law and US domestic political forces, it would seem impossible to achieve such an outcome. Interestingly, for one area of international policy the United States has developed policy tools to permit domestic legislation to be forced by international agreement. This area is trade policy. In light of the structural difficulties of the US political systems, it would appear impossible ever to conclude an international trade agreement with its characteristic mix of general rules and specific concessions affecting individual industries (not unlike prospective climate agreements). US negotiators would always be upstaged by subsequent congressional restructuring of agreements reached internationally – and no country would be willing to negotiate under such circumstances.

To meet this need, the US Congress has developed the tool of fast track authority. Essentially this is a device which provides Congress with some input into the definition of negotiating objectives, creates an opportunity for the House of Representatives to participate in the process more actively than envisaged by the Constitution, and provides for a continuous process of exchange between Administration and Congress during the course of negotiations and in the preparation of implementing legislation. In return Congress commits itself to vote on the entire resulting package, including all necessary domestic implementing legislation, without amendment and with a simple majority of all votes cast required for passage, rather than the prohibitive requirement for two-thirds in the Senate which affords almost every special interest a blocking minority.

The parallels between trade policy and climate change policy run deep. Both have, as their ultimate economic goal, structural change to increase efficiency, both have an inescapable international dimension, and both require international agreements leading to highly detailed changes to domestic legislation. Ultimately it is unlikely that the United States will meet the challenge of climate change without resort to fast track authority for its international climate change negotiators. But that will only come when the US becomes truly internationalist in its handling of the post-Rio agenda.

CONTRACT WITH AMERICA AND US CLIMATE CHANGE POLICY

In November 1994 just under half the US electorate went to the polls to vote in a Republican majority in Congress for the first time in over 40 years. This was widely regarded as a politically seismic event, because the mandate to the new Congress was seen as a popular backlash against big government, interference in personal freedoms generally, and the setting of an agenda for deregulation and unfettered youth. This in turn created a form of anti-environmentalism, known as the 'wise use coalition'. The phrase is misleading. The motive is to cut spending in federal programmes of pollution control and environmental pollution generally where Congress had not specifically voted the funds to the states for implementation. In effect there could be a federal initiative, but no supporting budget.

The rhetoric of the wise use, moreover, is angry and anti-government; as Brick (1995, 5) quotes, the adherents accuse environmentalists of 'putting rats ahead of family wage jobs, impeding economic progress, and drowning individual rights with big governmental rules and regulations'. This group wants to rescind key environmental statutes such as the Endangered Species Act. In addition it is looking for environmental regulations to be cut, or removed, and at the very least subjected to a cost benefit analysis. This was designed as a euphemism for giving industry and developers generally powers in law to challenge proposed permits that appeared to be too financially onerous and when the benefits could not obviously be quantified.

This combination of reduced budgets and possibly very restrictive conditions in future environmental regulation panicked environmental NGOs and alarmed the more responsible and environmentally committed sectors of US industry. Such groups were in a minority, but sufficiently influential as to form important, if informed, alliances. By mid-1995, these two highly contentious issues had not been resolved. The outcome is likely to be a messy compromise, with cost benefit to a slightly less rigid form than being applied for all new regulatory cuts, and widened scope for appeal by regulated parties when they feel that risk assessments have been too strictly applied.

This suggests that the new Congress is unlikely to pass new laws promoting the cause of CO_2 reduction. More probable will be a modest switch to gas, and a series of voluntary agreements in industry and at state and local level to reduce energy wastage. The crucial transport sector is unlikely to be touched. In 1993 the American people protested very vociferously over proposed new taxes on gasoline, and won their alleged right to purchase gasoline at prices less than a third of European prices. In mid-1995 Congress voted to relax the long standing 55 mph highway speed limit, created in the wake of the oil crisis.

Europe will clearly have to play its full part in the greenhouse gas reduction programme before the Americans move. The European negotiating position in future Conferences of Parties will depend enormously on a clear purpose of going beyond 2000 stabilization. Until this can be seen to be deliverable, it is most unlikely that the United States will take any specific action. How Europe handles the critical next few years will undoubtedly have profound repercussions for the whole of the Convention. It is as simple as that.

A VIEW FROM BANGLADESH (by Atiq Rahman)

THE KEY ISSUES

Introduction: developing country views

Bangladesh's perspective on climate change is based on a set of views common amongst developing countries. The dominant amongst these is the belief that the impacts of global climate change are going to have disastrous effects on the development pathways of several developing countries and that their people and ecosystems are the victims of a global phenomenon to which their contribution is insignificant. As is well established by the IPCC reports, outlined in the introductory chapter, the primary responsibility for global climate change lies with the OECD (Annex 1 countries) dominated by North America and Europe.

The developing countries have to express solidarity with the Group of 77 (the most important political block in the UN). Yet the G77 represents a set of heterogeneous and even conflicting interest groups as far as climate change is concerned. Accordingly, the G77 may not be the adequate mechanism to express the respective positions and frustrations of the developing world as a whole within the climate negotiations process. But the G77 routeway is the best available. In general, the Third World countries tend to be most supportive of the AOSIS group of small island states. Yet AOSIS does not adequately represent the highly populated and threatened delta regions of Bangladesh or Egypt.

The responsibility for climate change past, present and in the near future lies with the north which should compensate countries such as Bangladesh through new and additional funding. But an emerging donor fatigue makes this ecologically just solution even less likely. There appears to be no real seriousness amongst the OECD countries to change the global institutional regime, or to provide requisite resources, or to change their consumption behaviour. Yet all of this is so badly needed to address the threats of global climate change.

From the Third World perspective, the north takes convenient joint positions as far as the south is concerned to protect the interests of the north. The EU and the United States selectively express their differences as a vehicle for protecting northern interests when it comes to the question of targets, timetables and specific commitments.

The developing countries have made significant concessions during the different stages of negotiations leading up to the signing of the Climate Convention in Rio and CoP I in Berlin, particularly on the matter of additional funding and technology transfer. But the northern countries continued to set priorities suitable to their needs in the recent negotiations leading up to the Berlin conference. For example, they are seeking to enshrine the tactic of joint implementation as a major mechanism without addressing the prerequisite and fundamental issues of criteria for credits, concepts of per capita allocations and entitlements. Indeed, some EU members such as Germany and the Netherlands are willing to make public announcements about national commitments but are unwilling to enforce them in common EU positions.

Perceptions about the EU and its member countries

The perception amongst the developing countries about EU Member States was that the latter are more progressive than the United States within the OECD bloc and marginally more serious about achieving a functional climate convention with limited or minimum prior commitments. Public opinion in Europe is generally seen as more supportive of a more serious commitment than is the case in the United States. While the industrial, coal and oil lobbies of the United States were most influential in assisting or dictating to the US government regarding its positions before and even more consistently after Rio, European industries were less vocal or public in their pronouncements.

The issue of climate change has exposed the biases of each Member State of the EU according to its particular economic conditions, political perceptions and public preferences. Each subset of issues of climate change resulted in a separate combination or clustering of European countries. For example, the Toronto target which formed the basis of the AOSIS Protocol proposal in CoP I included countries such as Germany, the Netherlands and Denmark. The UK was perceived to have positions closer to the United States than its European partners. France and Germany jointly pushed to make the Global Environment Facility (GEF) the sole financial mechanism under the UN FCCC but in its implementation phase France chose to invest some of its money in a 'French–GEF' modality. Norway and the Netherlands were the major players in promoting the cause of joint implementation and initiating bilateral initiatives. These two countries apparently have limited possibilities for reducing greenhouse gas emissions at home and hence need measures in other countries to meet their own climate commitments under the UN FCCC. While Germany, France, the UK and the Netherlands played a leading role, Portugal, Greece and Ireland remained on the periphery of the debate.

Bangladesh's role and perspectives: crisis and helplessness

A recent study undertaken by the government of Bangladesh, following the IPCC methodology and conducted jointly by the Bangladesh Centre for Advanced Studies (1994) of Dhaka and Resource Analysis of the Netherlands, clearly demonstrated the severity of the threat of climate change for Bangladesh. A 1 metre or 30 cm rise in sea level will inundate over 17 or 13 per cent respectively of the country and most of its highly populated and productive coastal areas. Furthermore any significant sea-level rise would affect the world's largest mangrove forest – the Sundsbans. The rate at which sea-level rise is expected to take place will be much faster than the rate at which the mangrove system can migrate and re-establish. Further, enormous demographic pressures make any such ecological migration impossible. The associated loss of biodiversity is also extremely high, irrespective of the very considerable social and economic costs.

The use of a scenario of a 2 °C and 4 °C temperature rise on a set of nine vulnerability zones of Bangladesh revealed the astounding prediction that the severity of drought in the dry season in Northwestern Bangladesh will have as great an impact on agriculture, i.e. rice productivity, as the impact of sea-level rise. Further, the impact on existing floods will result in greater monsoon flow and more severe droughts at different seasonal periods.

This limiting combination, further analysed by using multicriteria analysis, shows that if the threat of sea-level rise is real and if Bangladesh does not get cooperation from its neighbours on equitable water sharing of its main rivers, then all the development efforts undertaken by Bangladesh in the next 30 years will be nullified by these two exogenous events (global climate change and non-cooperation on water issues). Bangladesh has little political control over either of these two threats.

Given endemic poverty, rampant and increasing disaster of floods, cyclones, drought and environmental degradation, Bangladesh cannot afford the luxury of giving any significant priority to the mitigation of climate change. On the other hand, it is doing its best in reporting, studying and undertaking what vitally necessary measures it can afford. Bangladesh has already developed its Ozone Depleting Substances Phase-out Plan under the Montreal Protocol, undertaking its Climate Country Plan and developing a least-cost abatement strategy for climate change.

Unfortunately, the Bangladesh government was not a major player in the negotiations leading to the UN FCCC, or in the IPCC, or in the GEF. But wherever possible it has highlighted the most major concerns of sea-level rise and the very real possibility of intensified and unpredictable extreme events to various global fora. At home it has highlighted climate concerns at all levels, set up an interministerial group and supported research groups and NGOs.

The non-governmental organizations of Bangladesh, on the other hand, have played a key role in the UN FCCC negotiations and all its agencies including IPCC and GEF. A Bangladeshi NGO representative was elected by the world NGOs participating in the UN FCCC negotiations to address the UN both at the first and the last INCs (i.e. INC 1 and INC 11) at their formal deliberations. The NGOs have succeeded in making the concerns of Bangladesh a major reminder and a moral and ethical milepost in the negotiation process. Further, in the IPCC, Bangladeshi NGO representatives have highlighted the concerns of equity, social considerations, and the special issues facing most affected social groups. In the GEF process Bangladeshi NGOs played a leadership role in its transformation from the pilot phase to a permanent UN agency.

The government of Bangladesh has mostly worked through G77 where the dominant voices have been China, India and Brazil together with the countervailing interest group of fossil fuel exporting countries represented by Saudi Arabia and Kuwait. The absence of a group representing the low lying vulnerable delta or coastal states has limited the role of the Bangladesh government to influence the UN FCCC and the subsequent global climate change debate. Thus Bangladesh's sense of crisis and helplessness continues.

THE ETHICAL ISSUES

The three moral principles of global governance which have the highest relevance to climate change have dominated the argument of some of the developing countries and NGO politics. These are as follows:

1 All human beings are equal (as enshrined in the UN Charter).
2 The global commons such as the atmosphere and oceans belong to all human beings equally. This principle was supported by thousands at Rio under the Declaration of the Global Forum on Environment and Poverty (GFEP).
3 The polluter pays principle, accepted increasingly in the north.

A juxtaposition of the above three principles gives rise to an argument which makes the transfer of resources to developing countries from Annex 1 countries under the UN FCCC quite evident. But there is very little recognition in northern thinking of these southern positions. The globally concerned international NGO community has been far more amenable to such a moral imperative. Thus the civil society participating in the climate change negotiation process has been far more coherent in working towards achieving the objective of the UN FCCC than their governments who purport to represent them.

The view from the developing world is that until the industrialized nations show real progress in reducing their greenhouse gas emissions, the UN FCCC process is invalid. A study by Warrick and Rahman (1992) concluded that between 1800 and 1990, the rich nations have cumulatively accounted for over 84 per cent of all CO_2 emissions caused by fossil fuel burning, and over 75 per cent of CO_2 emissions associated with deforestation. At the very least these countries have a 'polluter pays' responsibility for taking a lead. Europe simply has to play more than its part by putting its own house in order. Until that is the case there can be no question of entering into serious negotiations over joint implementation or multinational development bank preferential loans, or even GEF-inspired deals. The north has got to deliver.

More serious still is the need for a different perception of the causes of greenhouse gas warming and the relative future responsibilities of the north and the south. The IPCC response scenarios published in both 1990 and

1992 failed to reflect the principle of common and differentiated responsibility for climate change. This is because the scenarios build in a world view that actually widens the inequality gap between north and south. The assumption in the 1990 reference scenario is that per capita CO_2 emissions in North America will rise from 5.08 tC to 7.12 tC by 2025. According to this scenario, the emissions of one American are equivalent to over seven Latin Americans, 11 Asians and 13 Africans. Set against this reference scenario were four alternatives, of which one was the stabilization of global CO_2 emissions at the constant 1985 level of 5.12 billion tonnes of carbon per capita by 2025. This would mean a cut of over 7 billion tC from the reference scenarios of 12.43 billion tC. To get this, the IPCC considers stabilizing the life styles of the rich and shifting the legitimate development aspirations of the poor.

There is far too facile a case that future CO_2 emissions will be driven more by population growth and subsequent energy demands and land use practices than by overconsumption in the north. The climatic problem is caused by overconsumption in the north and by underconsumption in the south. The linkages between poverty and environmental stress have been highlighted ever since the Stockholm Conference on the Human Environment in 1972. The point was picked up in the Brundtland Report (1987), which decried a 'downward spiral of poverty and degradation', whereby the poor are forced to draw unsustainably on available natural resources to satisfy immediate survival needs.

But to blame the poor for their environmental inputs would be both immoral and ineffectual. Harrison (1992, 121) observed that 'the poor probably tread lightest upon the earth, and do less damage to the environment than any other group. They are victims, not perpetrators.' What does degrade both the environment and humanity are explorative relationships which marginalize whole sections of society and extract wealth from the earth without concern for equity or sustainability.

The real 'climate bomb' is overconsumption in the north (Rahman *et al.*, 1992). A study by Parikh (1992) shows that though the developed countries constitute just under a quarter of the world's population, their share in the global consumption of the various commodities ranges from 50 per cent to 90 per cent. Despite the much publicized improvements in energy efficiency since the oil shocks of 1973, these countries still consume three-quarters of all commercially produced energy. The scale of this gulf can be measured by calculating the average disparity ratio between the per capita consumption levels of developed and developing countries. This ranges from around 5 for food to over 20 for chemicals and cars, and around 10 for energy. These figures are distorted a little by including the former Soviet countries in the developing world, and the newly industrialized nations in the developed world. If the extreme ratio of the per capita consumption levels between the richest and the poorest nations is used, as exemplified by the United States and India, then the ratio widens from 4 for basic foods to 50 for meat, to over 200 for metals, 30 for chemicals, and 320 for cars.

Clearly there are central and most fundamental equity issues at stake here. These would suggest that there should be a convergence in the distribution of goods and services to levels that are both humanly necessary and environmentally sustainable. The national report of India explained that

> while to a large extent the lower per capita consumption is due to poverty, an important reason is a majority of the population of India follows a sustainable lifestyle which is based on environmentally sound practices evolved over thousands of years.
>
> (Quoted in Rahman *et al.*, 1992, 43)

Agarwal and Narain (1991) have shown that the resource transfer from north to south in a carbon regime would be higher than total overseas development aid.

Future development patterns for China and India envisage a potentially huge expansion for coal, with profound implications for further CO_2 emissions. But on a per capita basis even these awesome increases will result in very low levels of energy consumption compared with northern consumption patterns today. So the north, and that certainly includes Europe, has to reduce its projected consumption patterns, so as to allow the 'ecological space' for the south to develop, and the south should be allowed to develop along energy-efficient and resource use minimizing pathways. For this to happen technology and financial transfers on a large scale will be required.

If joint implementation is to become a reality, it must occur only in those sectors clearly affecting the poor, or where there is a substantial reduction of emissions in the energy sector. For the time being at least, all joint implementation agreements must be bilateral, with no greenhouse gas (GHG) credit to the industrial partner, and where the funding must be additional to normal trade or aid flows. All projects must also result in verifiable GHG emission reductions. Any deal must also be backed by clear evidence of an effective climate GHG reduction programme, backed by independent verification. The packaging of any joint implementation scheme must be decentralized, involving genuine social and economic gains to the poor, and handled by NGOs, the private sector, and by governments in this order, as indicated by communities themselves.

This means that the CoP will have to fund and build up national teams for joint implementation pilot schemes and for unification. This in turn requires that the carbon credit donor countries must be satisfied that real GHG reductions are taking place. The funding for this should come from central taxes or other fiscal measures undertaken in the creditor countries. The CoP will therefore have to develop institutional arrangements along with special standards and methodologies to establish suitable implementing and monitoring teams. The criteria of accreditation of any credit must be established and the attribution of credit rights will have to be decided. The southern countries would definitely prefer a per capita entitlement principle as a basis. As noted in Chapter 3, all this has to take place in the context of verifiable national GHG emission monitoring programmes in every recipient country.

Europe could take a lead in all this. But it has to show willing to recognize its own microcosm of similar arrangements from its own north to south before it can seriously address the global equivalent. Southern eyes will be gazing intently upon the EU Member States to take that lead. That in turn will require institutional innovation of a high order. The Bonn Secretariat is well located. Germany has as much to do for the EU as it has in partnership with the other wealthy countries for the world stage.

CONCLUDING REMARKS

Achieving the scientific requirement for a reduction of 60 per cent of greenhouse gas emissions as stated by IPCC as a prerequisite for a sustainable planet, free from the threat of climate change, seems an impossibility. The imperative for economic development in both the south and the north, the incapacity of short term national planning horizons to address long term climate change, the complacency of the northern leadership and electorate in the face of an extremely iniquitous world, and increasing GHG emissions make it almost impossible to find a consensus for adequate measures to combat the threat of climate change. So, what does the future hold?

Two overlapping but separate long term visions offer a possible way out of this ensuing global crisis despite the tremendous opposition and difficulties. If we believe in an equitable and fairer and yet safer world in terms of global climate change, then hard choices will have to be made.

First is the 'concept of a low carbon future' which does not compromise the quality of life of the population of the northern countries. This can be achieved by reducing the flow of energy and materials in the production and consumption processes at all levels, from individuals to the processes. Technology will probably offer part of the solution. So too must come societal and behavioural changes in terms of respect for frugality and a commitment towards a higher quality of life rather than a so-called high living standard loaded as it is by high energy, usage, and high material demand, and consequently high carbon content. This will require a fundamental shift in thinking and modalities which will be difficult but not impossible.

The second is the 'concept of convergence'. There is a well-founded fear that developing countries will continue to emit greenhouse gases and in the next decades will be responsible for increasingly greater quantities of emissions. But the truth remains that if today all developing country human beings were to stop breathing, and the north continues to emit greenhouse gases without reduction, the result would not be sustainable in terms of tolerable climate change and stability of the atmosphere. Thus a reduction of emissions from the north is a must while the south must become more and more carbon responsive in its growth and development. It is possible to foresee a future where the north will continue to reduce emissions and the south will increase in a frugal and carbon-responsible world. These two trends will converge to a point where all countries will have the same per capita carbon entitlement. That is more likely to be in the 22nd rather than the 21st century if everyone is carbon responsible. Beyond 'the point of convergence' of the per capita carbon entitlement, with its decreasing per capita trend of the north and increasing per capita trend of the south, all countries, i.e. the global population, will decide to reduce further GHG emissions congruently. That future might look far fetched, but that is the future where today's north and south must work together. The UN FCCC may provide the first real opportunity for an equitable global governance regime and a global achievement of a high quality of life for all its citizens, achieved with a significantly lower energy content and material flow and consequent dwindling carbon emissions.

REFERENCES

Agarwal, A. and Narain, S. (1991) *Global Warming in an Unequal World*, Delhi: Centre for Science and Environment.

Bangladesh Centre for Advanced Studies (1994) *Vulnerability of Bangladesh to Climate Change and Sea Level Rise*, Dhaka: Bangladesh Centre for Advanced Studies.

Brick, P. (1995) Determined opposition: the wise use movement challenges environmentalism, *Environment* 37(8), 16–20, 36–41.

Brundtland, H. G. (Chair) (1987) *Our Common Future*, Oxford: Oxford University Press.

Harrison, P. (1992) *The Third Revolution*, London: Tauris.

Manne, A.S. and Richels, R. (1992) *Buying Greenhouse Insurance: The Economic Costs of Carbon Dioxide Emission Limits*, Cambridge, MA: MIT Press.

National Academy of Sciences (1991) *Policy Implications of Greenhouse Warming – A Synthesis Panel*, Washington, DC: National Academy Press.

Nordhaus, W. (1991) Economic approaches to greenhouse warming, pp. 71–89 in R. Dornbusch and J. M. Poterba (eds) *Global Warming*, Cambridge, MA: MIT Press.

Parikh, J. (1992) *Consumption Patterns: the Driving Force of Environmental Stress*, New Delhi: Indira Ghandi Institute of Development Research.

Rahman, A., Robbins, N. and Roncerel, A. (eds) (1992) *Consumption vs Population: Which is the Climate Bomb?*, Brussels: Climate Network Europe.

Vernon, R. (1993) Behind the scenes: how policymaking in the European Community, Japan and the United States affects global negotiations, *Environment* 35(5), 12–20, 35–42.

Warrick, R. and Rahman, A. (1992) Future sea level rise: environmental and socio-political considerations, pp. 97–112, in I. Mintzer (ed.) *Confronting Climate Change: Risks, Implications and Responses*, Cambridge: Cambridge University Press.

12

BEYOND CLIMATE CHANGE
SCIENCE AND POLITICS

Tim O'Riordan and Jill Jäger

PERSPECTIVE

The purpose of this exercise was to evaluate institutional adaptation in the European response to the perceived threat of global warming. The idea was to assess how far there are 'climate change politics' that are separately identifiable, purposeful, and capable of changing political relationships, organizational procedures, individual outlooks, and mutually beneficial behaviour. In short, was the debate and commitment to a climate convention sufficiently focused and powerful to alter what would otherwise have occurred in its absence?

There is an emerging literature on the effectiveness of international environmental regimes. Regimes are institutions in their own right, namely constellations of governmental and intergovernmental structures organized around a set of shared roles and procedures to achieve a common purpose. In Chapter 1 we suggested that there may in the UN FCCC be three loosely linked subregimes, namely scientific investigation and analysis; intergovernmental negotiating, monitoring and reviewing structures; and national governmental interpretation and responses. It would, in our view, be unwise to lump these together into an amorphous whole. They are certainly different, they involve importantly separate rules and responsibilities, and the institutional change triggers, outlined in Figure 4.1, will act on them in separately identifiable ways.

The science of climatic change, as we discussed in Chapter 1, is increasingly being drawn into politically supported analytical structures, to the point where 'climate change science' is not always separately identifiable from the political process that shapes it. But there is still a recognizable climate change science, and this remains one organizing framework. Similarly, as we have shown throughout the case studies, though climate change politics and organizational structures are undoubtedly linked to many other policy motives, nevertheless there is an identifiable national response that can be attributed to climate change *per se*.

346

We suggest that over the course of time, one possibly very important institutional innovation will be the merger of these three subregimes into a more coherent climate change regime structure. This will be an amalgamation of a series of institutional innovations, outlined in Figure 4.1 and summarized below. The scientific and evaluative component is steadily becoming enmeshed in both the CoP process and national responses. The politicization of the social science agenda on climate change analysis has barely begun. It is likely that any future evaluation of what is driving climate change, for example who should be responsible for taking the lead, how far joint implementation will be promoted, and the extent to which all this becomes engulfed in a much wider north–south dispute over aid and targeted development, will be enmeshed in political rhetoric of equity rather than efficiency. In Chapter 11 Atiq Rahman clearly stated the views of the southern NGO community on such issues. And in Chapter 1 we reported on the increasingly vitriolic dispute over the equity–ethical interpretations of cost benefit analysis of response options.

We also concluded both in Chapters 1, 3 and in Chapter 6 that the CoP-based subregimes around monitoring, verification and supportive investments are highly active and immensely important for building trust in the process, and guaranteeing compliance. These subregimes are also articulating a weighty role for scientific appraisal of both greenhouse gas forcing comparisons and improved interdisciplinary and international approaches to measuring national emissions. We regard this as a significant breakthrough in the linkage of science and policy, a breakthrough which is being achieved with a remarkable amount of consensus.

Within this science monitoring mode, we observe another important institutional innovation. The United States and the United Kingdom led the way with a view that the greenhouse gas inventory should be regarded as a totality, or a *basket of gases*. This is the crucial basis of the British position for 2000 and beyond, as outlined in Table 8.4. The UK also wants to take a lead in guiding the European Environment Agency into a viable organization for comparing greenhouse gas inventories, as well as a host of other pollutants. In this way the UK, apparently followed by Germany, Norway and, we believe, Italy, is pushing for a monitoring approach that is comprehensive, and complementary to policy mixing – yet still controversial. This is because the method of consigning CO_2 equivalence remains unresolved. The controversy will be even more the case on the north–south agenda, as doubtless future CoPs will discover. This in turn is linked to a strong national desire to ensure that there is reasonable comparability over emissions data, and that any burden sharing in intergovernmental trades should take place in the open and according to agreed rules of monitoring and compliance. So even here, the quality and extent of any national response is a function of a more profound willingness to generate better data, to guarantee a clearer interpretation of trends and performance achievements, and to legitimate inter-Member State deals over differentiated responsibility according to national needs and economic means. This is why we have placed so much emphasis on the European community monitoring mechanism as an innovative institutional precursor to its global equivalent.

But on the third subregime, namely the trigger effect of the UN FCCC on national capacity and response, we remain more circumspect. This is the central element of our analysis, though we believe that it is appropriate to place it in its wider institutional perspective. The most widely cited evaluation of regime effectiveness in terms of national responses is that of Marc Levy and his collegues (1993) as summarized in Table 12.1. They posit that international institutions geared to global environmental agreements may be judged on the basis of three parameters:

- the degree to which governmental *concern* was increased directly as a result of the convention, or treaty or protocol;
- the degree to which the *contractual* conditions of compliance through monitoring and deciding was made easier and more explicit as a consequence;
- the extent to which national *capacity* to deliver the agreement was improved, together with supporting policies, advocacy coalitions and focused public support.

Table 12.1 Paths to effectiveness: how international environmental institutions boost the three Cs

Role of institution	Representative institutional activities
Increase governmental concern	Create, collect and disseminate scientific knowledge
	Create opportunities to magnify domestic public pressure
Enhance contractual environment	Provide bargaining forums that
	• reduce transaction costs
	• create an iterated decision making process
	Conduct monitoring of
	• environmental quality
	• national environmental performance
	• national environmental policies
	Increase national and international accountability
Building national capacity	Create interorganizational networks with operational organizations to transfer technical and management expertise
	Transfer financial assistance
	Transfer policy-relevant information and expertise
	Boost bureaucratic power of domestic allies

Source: Levy *et al.* (1993, 406)

These are normative conditions, arguably very broad based, and reasonably self-evident. Oran Young and Konrad von Moltke (1994) diplomatically commented that such criteria were too vague to be reliably testable, and too generic to provide any real insight into the convoluted dynamics of response. We propose to develop this in another way, taking as our line the notion of policy open-endedness introduced by Rhodes (1994) and touched on in Chapter 4. Rhodes (1994) describes this a 'non correspondence of policy systems and policy processes'. This is because the transcendental return of certain policy arenas, of which climate change is a marvellous example, incorporates both a vertical dimension of interest group coalitions and a horizontal dimension of issue networks. There may be no articulation of common interest between these two dimensions. The state may be sufficiently incapacitated to bring such a disparate collection of influences together. Furthermore, the diffuse nature of the policy problem suggests that it does not require a specifically thought-out coordinated response to become an active area for state initiative.

In such a situation there are severe problems of policy non-coordination, conflicts between government departments and their supporter clients, and crises of legitimation, or policy justification. This was clearly evident on the UK VAT proposals, the Norwegian carbon tax, and the failed German efforts to come to terms with the transport sector. In Italy, the legitimation crisis is rooted in the uncertain role of new administrative structures and a legacy of deep political suspicion over the bureaucratic incompetence of any administrative arrangement, new or established.

Yet in the circumstances of policy open-endedness, policy communities can often work successfully at cross purposes, even though the outcome may still be favourable for implementation. For example, the UK Treasury remains implacably opposed to a pre-determined commitment for any tax gathering exercise, known as hypothecation, yet this principle was partly breached in the aftermath of the VAT row when the poor received the benefit of various income support measures, and the Home Energy Efficiency Scheme obtained new money at a time of severe public expenditure restraint. In Italy, local government has begun to adopt a more coordinated and community centred role for the climate change agenda, but is thwarted by bureaucratic inertia higher up. In Germany, likewise, local climate action is lively and promotional within democratic politics, though also impeded because of ill fitting and unsupportive national and regional policies and structures.

This has led us to examine how policy coordination of contradictory policy alliances can actually succeed in 'wide net' policy arenas such as climate change, or, indeed, sustainable development generally. We conclude that implementation takes place in a host of directions, and that neither 'top-down' nor 'bottom-up' processes are sufficient. Effective implementation is supported by the more open, negotiated process of policy bargaining.

This is neither 'top' nor 'bottom'. Rather interest coalitions form opportunistically or strategically to obtain leverage on a process that is sufficiently open and accommodative. In any case such coalitions can create information, and hence policy biases, by simply combining forces and expertise. In addition, through traditional means of campaigning and lobbying they can command public attention and support in ways that neither 'top-down' nor 'bottom-up' approaches can ensure. (For a discussion, see Barrett and Fudge, 1981; Ingram, 1990; Ham and Hill, 1993.)

Indeed, implementation may not even need a specific role for 'the state', weakened as it is by deregulation, privatization, a wider European agenda, and interest group capture of key policy areas through innovative alliances. The process of implementation in these 'open-ended' policy arenas may therefore be far more decentralized, disjointed and opportunistic than has been interpreted by many analysts so far. In such 'wide net' policy arenas there may be many 'chaotic' ways in which policies emerge, unite, coalesce and submerge in political arrangements of competing and cooperative parts. The key, therefore, is to look for interesting alliances, fresh forms of evaluation, cultural trends, and active social movements which combine and fragment in intriguing ways. Regimes play their role in providing the all important basis for a set of potentially coordinating alliances and actions. But they are by no means the driving force for subsequent action. What we therefore may be witnessing is a more chance process of alliance building and fragmentation, within which the Climate Convention plays its part, but only as it is allowed to do by circumstances that may have little to do with global warming. Conventions provide a vitally important focus for action and evaluation, so one must not underestimate their significance. But their stage may be peopled by many 'walk on' actors.

So we conclude that the answer to the opening question in the first paragraph, namely whether the UN FCCC has altered the course of events in European politics, is 'yes', so long as 'climate change' is broadly interpreted as part of a general institutional dynamic around restructuring economies, social rehabilitation, international realignments of influence and trade, and the amalgamation of policy into coupled objectives and programmes. Climate change politics are increasingly immersed in a ferment of institutional change that is a feature of a European Union, bent on greater integration, pushed by global economic forces not always in its control, the post cold war era of freer trade and troublesome regional conflicts, and the relentless progress of information technology and ubiquitous communication. To try to assess a specific 'distinctiveness' for climate change politics would therefore be misguided and foolhardy. Misguided because the focus is more blurred, foolhardy because the pattern of influence would be misunderstood.

CIVIC SCIENCE AND CLIMATE POLITICS

One indicator of a climate-change-driven response will be the gradual incorporation of a 'civic science' into the process. Civic science, in a term coined by Kai Lee (1994), is an extension of 'conventional' science through which both data and projections are subject to open and more trusting negotiations amongst a wide range of stakeholders. Here is where the infusion of the NGO communities, including the nine stakeholding groups formally represented in the post-Rio process via Agenda 21 special 'chapters', might eventually play an increasingly significant role. These groups are: indigenous people; women; youth; business; academia; agriculture; trade unions; local government; and non-governmental environmental and developmental organizations. Because they represent a cross-section of global interests, because they are heavily networked via communications technology, and because they hold the key to local support and action, the CoP process would benefit by absorbing both a participatory mode of working with strong decentralizing links and a progressive civic science mode of analysis that stealthily infiltrates the IPCC process. As a result the IPCC may well become more distanced from policy making. Its significance in the future could depend upon this wider democratization of the science policy process. This would in turn allow scientific uncertainty to be channelled into workable 'bite-sized' programmes of progressive action, and cost benefit analyses to be made both more equitable and locationally specific. Awkward political choices, having to be determined ahead of apparently pressing need, might then be better translated into more tolerated approval. NGO participation in its widest possible interpretation could act as educator, advance intelligence, mediator, and implementer at various scales between local action and global consensus. This would be a most important UN FCCC-inspired institutional adaptation.

COMMITMENT

In Chapter 4 we concluded that institutions, for all their mistiness, are vital mechanisms for holding societies together, and for enabling them to adjust peacefully to threats and opportunities. We sought an approach that visualizes institutions as combinations of formal structures and informal relationships that link individuals to the social world of politics, peer groups and customary norms. Climate change is posing sufficiently serious a challenge that it has brought about change which is institutionally mediated. Despite its frustratingly tortuous and tardy progress, a process to respond to the issue of climate change has begun. Nations are obliged to publish plans and emissions data, and a pilot phase of joint implementation will take place. The precautionary principle may annoy many politicians, not a few economists and plenty of traditionalist scientists, but it, too, is now embedded in the international machinery of global environmental change. All countries have done *something*, no matter how small or how beneficial to other political objectives such as energy use efficiency or controlling the environmental and social costs of the automobile. But that something is only partly *precautionary*. Though climate change has not yet exerted any separable 'cost' in an anxious but easily distracted world, there is no consensus that any long term costs are going to be regarded as being so great that profoundly expensive and unpopular short term action, with climate change as the sole driving force, is justified.

The Convention is taken seriously in at least some parts of all the governments studied, so there is a 'national view' on its purpose and significance. But because of fragmented issue networks and interweaving policy communities there is no integrating focus within governments as to their future role in reducing greenhouse gases. For the most part only environment and foreign ministries are involved in CoP, not finance or trade and industry ministries. The link to employment, social security and education is yet to be made in any country studied. Nevertheless all countries have abided by the task of producing plans, policies and data for the UN FCCC Secretariat, and they all recognize the moral as well as political significance of at least adhering to the basic responsibilities of embarking on measurable and purposeful action.

To do this, all countries have created some form of coordinating machinery. This may be in the form of an interdepartmental committee, or a ministerial working group, or even, as in Italy, a range of national, regional and local structures that are wholly new. At the very least these bodies are responsible for coordinating data, evaluating possible responses, and advising mostly environment (and some foreign affairs) ministers on what to take to Cabinet. These structures in turn have spawned or are connected to scientific research organizations, regulatory agencies, and to a wide range of NGO interests. The actual record is patchy here, because the competence and bureaucratic administration of these connecting parts vary so widely.

We can further conclude that much depends on the support and on the vision of the prime minister or president, of his or her policy office, on the stance taken by finance, employment and industry ministers, on the level and significance of the public debate over transport and climate change and other environmental problems, and on the degree to which a formal 'social and environmental audit' is made of current and possibly future policies which bear on greenhouse gas emissions and sinks. In no country we studied is there noticeable prime ministerial support beyond the lip service of signing preferences and attending the occasional international function. The one exception is Norway's Brundtland, the promoter of sustainable development, who has recently chosen to turn her articulated attention onto climate change issues. She has led the Norwegian 'White Paper' debate following the publication of the Strategy, and has championed its cause in the press.

This is partly because Norway is in some sense ahead of the pack. Because Norwegian power generation is not fossil fuel based, Norway has reached the post 'no regrets' phase where the hard political and economic pounding starts. How Norway handles the mix of domestic hard grind and joint implementation may prove to be a marker for those countries, like Germany and the UK, beginning to emerge from the comfort of 'no regret' energy management economic strategies.

RESPONSE

Institutional response comes in many forms – policy shift, organizational change, collective outlook, informal relationships, personal commitment. It is not easy to provide a coherent picture of all these patterns, but in Chapter 4 we did suggest that such response could be portrayed along the lines summarized in Figure 4.1.

At the core of the diagram are the two principal conclusions of this book:

1 If you want to implement a climate policy, it is wise not to call it that. Call it an economic policy, a fiscal policy, an employment policy, a social policy, or an international relations policy. Climate policy on its own does not command either widespread public support or effective political weight. But a climate policy that resonates with or reinforces other arenas of policy, where there is higher public and political profile, can help give it strength and durability.

2 Global commitment through the Convention process is vital to create a framework for common action that is ultimately mutually beneficial. As Levy and his colleagues (1993) concluded, the national task is to co-ordinate a range of actions across a broad sweep of policy, to ensure that the reporting mechanisms are sound and verifiable, and to create an atmosphere of flexible interpretation of costs and benefits of policy options. But it is at the local level that real response will be found, for here is where the climate issue touches the lives of every city and household.

   This is why, closest to the central box, are four key outcomes:

   - *policy integration* involving new mixes of issues, and new constellations of stakeholders;
   - *recognition of mutual advantage* by cooperation and the coupling of policy mixes;
   - *realignment of power* between nations, within nations, amongst actors;
   - *progressive change* in small steps, always through consensus, and often by conscious trial and error.

No matter how aggravating to many, climate change politics cannot be hurried. Embedding these politics in wider structures of power, education, democracy and authority will enable a more successful and reliable process to be worked out, a process that intriguingly may become increasingly distinguishable as climate change driven. No matter how much the advocates of strong and single-minded climate change measures may dislike the fuzziness of the response process, interlocking that response into cohesive new relationships and outlooks remains the most likely way of producing the outcome that the majority should want, namely a peaceful, prosperous, precautionary and power sharing world that will both create and equitably respond to climate change.

INSTITUTIONAL ADAPTATION

The most significant institutionally adaptive measures from all the countries studied here include tax policies, policy integration arrangements, modest initiatives on the transport front, improved consultative and coordination procedures, and the momentum towards local Agenda 21.

Tax policies

Climate policy inevitably becomes entangled in tax policy, regional development strategies and social well-being issues. The most adventurous attempt to promote this mix was the proposal by DG XI to promote a powerful but competitively equalizing tool in the form of a union-wide carbon/energy tax. With a potential revenue of around 30 billion ECU annually, this could have generated around 2 per cent of tax revenue for the whole Community. The aim of the Commission was to make this revenue fiscally neutral, so that it would not actually add to the tax burden. It would have been redistributed in various ways, including social welfare programmes for those most disadvantaged, labour cost reductions to industry to help keep jobs or create new ones, and regional development expenditures geared to energy efficiency and reduced road and air transport requirements. These proposals are indicated clearly in the Commission's White Paper on Growth Competitiveness and Employment published in December 1993.

The carbon/energy tax would have been real institutional reform. But it was promoted too rapidly, and too forcefully for a non-fiscal Directorate-General to be tolerated in the turf-conscious Brussels administrative machinery. In any case any fiscal strategy requires unanimity amongst Member States, and the UK for one was not in a political position domestically to accept it. In addition the many complicated details of this truly radical programme were not thought through. For example, the pattern of energy subsidies and excise taxes varies enormously across Member States, as does the structure and industrial role of the energy industry. It is simply not politically possible to impose a fixed tax into such an amorphous and non-comparable set of arrangements. The Commission has not abandoned the idea. The germ of revolutionary fiscal reform is in place. Its time will surely come as further economic and social restructuring will need new income and a justification for a tax reform that will be socially tolerated. So certain preparatory institutional structures are now in place, even if not yet at the Community level.

For example, in the UK the fuel tax escalator and the proposal, advanced in November 1994, for a landfill tax on industry, the proceeds of which would be earmarked for offsetting labour costs, are true innovations. Admittedly both are connected to other fiscal policies, and to political concern over popular protest regarding new roads, the growing alarm over the health effects of vehicle-linked air pollution, and the rumbling planning disputes over landfill sites and incinerators. Both show the significance of issue linkage and interest group realignments, and both give hope to tax reformers and environmental economists that this small shift will become progressive. The investment from VAT on fuel and excise tax on petrol in the Energy Saving Trust and in domestic insulation is a step in this direction.

The Norwegian carbon tax, though modified through political reaction, remains the Norwegians' only effective weapon in their carbon war. Joint implementation of emissions trading measures is being tried out as part of the pilot phase promoted by the Berlin Mandate. Norway will not get any carbon credits in the interim. One might expect the Norwegians to push for some version of tradable permits in the future; this is hinted at in current Norwegian debates.

The Germans used their tax system to finance restructuring of the new eastern *Länder*. This is proving to be costly and unpopular in the old western *Länder*. But it has helped to carry the German carbon reduction load in a form of an internal joint implementation strategy. In the future Germans may look towards more technologically based initiatives in industrial materials use efficiency, and thus may be disposed to nudging prices in an environmentally friendly direction even in the absence of an EU carbon/energy tax. This would be a crafty move as materials and energy efficiency technologies are bound to sell in the open European markets both within the expanded EU and across the borders to possible new members, especially those in Central and Eastern Europe. Here is where trade, social and industrial policies may intertwine to the benefit of long run carbon reduction.

The Italians are becoming tax resistant. Intriguingly there may be a form of 'tax fatigue' through which Italians seek devious ways of offsetting further taxes on gasoline or energy generally by such devices as company income credits or some social security adjustments. The Italian experience could prove very interesting. Income neutrality may be politically popular, but it may not cut carbon emissions as much as intended. This is, as we shall see, where active local Agenda 21 politics may come into play.

Policy integration

Policy integration is long called for but rarely takes place. This should not be a wonder when both administrative departments and also parliamentary procedure usually focus on departmental separation for budgeting and political scrutiny. The interdepartmental coordinating machinery is much more likely to become effective via cost reduction measures and multiple objective planning than it is through formal political coordination of ministerial whims. The British interest in programme coordination aimed at least-cost combinations of departmental measures is also a powerful innovation, again carried out for fiscal reasons as much as environmental. Admittedly these are very early days: the Treasury is not yet convinced, the integrative economic techniques are still being developed, the policy coordinating units are small with little clout, and the whole process has negligible public visibility. Nevertheless the aims of systematic, cross-departmental cost benefit analyses using environmental economic tools aimed at minimizing social costs in the context of multiple policy combinations are a huge step forward, if only conceptually. In all probability this will be the basis of future political coordination of policies elsewhere. Just as privatization gripped the world, so eventually will multiobjective cost minimization – though it will need a catchy title. One can detect the elements of this in Norway certainly, Italy to a lesser extent because of its peculiar bureaucratic politics, and in Germany where administrative efficiency is highly popular. In the European Commission, however, it is a distant prospect – at least until it attracts the attention of the President's Office and the key finance ministers.

Consultation and outreach

All the governments studied explicitly recognized that any move in a precautionary direction requires stakeholder support. A common innovation was staged participation in a more open manner than has been the case before in such exercises, and the emergence of two developments. One was the parliamentary commission such as took place in Germany and Italy, involving distinguished experts as well as legislators. The other was the round table, notably drawing in business, but increasingly involving environmental groups. This process has helped the various interests share each other's viewpoints and assumptions, it has begun the collegiate approach to problem solving based on mutual interest, and it has created a sense of trust in political arrangements that was fast disappearing.

Again these are early days. One should be careful of overemphasizing a necessary condition for any precautionary political commitment. The fact that the various stakeholders actually recognize their shared interests is an important step forward in creating both effective response to international agreements and the basis for local action backed by national consensus.

Cross-party support would also help this process, but this is very unlikely given the fact that climate change politics are enmeshed in so many other policy arenas where party ideologies conflict. So the concept of mutual advantage via policy networks and interest realignments beyond the party political battleground is the basis for further development. We regard this as a much more significant institutional change than Icharynthian efforts to create formal integrative bureaucratic structures.

The transport crunch

Everywhere we look transport-based CO_2 emissions are rising and show little indication of slowing down. Both air and land transport demands show no sign of abating, yet both create environmental and social costs that are alarmingly high. Admittedly the calculations need refinement, but even with a heavy carbon tax income, these aggregated external costs could still be greater. At present, it is this underfinanced social burden that strikes at the heart of the schizophrenia over transport. The modern age has all but made air and road transport indispensable for many people in industrial countries. This is partly a function of specializing and globalizing economies where it still pays to ship components and final products over great distances rather than to retain locally provided distribution networks. Because of this perceived, yet politically demanding, transport dependency, the clash between transport emissions and climate change reduction measures will be most severe and painful.

Here is where the model summarized in Figure 4.1 is perhaps most appropriate. Technological innovation, spurred on by price signals, can make the car lighter, more recyclable and hence much more fuel efficient. Similarly a combination of regulations and pricing might shift the growing amounts of short haul air traffic towards high speed trains linking all European capitals and other centres. But such technological fixes in themselves are nothing like enough. Local climate action plans may encourage businesses to permit teleworking, to decentralize offices, to encourage cycling and car pooling and to redeploy their supplier–distributor geographies. The German experience shows that localization of climate action within Local Agenda 21 politics has helped to coordinate efforts at reducing individual private travel in favour of more collective modes or cycling, even though the national picture still reveals growing transport-related emissions.

For transport to be tackled on a grand scale, all the forces depicted on Figure 4.1 will need to be put in place, including the fiscal, multiobjective coordination, health–social gains, interest realignments and international cooperative arrangements. This will test the nerve of governments and the basis of our conclusions. Informal commitments to fresh socially supportive norms, coupled to local activism, could become the way out. The Germans are most advanced in this respect, but the British are catching up. The

Norwegians and the Italians are harnessing their business communities in this area too. This is already proving to be fertile ground for institutional adaptation.

Local Agenda 21

It is possible that the bulk of climate change politics will have to devolve to the local level if it is to become effective in the informal institutional dynamics of individuals and households. The rise of informal networks of cooperation is an important development here, spurred on via schools and colleges, various social groupings, and local business. This is by far the most exciting arena for innovation in climate change politics. For this transition to succeed, it will require:

1 An effective mechanism through which the international message can be converted to the local level, so that local action is seen as resonating with global action. This will mean a fresh role for those NGOs that straddle the global and the local in their lobbying and educational roles, as well as clearly coordinated civic initiative and school teachers.
2 Visionary leadership by local authority politicians and officers who begin with their own official life styles and their personal commitments. This will be a demanding task, requiring dedication and careful publicity, helped through media and schools support.
3 Constructive national policies, in such matters as tax, housing, employment, education, health care and transport, to help local authorities find their own level in Local Agenda 21 without being impeded by unnecessary and unhelpful restrictions.
4 Collaboration by local business, both in backing local strategies as well as in putting their own greenhouse gas emission reductions in order. Much can be done via regional, interbusiness collaboration, particularly amongst the smaller firms, as well as through careful audits of supplier and distributor usage of climate change inducing emissions.
5 Some central unit of advice, guidance and intelligence gathering for all local initiatives. This unit should be electronically networked to allow for speedy and effective communication. It should be prepared to review local action but in a helpful and encouraging manner. And it should be able to pinpoint blockages and suggests ways forward, even if a range of non-local activities have also to be harnessed.

CONCLUSION

All that has been described above is only just beginning in climate change politics. Nevertheless a real start has been made, in the case studies reviewed. Just as we suggested that climate change will metamorphose into social and economic change generally, so local climate action strategies should evolve into the broader and more democratic compass of Local Agenda 21.

We conclude where we began. Climate change is an identifiable arena of concern and action and will remain so, despite the push towards 'wide policy' arenas. It provides an organizing focus for public interest in the future of a healthy globe, as well as a target for a wide array of actions at every conceivable scale. But the institutional dynamics that will deliver a more coordinated and effective response to climate change lie far beyond the realm of climate change science and politics. Some progress will be made when climate change science and politics will become part of a pattern of social and economic activity that will capture their message regarding the comprehensive analysis, evaluation and supportive enforcement of response strategies.

Progress will also be made as climate change generates, or reinforces, realignments of power, patterns of cost benefit analysis and democratic involvement, issue networking and interest group reconnection that are always taking place, but which can be loosely organized in the larger stage of wide policy momentum. Taking a final look at Figure 4.1, we can see that institutional innovation has centred on a series of vital themes:

- The European Community has created a coherent, response sharing perspective that it has backed by the monitoring mechanism, incipient policy integration, the beginnings of an awareness of the need for ecological tax reform, and the organizing focus of a regular response to the UN FCCC Secretariat. This is global commitment at the multinational level.
- National cohesion is incipient, but at least in every country structures are in place that should maintain the momentum of integrated climate change science and a depth of coordinated national response, if for no other reason than the reporting mechanisms are now there, and some form of identifiable reaction is expected by the NGO community in its loosest sense. But the UN FCCC Secretariat has to enjoy full CoP backing for this to prove effective. So far that backing has not been guaranteed. This is the most important single innovation if the changes outlined below are to prove truly effective. Local action has begun. It may be faltering and it is inhibited, but it carries the seeds of democratic involvement through learning, doing and cooperating that cannot effectively take place in any other arena.
- Policy integration has occurred because of policy width, and the opportunistic scope for interesting suballiance within an amorphous policy field. This in turn has enabled valuable new approaches to policy instruments of informing, consulting, mediating and distributing that occur anyway but can cooperate around climate change without being too obvious or subversive.

- This in turn is creating social learning, both in interministerial arrangements and at the local community level, in a host of ways. Again, the learning is by no means only climate change driven, but it is a productive area of ferment.
- The precautionary principle has its place, but action remains most acceptable in the 'no regrets' stages of the response cycle. The precautionary principle will be more properly tested when that phase has run its course. By then, only if the points made above are in place will the precautionary principle prove effective.
- The connections of scientific evaluation, technological advance, organizational repositioning, and the shift in power caused by new amalgamations of policy communities, provide a vital trigger for all these changes. These connections are the fluid of institutional adaptation to climate change in a *paradoxical world* where climate change politics are both separately identifiable and comfortingly obscure.

REFERENCES

Barrett, B. and Fudge, S. (1981) Examining the policy-action relationship, pp. 210–240 in S. Barrett and C. Fudge (eds) *Policy and Action*, Methuen: London.

Ham, C. and Hill, M. (1993) *The Policy Process in the Modern Capitalist State*, London: Harvester Wheatsheaf.

Ingram, H. (1990) Implementation: a review and a suggested framework, pp. 70–96 in A. Lynn and A. Wildawsky (eds) *Public Administration*, New York: Chatham House.

Lee, K. (1994) *Compass and Gyroscope: Integrating Science and Politics for the Environment*, New York: Island Press.

Levy, M.A., Keohane, R. O. and Haas, P.M. (1993) Improving the effectiveness of international environmental institutions, pp. 397–426 in P.M. Haas, R. O. Keohane and M. A. Levy (eds) *Institutions for the Earth: Sources of Effective Environmental Protection*, Cambridge, MA: MIT Press.

Rhodes, R.A.W. (1994) Interorganizational networks and control: a critical conclusion, pp. 525–534 in F. X. Kaufmann (ed.) *The Public Sector: Challenge for Coordination and Learning*, Berlin and New York: De Gruyter.

Unterdahl, A. (1992) *The Concept of Effectiveness*, writing paper 2, Oslo: Department of Political Science, University of Oslo.

Young, O. and von Moltke, K. (1994) The consequences of international environmental regimes: lessons from the Barcelona Workshop, *International Environmental Affairs* 6(4), 348–370.

APPENDIX I

UNITED NATIONS FRAMEWORK CONVENTION ON CLIMATE CHANGE*

INTRODUCTION

During the 1980s, scientific evidence about the possibility of global climate change led to growing public concern. By 1990, a series of international conferences had issued urgent calls for a global treaty to address the problem. The United Nations Environment Programme (UNEP) and the World Meteorological Organization (WMO) responded by establishing an intergovernmental working group to prepare for treaty negotiations. Rapid progress was made, in part because of work by the Intergovernmental Panel on Climate Change (IPCC) and by meetings such as the 1990 Second World Climate Conference.

In response to the working group's proposal, the United Nations General Assembly at its 1990 session set up the Intergovernmental Negotiating Committee for a Framework Convention on Climate Change (INC/FCCC). The INC/FCCC was given a mandate to draft a framework convention and any related legal instruments it considered necessary. Negotiators from over 150 States met during five sessions between February 1991 and May 1992. They adopted the United Nations Framework Convention on Climate Change on 9 May 1992 at UN Headquarters in New York.

Soon after, at the June 1992 United Nations Conference on Environment and Development (known as the Rio 'Earth Summit'), the Convention received 155 signatures. Other States have since signed, and a growing number have ratified. The Convention will enter into force 90 days after the 50th ratification. The first session of the Conference of the Parties must then be convened within the following year. This meeting of all ratifying States will be hosted by Germany, possibly in early 1995. The INC/FCCC, which is continuing with important preparatory work, will then be dissolved and the Conference of the Parties will take over responsibility for the lengthy process of implementing the Convention.

* From the UNEP/WMO Information Unit on Climate Change (IUCC) on behalf of the Interim Secretariat of the Convention.

361

UNITED NATIONS FRAMEWORK CONVENTION
ON CLIMATE CHANGE

The Parties to this Convention

Acknowledging that change in the Earth's climate and its adverse effects are a common concern of humankind,

Concerned that human activities have been substantially increasing the atmospheric concentrations of greenhouse gases, that these increases enhance the natural greenhouse effect, and that this will result on average in an additional warming of the Earth's surface and atmosphere and may adversely affect natural ecosystems and humankind,

Noting that the largest share of historical and current global emissions of greenhouse gases has originated in developed countries, that per capita emissions in developing countries are still relatively low and that the share of global emissions originating in developing countries will grow to meet their social and development needs,

Aware of the role and importance in terrestrial and marine ecosystems of sinks and reservoirs of greenhouse gases,

Noting that there are many uncertainties in predictions of climate change, particularly with regard to the timing, magnitude and regional patterns thereof,

Acknowledging that the global nature of climate change calls for the widest possible cooperation by all countries and their participation in an effective and appropriate international response, in accordance with their common but differentiated responsibilities and respective capabilities and their social and economic conditions,

Recalling the pertinent provisions of the Declaration of the United Nations Conference on the Human Environment, adopted at Stockholm on 16 June 1972,

Recalling also that States have, in accordance with the Charter of the United Nations and the principles of international law, the sovereign right to exploit their own resources pursuant to their own environmental and developmental policies, and the responsibility to ensure that activities within their jurisdiction or control do not cause damage to the environment of other States or of areas beyond the limits of national jurisdiction,

Reaffirming the principle of sovereignty of States in international cooperation to address climate change,

Recognizing that States should enact effective environmental legislation, that environmental standards, management objectives and priorities should reflect the environmental and developmental context to which they apply, and that standards applied by some countries may be inappropriate and of unwarranted economic and social cost to other countries, in particular developing countries,

Recalling the provisions of General Assembly resolution 44/228 of 22 December 1989 on the United Nations Conference on Environment and Development, and resolutions 43/53 of 6 December 1988, 44/207 of 22 December 1989, 45/212 of 21 December 1990 and 46/169 of 19 December 1991 on protection of global climate for present and future generations of mankind,

Recalling also the provisions of General Assembly resolution 44/206 of 22 December 1989 on the Possible adverse effects of sea-level rise on islands and coastal areas, particularly low-lying coastal areas and the pertinent provisions of General Assembly resolution 44/172 of 19 December 1989 on the implementation of the Plan of Action to Combat Desertification,

Recalling further the Vienna Convention for the Protection of the Ozone Layer, 1985, and the Montreal Protocol on Substances that Deplete the Ozone Layer, 1987, as adjusted and amended on 29 June 1990,

Noting the Ministerial Declaration of the Second World Climate Conference adopted on 7 November 1990,

Conscious of the valuable analytical work being conducted by many States on climate change and of the important contributions of the World Meteorological Organization, the United Nations Environment Programme and other organs, organizations and bodies of the United Nations system, as well as other international and intergovernmental bodies, to the exchange of results of scientific research and the coordination of research,

Recognizing that steps required to understand and address climate change will be environmentally, socially and economically most effective if they are based on relevant scientific, technical and economic considerations and continually re-evaluated in the light of new findings in these areas,

Recognizing that various actions to address climate change can be justified economically in their own right and can also help in solving other environmental problems,

Recognizing also the need for developed countries to take immediate action in a flexible manner on the basis of clear priorities, as a first step towards comprehensive response strategies at the global, national and, where agreed, regional levels that take into account all greenhouse gases, with due consideration of their relative contributions to the enhancement of the greenhouse effect,

Recognizing further that low-lying and other small island countries, countries with low-lying coastal, arid and semi-arid areas or areas liable to floods, drought and desertification, and developing countries with fragile mountainous ecosystems are particularly vulnerable to the adverse effects of climate change,

Recognizing the special difficulties of those countries, especially developing countries, whose economies are particularly dependent on fossil fuel production, use and exportation, as a consequence of action taken on limiting greenhouse gas emissions,

Affirming that responses to climate change should be coordinated with social and economic development in an integrated manner with a view to avoiding adverse impacts on the latter, taking into full account the legitimate priority needs of developing countries for the achievement of sustained economic growth and the eradication of poverty,

Recognizing that all countries, especially developing countries, need access to resources required to achieve sustainable social and economic development and that, in order for developing countries to progress towards that goal, their energy consumption will need to grow taking into account the possibilities for achieving greater energy efficiency and for controlling greenhouse gas emissions in general, including through the application of new technologies on terms which make such an application economically and socially beneficial,

Determined to protect the climate system for present and future generations,

Have agreed as follows:

ARTICLE 1: DEFINITIONS *

For the purposes of this Convention:

1 'Adverse effects of climate change' means changes in the physical environment or biota resulting from climate change which have significant deleterious effects on the composition, resilience or productivity of natural and managed ecosystems or on the operation of socio-economic systems or on human health and welfare.

2 'Climate change' means a change of climate which is attributed directly or indirectly to human activity that alters the composition of the global atmosphere and which is in addition to natural climate variability observed over comparable time periods.

3 'Climate system' means the totality of the atmosphere, hydrosphere, biosphere and geosphere and their interactions.

4 'Emissions' means the release of greenhouse gases and/or their precursors into the atmosphere over a specified area and period of time.

5 'Greenhouse gases' means those gaseous constituents of the atmosphere, both natural and anthropogenic, that absorb and re-emit infrared radiation.

6 'Regional economic integration organization' means an organization constituted by sovereign States of a given region which has competence in respect of matters governed by this Convention or its protocols and has been duly authorized, in accordance with its internal procedures, to sign, ratify, accept, approve or accede to the instruments concerned.

7 'Reservoir' means a component or components of the climate system where a greenhouse gas or a precursor of a greenhouse gas is stored.

8 'Sink' means any process, activity or mechanism which removes a greenhouse gas, an aerosol or a precursor of a greenhouse gas from the atmosphere.

9 'Source' means any process or activity which releases a greenhouse gas, an aerosol or a precursor of a greenhouse gas into the atmosphere.

ARTICLE 2: OBJECTIVE

The ultimate objective of this Convention and any related legal instruments that the Conference of the Parties may adopt is to achieve, in accordance with the relevant provisions of the Convention, stabilization of greenhouse gas concentrations in the atmosphere at a level that would prevent dangerous anthropogenic interference with the climate system. Such a level should be achieved within a time-frame sufficient to allow ecosystems to adapt naturally to climate change, to ensure that food production is not threatened and to enable economic development to proceed in a sustainable manner.

ARTICLE 3: PRINCIPLES

In their actions to achieve the objective of the Convention and to implement its provisions, the Parties shall be guided, *inter alia*, by the following:

* Titles of articles are included solely to assist the reader.

1   The Parties should protect the climate system for the benefit of present and future generations of humankind, on the basis of equity and in accordance with their common but differentiated responsibilities and respective capabilities. Accordingly, the developed country Parties should take the lead in combating climate change and the adverse effects thereof.

2   The specific needs and special circumstances of developing country Parties, especially those that are particularly vulnerable to the adverse effects of climate change, and of those Parties, especially developing country Parties, that would have to bear a disproportionate or abnormal burden under the Convention, should be given full consideration.

3   The Parties should take precautionary measures to anticipate, prevent or minimize the causes of climate change and mitigate its adverse effects. Where there are threats of serious or irreversible damage, lack of full scientific certainty should not be used as a reason for postponing such measures, taking into account that policies and measures to deal with climate change should be cost-effective so as to ensure global benefits at the lowest possible cost. To achieve this, such policies and measures should take into account different socio-economic contexts, be comprehensive, cover all relevant sources, sinks and reservoirs of greenhouse gases and adaptation, and comprise all economic sectors. Efforts to address climate change may he carried out cooperatively by interested Parties.

4   The Parties have a right to, and should, promote sustainable development. Policies and measures to protect the climate system against human-induced change should be appropriate for the specific conditions of each Party and should be integrated with national development programmes, taking into account that economic development is essential for adopting measures to address climate change.

5   The Parties should cooperate to promote a supportive and open international economic system that would lead to sustainable economic growth and development in all Parties, particularly developing country Parties, thus enabling them better to address the problems of climate change. Measures taken to combat climate change, including unilateral ones, should not constitute a means of arbitrary or unjustifiable discrimination or a disguised restriction on international trade.

ARTICLE 4: COMMITMENTS

1   All Parties, taking into account their common but differentiated responsibilities and their specific national and regional development priorities, objectives and circumstances, shall:

(a) Develop, periodically update, publish and make available to the Conference of the Parties, in accordance with Article 12, national inventories of anthropogenic emissions by sources and removals by sinks of all greenhouse gases not controlled by the Montreal Protocol, using comparable methodologies to be agreed upon by the Conference of the Parties;

(b) Formulate, implement, publish and regularly update national and, where appropriate, regional programmes containing measures to mitigate climate change by addressing anthropogenic emissions by sources and removals by sinks of all greenhouse gases not controlled by the Montreal Protocol, and measures to facilitate adequate adaptation to climate change;

(c) Promote and cooperate in the development, application and diffusion, includ-ing transfer, of technologies, practices and processes that control, reduce or prevent anthropogenic emissions of greenhouse gases not controlled by the Montreal Protocol in all relevant sectors, including the energy, transport, industry, agriculture, forestry and waste management sectors;

(d) Promote sustainable management, and promote and cooperate in the conserva-tion and enhancement, as appropriate, of sinks and reservoirs of all greenhouse gases not controlled by the Montreal Protocol, including biomass, forests and oceans as well as other terrestrial, coastal and marine ecosystems;

(e) Cooperate in preparing for adaptation to the impacts of climate change; develop and elaborate appropriate and integrated plans for coastal zone management, water resources and agriculture, and for the protection and rehabilitation of areas, particularly in Africa, affected by drought and desertifi-cation, as well as floods;

(f) Take climate change considerations into account, to the extent feasible, in their relevant social, economic and environmental policies and actions, and employ appropriate methods, for example impact assessments, formulated and deter-mined nationally, with a view to minimizing adverse effects on the economy, on public health and on the quality of the environment, of projects or measures undertaken by them to mitigate or adapt to climate change;

(g) Promote and cooperate in scientific, technological, technical, socio-economic and other research, systematic observation and development of data archives related to the climate system and intended to further the understanding and to reduce or eliminate the remaining uncertainties regarding the causes, effects, magnitude and timing of climate change and the economic and social conse-quences of various response strategies;

(h) Promote and cooperate in the full, open and prompt exchange of relevant scientific, technological, technical, socio-economic and legal information related to the climate system and climate change, and to the economic and social consequences of various response strategies;

(i) Promote and cooperate in education, training and public awareness related to climate change and encourage the widest participation in this process, including that of non-governmental organizations; and

(j) Communicate to the Conference of the Parties information related to imple-mentation, in accordance with Article 12.

2 The developed country Parties and other Parties included in Annex I commit them-selves specifically as provided for in the following:

(a) Each of these Parties shall adopt national† policies and take corresponding measures on the mitigation of climate change, by limiting its anthropogenic emissions of greenhouse gases and protecting and enhancing its greenhouse gas sinks and reservoirs. These policies and measures will demonstrate that developed countries are taking the lead in modifying longer-term trends in anthropogenic emissions consistent with the objective of the Convention, recognizing that the return by the end of the present decade to earlier levels of anthropogenic emissions of carbon dioxide and other greenhouse gases not

† This includes policies and measures adopted by regional economic integration organizations.

controlled by the Montreal Protocol would contribute to such modification, and taking into account the differences in these Parties' starting points and approaches, economic structures and resource bases, the need to maintain strong and sustainable economic growth, available technologies and other individual circumstances, as well as the need for equitable and appropriate contributions by each of these Parties to the global effort regarding that objective. These Parties may implement such policies and measures jointly with other Parties and may assist other Parties in contributing to the achievement of the objective of the Convention and, in particular, that of this subparagraph;

(b) In order to promote progress to this end, each of these Parties shall communicate, within six months of the entry into force of the Convention for it and periodically thereafter, and in accordance with Article 12, detailed information on its policies and measures referred to in subparagraph (a) above, as well as on its resulting projected anthropogenic emissions by sources and removals by sinks of greenhouse gases not controlled by the Montreal Protocol for the period referred to in subparagraph (a), with the aim of returning individually or jointly to their 1990 levels these anthropogenic emissions of carbon dioxide and other greenhouse gases not controlled by the Montreal Protocol. This information will be reviewed by the Conference of the Parties, at its first session and periodically thereafter, in accordance with Article 7;

(c) Calculations of emissions by sources and removals by sinks of greenhouse gases for the purposes of subparagraph (b) above should take into account the best available scientific knowledge, including of the effective capacity of sinks and the respective contributions of such gases to climate change. The Conference of the Parties shall consider and agree on methodologies for these calculations at its first session and review them regularly thereafter;

(d) The Conference of the Parties shall, at its first session, review the adequacy of subparagraphs (a) and (b) above. Such review shall be carried out in the light of the best available scientific information and assessment on climate change and its impacts, as well as relevant technical, social and economic information. Based on this review, the Conference of the Parties shall take appropriate action, which may include the adoption of amendments to the commitments in subparagraphs (a) and (b) above. The Conference of the Parties, at its first session, shall also take decisions regarding criteria for joint implementation as indicated in subparagraph (a) above. A second review of subparagraphs (a) and (b) shall take place not later than 31 December 1998, and thereafter at regular intervals determined by the Conference of the Parties, until the objective of the Convention is met;

(e) Each of these Parties shall:

(i) Coordinate as appropriate with other such Parties, relevant economic and administrative instruments developed to achieve the objective of the Convention; and

(ii) Identify and periodically review its own policies and practices which encourage activities that lead to greater levels of anthropogenic emissions of greenhouse gases not controlled by the Montreal Protocol than would otherwise occur;

(f) The Conference of the Parties shall review, not later than 31 December 1998, available information with a view to taking decisions regarding such amendments to the lists in Annexes I and II as may be appropriate, with the approval of the Party concerned;

(g) Any Party not included in Annex I may, in its instrument of ratification, acceptance, approval or accession, or at any time thereafter, notify the Depositary that it intends to be bound by subparagraphs (a) and (b) above. The Depositary shall inform the other signatories and Parties of any such notification.

3 The developed country Parties and other developed Parties included in Annex II shall provide new and additional financial resources to meet the agreed full costs incurred by developing country Parties in complying with their obligations under Article 12, paragraph 1. They shall also provide such financial resources, including for the transfer of technology, needed by the developing country Parties to meet the agreed full incremental costs of implementing measures that are covered by paragraph 1 of this Article and that are agreed between a developing country Party and the international entity or entities referred to in Article 11, in accordance with that Article.The implementation of these commitments shall take into account the need for adequacy and predictability in the flow of funds and the importance of appropriate burden sharing among the developed country Parties.

4 The developed country Parties and other developed Parties included in Annex II shall also assist the developing country Parties that are particularly vulnerable to the adverse effects of climate change in meeting costs of adaptation to those adverse effects.

5 The developed country Parties and other developed Parties included in Annex II shall take all practicable steps to promote, facilitate and finance, as appropriate, the transfer of, or access to, environmentally sound technologies and know-how to other Parties, particularly developing country Parties, to enable them to implement the provisions of the Convention. In this process, the developed country Parties shall support the development and enhancement of endogenous capacities and technologies of developing country Parties. Other Parties and organizations in a position to do so may also assist in facilitating the transfer of such technologies.

6 In the implementation of their commitments under paragraph 2 above, a certain degree of flexibility shall be allowed by the Conference of the Parties to the Parties included in Annex I undergoing the process of transition to a market economy, in order to enhance the ability of these Parties to address climate change, including with regard to the historical level of anthropogenic emissions of greenhouse gases not controlled by the Montreal Protocol chosen as a reference.

7 The extent to which developing country Parties will effectively implement their commitments under the Convention will depend on the effective implementation by developed country Parties of their commitments under the Convention related to financial resources and transfer of technology and will take fully into account that economic and social development and poverty eradication are the first and overriding priorities of the developing country Parties.

8 In the implementation of the commitments in this Article, the Parties shall give full consideration to what actions are necessary under the Convention, including actions related to funding, insurance and the transfer of technology, to meet the specific needs and concerns of developing country Parties arising from the adverse effects of climate change and/or the impact of the implementation of response measures, especially on:

(a) Small island countries;

(b) Countries with low-lying coastal areas;

(c) Countries with arid and semi-arid areas, forested areas and areas liable to forest decay;

(d) Countries with areas prone to natural disasters;

(e) Countries with areas liable to drought and desertification;

(f) Countries with areas of high urban atmospheric pollution;

(g) Countries with areas with fragile ecosystems, including mountainous ecosystems;

(h) Countries whose economies are highly dependent on income generated from the production, processing and export, and/or on consumption of fossil fuels and associated energy-intensive products; and

(i) Land-locked and transit countries.

Further, the Conference of the Parties may take actions, as appropriate, with respect to this paragraph.

9 The Parties shall take full account of the specific needs and special situations of the least developed countries in their actions with regard to funding and transfer of technology.

10 The Parties shall, in accordance with Article 10, take into consideration in the implementation of the commitments of the Convention the situation of Parties, particularly developing country Parties, with economies that are vulnerable to the adverse effects of the implementation of measures to respond to climate change. This applies notably to Parties with economies that are highly dependent on income generated from the production, processing and export, and/or consumption of fossil fuels and associated energy-intensive products and/or the use of fossil fuels for which such Parties have serious difficulties in switching to alternatives.

ARTICLE 5: RESEARCH AND SYSTEMATIC OBSERVATION

In carrying out their commitments under Article 4, paragraph 1(g), the Parties shall:

(a) Support and further develop, as appropriate, international and intergovernmental programmes and networks or organizations aimed at defining, conducting, assessing and financing research, data collection and systematic observation, taking into account the need to minimize duplication of effort;

(b) Support international and intergovernmental efforts to strengthen systematic observation and national scientific and technical research capacities and capabilities, particularly in developing countries, and to promote access to, and the exchange of, data and analyses thereof obtained from areas beyond national jurisdiction; and

(c) Take into account the particular concerns and needs of developing countries and cooperate in improving their endogenous capacities and capabilities to participate in the efforts referred to in subparagraphs (a) and (b) above.

ARTICLE 6: EDUCATION, TRAINING AND PUBLIC AWARENESS

In carrying out their commitments under Article 4, paragraph 1(i), the Parties shall:

(a) Promote and facilitate at the national and, as appropriate, subregional and regional levels, and in accordance with national laws and regulations, and within their respective capacities:

(i) The development and implementation of educational and public awareness programmes on climate change and its effects;

(ii) Public access to information on climate change and its effects;

(iii) Public participation in addressing climate change and its effects and developing adequate responses; and

(iv) Training of scientific, technical and managerial personnel.

(b) Cooperate in and promote, at the international level, and, where appropriate, using existing bodies:

(i) The development and exchange of educational and public awareness material on climate change and its effects; and

(ii) The development and implementation of education and training programmes, including the strengthening of national institutions and the exchange or secondment of personnel to train experts in this field, in particular for developing countries.

ARTICLE 7: CONFERENCE OF THE PARTIES

1 A Conference of the Parties is hereby established.

2 The Conference of the Parties, as the supreme body of this Convention, shall keep under regular review the implementation of the Convention and any related legal instruments that the Conference of the Parties may adopt, and shall make, within its mandate, the decisions necessary to promote the effective implementation of the Convention. To this end, it shall:

(a) Periodically examine the obligations of the Parties and the institutional arrangements under the Convention, in the light of the objective of the Convention, the experience gained in its implementation and the evolution of scientific and technological knowledge;

(b) Promote and facilitate the exchange of information on measures adopted by the Parties to address climate change and its effects, taking into account the differing circumstances, responsibilities and capabilities of the Parties and their respective commitments under the Convention;

(c) Facilitate, at the request of two or more Parties, the coordination of measures adopted by them to address climate change and its effects, taking into account the differing circumstances, responsibilities and capabilities of the Parties and their respective commitments under the Convention;

(d) Promote and guide, in accordance with the objective and provisions of the Convention, the development and periodic refinement of comparable methodologies, to be agreed on by the Conference of the Parties, *inter alia*, for preparing inventories of greenhouse gas emissions by sources and removals by sinks, and for evaluating the effectiveness of measures to limit the emissions and enhance the removals of these gases;

(e) Assess, on the basis of all information made available to it in accordance with the provisions of the Convention, the implementation of the Convention by the Parties, the overall effects of the measures taken pursuant to the Convention, in particular environmental, economic and social effects as well as their cumulative impacts and the extent to which progress towards the objective of the Convention is being achieved;

(f) Consider and adopt regular reports on the implementation of the Convention and ensure their publication;

(g) Make recommendations on any matters necessary for the implementation of the Convention;

(h) Seek to mobilize financial resources in accordance with Article 4, paragraphs 3, 4 and 5, and Article 11;

(i) Establish such subsidiary bodies as are deemed necessary for the implementation of the Convention;

(j) Review reports submitted by its subsidiary bodies and provide guidance to them;

(k) Agree upon and adopt, by consensus, rules of procedure and financial rules for itself and for any subsidiary bodies;

(l) Seek and utilize, where appropriate, the services and cooperation of, and information provided by, competent international organizations and intergovernmental and non-governmental bodies; and

(m) Exercise such other function as are required for the achievement of the objective of the Convention as well as all other functions assigned to it under the Convention.

3 The Conference of the Parties shall, at its first session, adopt its own rules of procedure as well as those of the subsidiary bodies established by the Convention, which shall include decision making procedures for matters not already covered by decision making procedures stipulated in the Convention. Such procedures may include specified majorities required for the adoption of particular decisions.

4 The first session of the Conference of the Parties shall be convened by the interim secretariat referred to in Article 21 and shall take place not later than one year after the date of entry into force of the Convention. Thereafter, ordinary sessions of the Conference of the Parties shall be held every year unless otherwise decided by the Conference of the Parties.

5 Extraordinary sessions of the Conference of the Parties shall be held at such other times as may be deemed necessary by the Conference, or at the written request of any Party, provided that, within six months of the request being communicated to the Parties by the secretariat, it is supported by at least one third of the Parties.

6 The United Nations, its specialized agencies and the International Atomic Energy Agency, as well as an State member thereof or observers thereto not Party to the Convention, may be represented at sessions of the Conference of the Parties as observers. Any body or agency, whether national or international, governmental or non-governmental, which is qualified in matters covered by the Convention, and which has informed the secretariat of its wish to be represented at a session of the Conference of the Parties as an observer, may be so admitted unless at least one third of the Parties present object. The admission and participation of observers shall be subject to the rules of procedure adopted by the Conference of the Parties.

ARTICLE 8: SECRETARIAT

1 A secretariat is hereby established.
2 The functions of the secretariat shall be:
   (a) To make arrangements for sessions of the Conference of the Parties and its
       subsidiary bodies established under the Convention and to provide them with
       services as required;
   (b) To compile and transmit reports submitted to it;
   (c) To facilitate assistance to the Parties, particularly developing country Parties,
       on request, in the compilation and communication of information required in
       accordance with the provisions of the Convention;
   (d) To prepare reports on its activities and present them to the Conference of the
       Parties;
   (e) To ensure the necessary coordination with the secretariats of other relevant
       international bodies;
   (f) To enter, under the overall guidance of the Conference of the Parties, into such
       administrative and contractual arrangements as may be required for the effec-
       tive discharge of its functions; and
   (g) To perform the other secretariat functions specified in the Convention and
       in any of its protocols and such other functions as may be determined by the
       Conference of the Parties.
3 The Conference of the Parties, at its first session, shall designate a permanent secretariat and make arrangements for its functioning.

ARTICLE 9: SUBSIDIARY BODY FOR SCIENTIFIC AND TECHNOLOGICAL ADVICE

1 A subsidiary body for scientific and technological advice is hereby established to
   provide the Conference of the Parties and, as appropriate, its other subsidiary
   bodies with timely information and advice on scientific and technological matters
   relating to the Convention. This body shall be open to participation by all Parties
   and shall be multidisciplinary. It shall comprise government representatives competent in the relevant field of expertise. It shall report regularly to the Conference of
   the Parties on all aspects of its work.
2 Under the guidance of the Conference of the Parties, and drawing upon existing
   competent international bodies, this body shall:

   (a) Provide assessments of the state of scientific knowledge relating to climate
       change and its effects;
   (b) Prepare scientific assessments on the effects of measures taken in the implemen-
       tation of the Convention;
   (c) Identify innovative, efficient and state-of-the-art technologies and know-how
       and advise on the ways and means of promoting development and/or transfer-
       ring such technologies;
   (d) Provide advice on scientific programmes, international cooperation in research
       and development related to climate change, as well as on ways and means of
       supporting endogenous capacity-building in developing countries; and

(e) Respond to scientific, technological and methodological questions that the Conference of the Parties and its subsidiary bodies may put to the body.

3 The functions and terms of reference of this body may be further elaborated by the Conference of the Parties.

ARTICLE 10: SUBSIDIARY BODY FOR IMPLEMENTATION

1 A subsidiary body for implementation is hereby established to assist the Conference of the Parties in the assessment and review of the effective implementation of the Convention. This body shall be open to participation by all Parties and comprise government representatives who are experts on matters related to climate change. It shall report regularly to the Conference of the Parties on all aspects of its work.

2 Under the guidance of the Conference of the Parties, this body shall:

(a) Consider the information communicated in accordance with Article 12, paragraph 1, to assess the overall aggregated effect of the steps taken by the Parties in the light of the latest scientific assessments concerning climate change;

(b) Consider the information communicated in accordance with Article 12, paragraph 2, in order to assist the Conference of the Parties in carrying out the reviews required by Article 4, paragraph 2(d); and

(c) Assist the Conference of the Parties, as appropriate, in the preparation and implementation of its decisions.

ARTICLE 11: FINANCIAL MECHANISM

1 A mechanism for the provision of financial resources on a grant or concessional basis, including for the transfer of technology, is hereby defined. It shall function under the guidance of and be accountable to the Conference of the Parties, which shall decide on its policies, programme priorities and eligibility criteria related to this Convention. Its operation shall be entrusted to one or more existing international entities.

2 The financial mechanism shall have an equitable and balanced representation of all Parties within a transparent system of governance.

3 The Conference of the Parties and the entity or entities entrusted with the operation of the financial mechanism shall agree upon arrangements to give effect to the above paragraphs, which shall include the following:

(a) Modalities to ensure that the funded projects to address climate change are in conformity with the policies, programme priorities and eligibility criteria established by the Conference of the Parties;

(b) Modalities by which a particular funding decision may be reconsidered in light of these policies, programme priorities and eligibility criteria;

(c) Provision by the entity or entities of regular reports to the Conference of the Parties on its funding operations, which is consistent with the requirement for accountability set out in paragraph 1 above; and

(d) Determination in a predictable and identifiable manner of the amount of funding necessary and available for the implementation of this Convention and the conditions under which that amount shall be periodically reviewed.

4 The Conference of the Parties shall make arrangements to implement the above-mentioned provisions at its first session, reviewing and taking into account the interim arrangements referred to in Article 21, paragraph 3, and shall decide whether these interim arrangements shall be maintained. Within four years, thereafter, the Conference of the Parties shall review the financial mechanism and take appropriate measures.

5 The developed country Parties may also provide and developing country Parties avail themselves of, financial resources related to the implementation of the Convention through bilateral, regional and other multilateral channels.

ARTICLE 12: COMMUNICATION OF INFORMATION RELATED TO IMPLEMENTATION

1 In accordance with Article 4, paragraph 1, each Party shall communicate to the Conference of the Parties, through the secretariat, the following elements of information:

(a) A national inventory of anthropogenic emissions by sources and removals by sinks of all greenhouse gases not controlled by the Montreal Protocol, to the extent its capacities permit, using comparable methodologies to be promoted and agreed upon by the Conference of the Parties;

(b) A general description of steps taken or envisaged by the Party to implement the Convention; and

(c) Any other information that the Party considers relevant to the achievement of the objective of the Convention and suitable for inclusion in its communication, including, if feasible, material relevant for calculations of global emission trends.

2 Each developed country Party and each other Party included in Annex I shall incorporate in its communication the following elements of information:

(a) A detailed description of the policies and measures that it has adopted to implement its commitment under Article 4, paragraphs 2(a) and 2(b); and

(b) A specific estimate of the effects that the policies and measures referred to in subparagraph (a) immediately above will have on anthropogenic emissions by its sources and removals by its sinks of greenhouse gases during the period referred to in Article 4, paragraph 2(a).

3 In addition, each developed country Party and each other developed Party included in Annex II shall incorporate details of measures taken in accordance with Article 4, paragraphs 3, 4 and 5.

4 Developing country Parties may, on a voluntary basis, propose projects for financing, including specific technologies, materials, equipment, techniques or practices that would be needed to implement such projects, along with, if possible, an estimate of all incremental costs, of the reductions of emissions and increments of removals of greenhouse gases, as well as an estimate of the consequent benefits.

5 Each developed country Party and each other Party included in Annex I shall make its initial communication within six months of the entry into force of the Convention for that Party. Each Party not so listed shall make its initial communication within three years of the entry into force of the Convention for that Party, or of the availability of financial resources in accordance with Article 4, paragraph 3.

Parties that are least developed countries may make their initial communication at their discretion. The frequency of subsequent communications by all Parties shall be determined by the Conference of the Parties, taking into account the differentiated timetable set by this paragraph.

6   Information communicated by Parties under this Article shall be transmitted by the secretariat as soon as possible to the Conference of the Parties and to any subsidiary bodies concerned. If necessary, the procedures for the communication of information may be further considered by the Conference of the Parties.

7   From its first session, the Conference of the Parties shall arrange for the provision to developing country Parties of technical and financial support, on request, in compiling and communicating information under this Article, as well as in identifying the technical and financial needs associated with proposed projects and response measures under Article 4. Such support may be provided by other Parties, by competent international organizations and by the secretariat, as appropriate.

8   Any group of Parties may, subject to guidelines adopted by the Conference of the Parties, and to prior notification to the Conference of the Parties, make a joint communication in fulfilment of their obligations under this Article, provided that such a communication includes information on the fulfilment by each of these Parties of its individual obligations under the Convention.

9   Information received by the secretariat that is designated by a Party as confidential, in accordance with criteria to be established by the Conference of the Parties, shall be aggregated by the secretariat to protect its confidentiality before being made available to any of the bodies involved in the communication and review of information.

10 Subject to paragraph 9 above, and without prejudice to the ability of any Party to make public its communication at any time, the secretariat shall make communications by Parties under this Article publicly available at the time they are submitted to the Conference of the Parties.

ARTICLE 13:
RESOLUTION OF QUESTIONS REGARDING IMPLEMENTATION

The Conference of the Parties shall, at its first session, consider the establishment of a multilateral consultative process, available to Parties on their request, for the resolution of questions regarding the implementation of the Convention.

ARTICLE 14: SETTLEMENT OF DISPUTES

1   In the event of a dispute between any two or more Parties concerning the interpretation or application of the Convention, the Parties concerned shall seek a settlement of the dispute through negotiation or any other peaceful means of their own choice.

2   When ratifying, accepting, approving or acceding to the Convention, or at any time thereafter, a Party which is not a regional economic integration organization may declare in a written instrument submitted to the Depositary that, in respect of any

dispute concerning the interpretation or application of the Convention, it recognizes as compulsory *ipso facto* and without special agreement, in relation to any Party accepting the same obligation:

(a) Submission of the dispute to the International Court of Justice, and/or
(b) Arbitration in accordance with procedures to be adopted by the Conference of the Parties as soon as practicable, in an annex on arbitration.

A Party which is a regional economic integration organization may make a declaration with like effect in relation to arbitration in accordance with the procedures referred to in subparagraph (b) above.

3   A declaration made under paragraph 2 above shall remain in force until it expires in accordance with its terms or until three months after written notice of its revocation has been deposited with the Depositary.

4   A new declaration, a notice of revocation or the expiry of a declaration shall not in any way affect proceedings pending before the International Court of Justice or the arbitral tribunal, unless the parties to the dispute otherwise agree.

5   Subject to the operation of paragraph 2 above, if after twelve months following notification by one Party to another that a dispute exists between them, the Parties concerned have not been able to settle their dispute through the means mentioned in paragraph 1 above, the dispute shall be submitted, at the request of any of the parties to the dispute, to conciliation.

6   A conciliation commission shall be created upon the request of one of the parties to the dispute. The commission shall be composed of an equal number of members appointed by each party concerned and a chairman chosen jointly by the members appointed by each party. The commission shall render a recommendatory award, which the parties shall consider in good faith.

7   Additional procedures relating to conciliation shall be adopted by the Conference of the Parties, as soon as practicable, in an annex on conciliation.

8   The provisions of this Article shall apply to any related legal instrument which the Conference of the Parties may adopt, unless the instrument provides otherwise.

ARTICLE 15: AMENDMENTS TO THE CONVENTION

1   Any Party may propose amendments to the Convention.

2   Amendments to the Convention shall be adopted at an ordinary session of the Conference of the Parties. The text of any proposed amendment to the Convention shall be communicated to the Parties by the secretariat at least six months before the meeting at which it is proposed for adoption. The secretariat shall also communicate proposed amendments to the signatories to the Convention and, for information, to the Depositary.

3   The Parties shall make every effort to reach agreement on any proposed amendment to the Convention by consensus. If all efforts at consensus have been exhausted, and no agreement reached, the amendment shall as a last resort be adopted by a three-fourths majority vote of the Parties present and voting at the meeting. The adopted amendment shall be communicated by the secretariat to the Depositary, who shall circulate it to all Parties for their acceptance.

4   Instruments of acceptance in respect of an amendment shall be deposited with the Depositary. An amendment adopted in accordance with paragraph 3 above shall enter into force for those Parties having accepted it on the ninetieth day after the date of receipt by the Depositary of an instrument of acceptance by at least three fourths of the Parties to the Convention.

5   The amendment shall enter into force for any other Party on the ninetieth day after the date on which that Party deposits with the Depositary its instrument of acceptance of the said amendment.

6   For the purposes of this Article, 'Parties present and voting' means Parties present and casting an affirmative or negative vote.

ARTICLE 16: ADOPTION AND AMENDMENT OF ANNEXES TO THE CONVENTION

1   Annexes to the Convention shall form an integral part thereof and, unless otherwise expressly provided, a reference to the Convention constitutes at the same time a reference to any annexes thereto. Without prejudice to the provisions of Article 14, paragraphs 2(b) and 7, such annexes shall be restricted to lists, forms and any other material of a descriptive nature that is of a scientific, technical, procedural or administrative character.

2   Annexes to the Convention shall be proposed and adopted in accordance with the procedure set forth in Article 15, paragraphs 2, 3 and 4.

3   An annex that has been adopted in accordance with paragraph 2 above shall enter into force for all Parties to the Convention six months after the date of the communication by the Depositary to such Parties of the adoption of the annex, except for those Parties that have notified the Depositary, in writing, within that period of their non-acceptance of the annex. The annex shall enter into force for Parties which withdraw their notification of non-acceptance on the ninetieth day after the date on which withdrawal of such notification has been received by the Depositary.

4   The proposal, adoption and entry into force of amendments to annexes to the Convention shall be subject to the same procedure as that for the proposal, adoption and entry into force of annexes to the Convention in accordance with paragraphs 2 and 3 above.

5   If the adoption of an annex or an amendment to an annex involves an amendment to the Convention, that annex or amendment to an annex shall not enter into force until such time as the amendment to the Convention enters into force.

ARTICLE 17: PROTOCOLS

1   The Conference of the Parties may, at any ordinary session, adopt protocols to the Convention.

2   The text of any proposed protocol shall be communicated to the Parties by the secretariat at least six months before such a session.

3   The requirements for the entry into force of any protocol shall be established by that instrument.

4   Only Parties to the Convention may be Parties to a protocol.

5   Decisions under any protocol shall be taken only by the Parties to the protocol concerned.

ARTICLE 18: RIGHT TO VOTE

1 Each Party to the Convention shall have one vote, except as provided for in paragraph 2 below.

2 Regional economic integration organizations, in matters within their competence, shall exercise their right to vote with a number of votes equal to the number of their Member States that are Parties to the Convention. Such an organization shall not exercise its right to vote if any of its Member States exercises its right, and vice versa.

ARTICLE 19: DEPOSITARY

The Secretary-General of the United Nations shall be the Depositary of the Convention and of protocols adopted in accordance with Article 17.

ARTICLE 20: SIGNATURE

This Convention shall be open for signature by States Members of the United Nations or of any of its specialized agencies or that are Parties to the Statute of the International Court of Justice and by regional economic integration organizations at Rio de Janeiro, during the United Nations Conference on Environment and Development, and thereafter at United Nations Headquarters in New York from 20 June 1992 to 19 June 1993.

ARTICLE 21: INTERIM ARRANGEMENTS

1 The secretariat functions referred to in Article 8 will be carried out on an interim basis by the secretariat established by the General Assembly of the United Nations in its resolution 45/212 of 21 December 1990, until the completion of the first session of the Conference of the Parties.

2 The head of the interim secretariat referred to in paragraph 1 above will cooperate closely with the Intergovernmental Panel on Climate Change to ensure that the Panel can respond to the need for objective scientific and technical advice. Other relevant scientific bodies could also be consulted.

3 The Global Environment Facility of the United Nations Development Programme, the United Nations Environment Programme and the International Bank for Reconstruction and Development shall be the international entity entrusted with the operation of the financial mechanism referred to in Article 11 on an interim basis. In this connection, the Global Environment Facility should be appropriately restructured and its membership made universal to enable it to fulfil the requirements of Article 11.

ARTICLE 22: RATIFICATION, ACCEPTANCE, APPROVAL OR ACCESSION

1 The Convention shall be subject to ratification, acceptance, approval or accession by States and by regional economic integration organizations. It shall be open

for accession from the day after the date on which the Convention is closed for signature. Instruments of ratification, acceptance, approval or accession shall be deposited with the Depositary.

2   Any regional economic integration organization which becomes a Party to the Convention without any of its Member States being a Party shall be bound by all the obligations under the Convention. In the case of such organizations, one or more of whose Member States is a Party to the Convention, the organization and its Member States shall decide on their respective responsibilities for the performance of their obligations under the Convention. In such cases, the organization and the Member States shall not be entitled to exercise rights under the Convention concurrently.

3   In their instruments of ratification, acceptance, approval or accession, regional economic integration organizations shall declare the extent of their competence with respect to the matters governed by the Convention. These organizations shall also inform the Depositary, who shall in turn inform the Parties, of any substantial modification in the extent of their competence.

ARTICLE 23: ENTRY INTO FORCE

1   The Convention shall enter into force on the ninetieth day after the date of deposit of the fiftieth instrument of ratification, acceptance, approval or accession.

2   For each State or regional economic integration organization that ratifies, accepts or approves the Convention or accedes thereto after the deposit of the fiftieth instrument of ratification, acceptance, approval or accession, the Convention shall enter into force on the ninetieth day after the date of deposit by such State or regional economic integration organization of its instrument of ratification, acceptance, approval or accession.

3   For the purposes of paragraphs 1 and 2 above, any instrument deposited by a regional economic integration organization shall not be counted as additional to those deposited by Member States of the organization.

ARTICLE 24: RESERVATIONS

No reservations may be made to the Convention.

ARTICLE 25: WITHDRAWAL

1   At any time after three years from the date on which the Convention has entered into force for a Party, that Party may withdraw from the Convention by giving written notification to the Depositary.

2   Any such withdrawal shall take effect upon expiry of one year from the date of receipt by the Depositary of the notification of withdrawal, or on such later date as may be specified in the notification of withdrawal.

3   Any Party that withdraws from the Convention shall be considered as also having withdrawn from any protocol to which it is a Party.

ARTICLE 26: AUTHENTIC TEXTS

The original of this Convention, of which the Arabic, Chinese, English, French, Russian and Spanish texts are equally authentic, shall be deposited with the Secretary-General of the United Nations.

IN WITNESS WHEREOF the undersigned, being duly authorized to that effect, have signed this Convention.

DONE at New York this ninth day of May one thousand nine hundred and ninety-two.

ANNEX I

Australia
Austria
Belarus‡
Belgium
Bulgaria‡
Canada
Czechoslovakia‡
Denmark
European Economic Community
Estonia‡
Finland
France
Germany
Greece
Hungary‡
Iceland
Ireland
Italy
Japan
Latvia‡
Lithuania‡
Luxembourg
Netherlands
New Zealand
Norway
Poland‡
Portugal
Romania‡
Russian Federation‡
Spain
Sweden
Switzerland
Turkey
Ukraine‡
United Kingdom of Great Britain and Northern Ireland
United States of America

‡ Countries that are undergoing the process of transition to a market economy.

APPENDIX I

ANNEX II

Australia
Austria
Belgium
Canada
Denmark
European Economic Community
Finland
France
Germany
Greece
Iceland
Ireland
Italy
Japan
Luxembourg
Netherlands
New Zealand
Norway
Portugal
Spain
Sweden
Switzerland
Turkey
United Kingdom of Great Britain and Northern Ireland
United States of America

APPENDIX II

THE BERLIN MANDATE

CONCLUSION OF OUTSTANDING ISSUES AND ADOPTION OF DECISIONS

Proposal on agenda item 5(a)(iii) submitted by the President of the Conference

Review of the adequacy of Article 4, paragraph 2(a) and (b) of the Convention, including proposals related to a protocol and decisions on follow-up

The Conference of the Parties, at its first session, having reviewed Article 4, paragraph 2(a) and (b) and concluded that these are not adequate, agrees to begin a process to enable it to take appropriate action for the period beyond 2000, including the strengthening of the commitments of Annex I Parties in Article 4, paragraph 2(a) and (b), through the adoption of a protocol or another legal instrument.

I

1   The process shall be guided, *inter alia*, by the following:

(a) The provisions of the Convention, including Article 3, in particular the principles in Article 3.1 that reads as follows: 'The Parties should protect the climate system for the benefit of present and future generations of humankind, on the basis of equity and in accordance with their common but differentiated responsibilities and respective capabilities. Accordingly, the developed country Parties should take the lead in combating climate change and the adverse effects thereof';

(b) The specific needs and concerns of developing country Parties referred to in Article 4.8; the specific needs and special situations of least developed countries referred to in Article 4.9; and the situation of Parties, particularly developing country Parties referred to in Article 4.10 of the Convention;

(c) The legitimate needs of the developing countries for the achievement of sustained economic growth and the eradication of poverty, recognizing also that all Parties have a right to, and should, promote sustainable development;

(d) The fact that the largest share of historical and current global emissions of greenhouse gases has originated in developed countries, that the per capita emissions in developing countries are still relatively low and that the share of global emissions originating in developing countries will grow to meet their social and development needs;

(e) The fact that the global nature of climate change calls for the widest possible cooperation by all countries and their participation in an effective and appropriate international response, in accordance with their common but differentiated responsibilities and respective capabilities and their social and economic conditions;

(f) Coverage of all greenhouse gases, their emissions by sources and removals by sinks and all relevant sectors;

(g) The need for all Parties to cooperate in good faith and to participate in this process.

II

2   The process will, *inter alia*:

(a) Aim, as the priority in the process of strengthening the commitments in Article 4.2(a) and (b) of the Convention, for developed country/other Parties included in Annex I, both

- to elaborate policies and measures, as well as
- to set quantified limitation and reduction objectives within specified timeframes, such as 2005, 2010 and 2020, for their anthropogenic emissions by sources and removals by sinks of greenhouse gases not controlled by the Montreal Protocol

taking into account the differences in starting points and approaches, economic structures and resource bases, the need to maintain strong and sustainable economic growth, available technologies and other individual circumstances, as well as the need for equitable and appropriate contributions by each of these Parties to the global effort, and also the process of assessment and analysis referred to in section III, paragraph 4, below;

(b) Not introduce any new commitments for Parties not included in Annex I, but reaffirm existing commitments in Article 4.1 and continue to advance the implementation of these commitments in order to achieve sustainable development, taking into account Article 4.3, 4.5 and 4.7;

(c) Take into account any result from the review as referred to Article 4.2(f), if available, and any notification as referred to in Article 4.2(g);

(d) Consider, as provided in Article 4.2(e), the coordination among Annex I Parties, as appropriate, of relevant economic and administrative instruments, taking into account Article 3.5;

(e) Provide for the exchange of experience on national activities in areas of interest, particularly those identified in the review and synthesis of available national communications; and

(f) Provide for a review mechanism.

III

3   The process will be carried out in the light of the best available scientific information and assessment on climate change and its impacts, as well as relevant technical, social and economic information, including, *inter alia*, IPCC reports. It will also make use of other available expertise.

4   The process will include in its early stages an analysis and assessment, to identify possible policies and measures for Annex I Parties which could contribute to limiting and reducing emissions by sources and protecting and enhancing sinks and reservoirs of greenhouse gases. This process could identify environmental and economic impacts and the results that could be achieved with regard to time horizons such as 2005, 2010, and 2020.

5   The protocol proposl of the Alliance of Small Island States (AOSIS), which contains specific reduction targets and was formally submitted in accordance with Article 17 of the Convention, along with other proposals and pertinent documents, should be included for consideration in the process.

6   The process should begin without delay and be conducted as a matter of urgency, in an open-ended *ad hoc* group of Parties hereby established, which will report to the second session of the Conference of the Parties on the status of this process. The sessions of this group should be scheduled to ensure completion of the work as early as possible in 1997 with a view to adopting the results at the third session of the Conference of the Parties.

INDEX

Note: Emboldened numbers indicate a table or figure on the page.